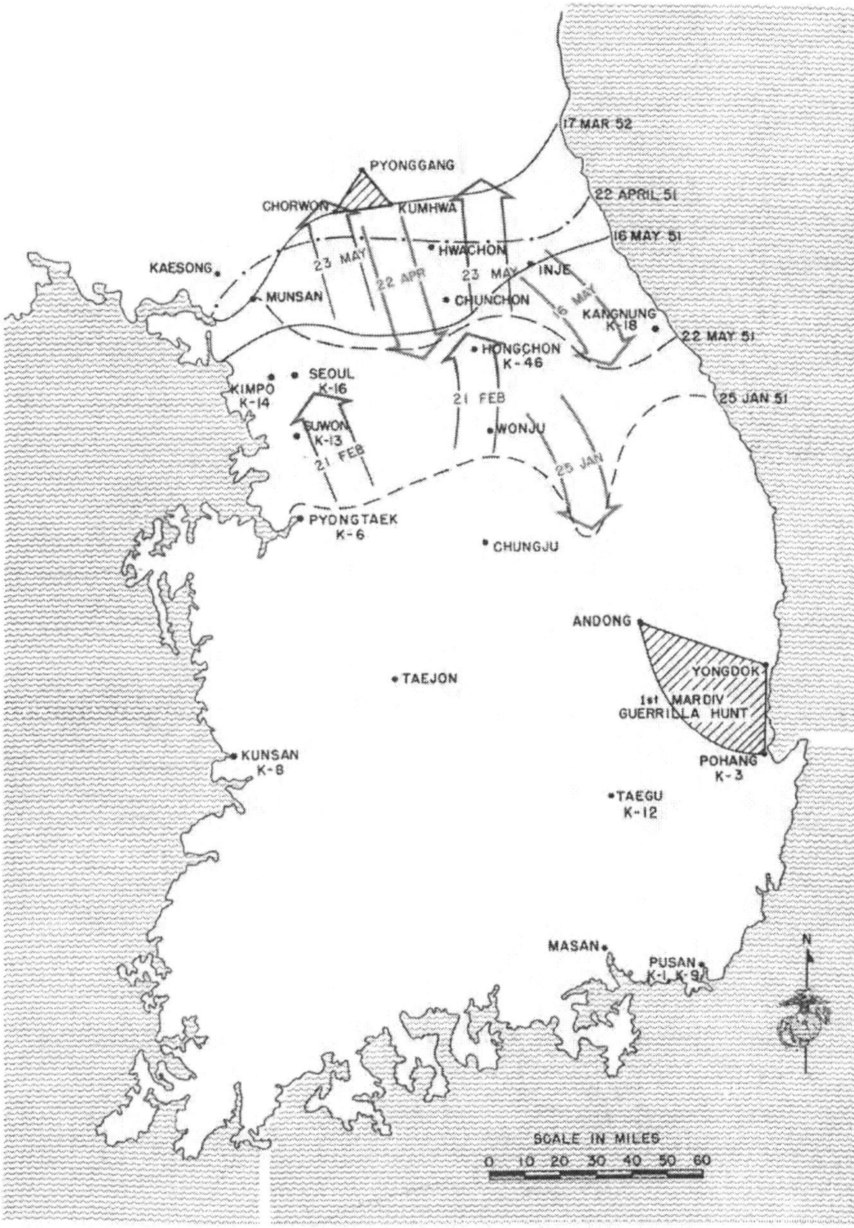
17 MAR 52
PYONGGANG
CHORWON
KUMHWA
22 APRIL 51
16 MAY 51
KAESONG
HWACHON
23 MAY
22 APR
23 MAY
INJE
MUNSAN
CHUNCHON
16 MAY
KANGNUNG
K-18
22 MAY 51
HONGCHON
K-46
KIMPO
K-14
SEOUL
K-16
SUWON
K-13
21 FEB
21 FEB
25 JAN 51
WONJU
25 JAN
PYONGTAEK
K-6
CHUNGJU
ANDONG
YONGDOK
1st MARDIV
GUERRILLA HUNT
TAEJON
KUNSAN
K-8
POHANG
K-3
TAEGU
K-12
MASAN
PUSAN
K-1, K-9
N
SCALE IN MILES
0 10 20 30 40 50 60

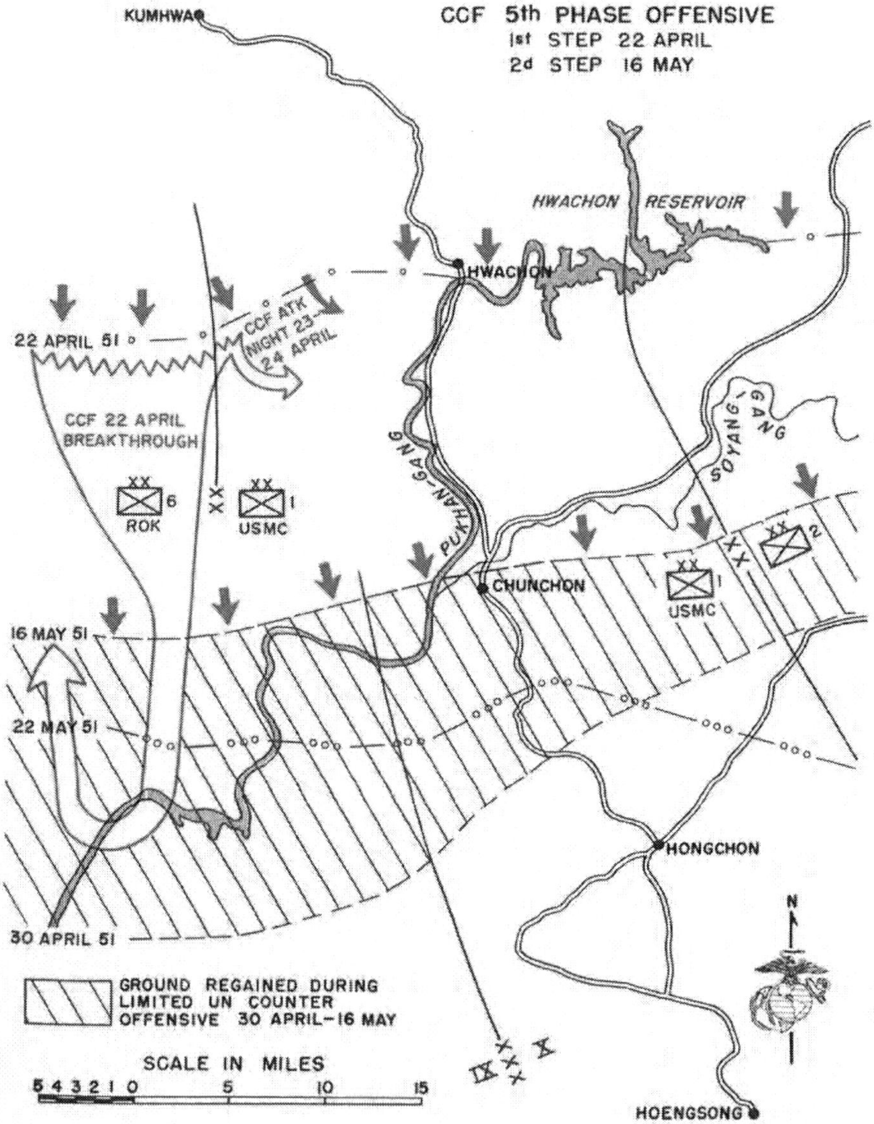
CCF 5th PHASE OFFENSIVE
1st STEP 22 APRIL
2d STEP 16 MAY
KUMHWA
HWACHON RESERVOIR
HWACHON
22 APRIL 51
CCF ATK NIGHT 23-24 APRIL
CCF 22 APRIL BREAKTHROUGH
ROK
USMC
PUKHAN-GANG
SOYANG-GANG
CHUNCHON
16 MAY 51
22 MAY 51
30 APRIL 51
HONGCHON
HOENGSONG
GROUND REGAINED DURING LIMITED UN COUNTER OFFENSIVE 30 APRIL–16 MAY
SCALE IN MILES
5 4 3 2 1 0 5 10 15
N

U. S. MARINE OPERATIONS IN KOREA

1950–1953

VOLUME IV

The East-Central Front

by

LYNN MONTROSS

MAJOR HUBARD D. KUOKKA, USMC

and

MAJOR NORMAN W. HICKS, USMC

Historical Branch, G–3

Headquarters, U. S. Marine Corps

Washington, D. C., 1962

Preceding Volumes of
U. S. Marine Operations in Korea

Volume I, "The Pusan Perimeter"

Volume II, "The Inchon-Seoul Operation"

Volume III, "The Chosin Reservoir Campaign"

Foreword

AMERICANS everywhere will remember the inspiring conduct of Marines during Korean operations in 1950. As the fire brigade of the Pusan Perimeter, the assault troops at Inchon, and the heroic fighters of the Chosin Reservoir campaign, they established a record in keeping with the highest traditions of their Corps. No less praiseworthy were the Marine actions during the protracted land battles of 1951, the second year of the Korean "police action."

The 1st Marine Division, supported wherever possible by the 1st Marine Aircraft Wing, helped stem the flood of the Chinese offensive in April. Then lashing back in vigorous and successful counterattack, the Marines fought around the Hwachon Reservoir to the mighty fastness of the Punchbowl. The Punchbowl became familiar terrain to Marines during the summer of 1951, and the Division suffered its heaviest casualties of the year fighting in the vicinity of that aptly named circular depression.

The fighting waxed hot, then cold, as the truce teams negotiated. They reached no satisfactory agreement, and the fighting again intensified. Finally, after a year of active campaigning on Korea's east-central front, the Marines moved west to occupy positions defending the approaches to the Korean capital, Seoul.

The year of desperate fighting, uneasy truce, and renewed combat covered by this volume saw the operational employment of a Marine-developed technique—assault by helicopter-borne troops. Tactics were continually being refined to meet the ever changing battle situation. However, throughout the period, the one constant factor on which United Nations commanders could rely was the spirit and professional attitude of Marines, both regular and reserve. This is their hallmark as fighting men.

DAVID M. SHOUP
General, U. S. Marine Corps,
Commandant of the Marine Corps.

Reviewed and Approved 20 Nov 1961.

Preface

THIS IS THE FOURTH in a series of five volumes dealing with the operations of United States Marines in Korea during the period 2 August 1950 to 27 July 1953. Volume IV presents in detail the operations of the 1st Marine Division and 1st Marine Aircraft Wing, the former while operating under Eighth Army control and also as part of IX Corps and X Corps, USA, and the latter while controlled by the Fifth Air Force.

The period covered in this volume begins in the latter part of December 1950, when the Division rested in the Masan "bean patch," and continues through the guerrilla hunt, the Punchbowl fighting, and all other operations during 1951. The account ends when the Marines move to positions in the west during March 1952.

Marines did not fight this war alone; they were a part of the huge Eighth United States Army in Korea. But since this is primarily a Marine history, the actions of the U. S. Army, Navy, and Air Force are presented only sufficiently to place Marine operations in their proper perspective.

Many participants in the fighting during this period have generously contributed to the book by granting interviews, answering inquiries, and commenting on first draft manuscripts. Their assistance was invaluable. Although it was not possible to use all the plethora of detailed comments and information received, the material will go into Marine Corps archives for possible use and benefit of future historians.

The manuscript of this volume was prepared during the tenure of Colonel Charles W. Harrison, Major Gerald Fink, and Colonel William M. Miller as successive Heads of the Historical Branch. Production was accomplished under the direction of Colonel Thomas G. Roe. Major William T. Hickman wrote some of the preliminary drafts and did much valuable research and map sketching. Dr. K. Jack Bauer and Mrs. Elizabeth Tierney assisted the authors in research, and Mr. Truman R. Strobridge assisted in proofreading and preparing the index.

To the Army, Navy, and Air Force officers, as well as Marine officers and NCOs, who submitted valuable comments and criticisms of preliminary drafts, thanks are also extended. These suggestions added to the accuracy and details of the text. Additional assistance was rendered by personnel of the Office of the Chief of Military History, Department of the Army; the Division of Naval History, Department of the Navy; and the Historical Division, Department of the Air Force.

The exacting administrative duties involved in processing the volume from first draft manuscripts through the final printed form were ably managed by Miss Kay P. Sue. All manuscript typing was done expertly by Mrs. Miriam R. Smallwood.

The maps contained in this volume were prepared by the Reproduction Section, Marine Corps Schools, Quantico, Virginia, and the Historical Branch, Headquarters Marine Corps. Official Department of Defense photographs were used.

The Marine Corps mourns the passing of the prime author of this series and other admirable works of Marine Corps and military history. Lynn Montross, after a lengthy illness, died on 28 January 1961.

H. W. Buse, Jr.
Brigadier General, U. S. Marine Corps,
Assistant Chief of Staff, G–3.

Contents

		Page
I	Interlude at Masan . . . . Return to the Bean Patch—1st Marine Division in EUSAK Reserve—General Ridgway New EUSAK Commander—Ridgway's Declaration of Faith—Marine Personnel and Equipment Shortages—Marine Air Squadrons in Action—The Air Force System of Control—X Corps Conference at Kyongju	1
II	The CCF January Offensive . . . . UN Forces Give Ground—Further Eighth Army Withdrawals—Marine Aircraft in the Battle—1st Marine Division Assigned Mission—Replacements by Air and Sea—The Move to Japanese Airfields—Red China's "Hate America" Campaign—A Tactical Formula for Victory	21
III	The Pohang Guerrilla Hunt . . . . The New Marine Zone of Operations—1st MAW Moves to Bofu—Marine Rice Paddy Patrols—Operations THUNDERBOLT and ROUNDUP—Action in the Pohang-Andong Zone—KMC Regiment Joins 1st Marine Division—10th NKPA Division Scattered—New Mission for the Marines	41
IV	Operation KILLER . . . . The Move to the Chungju Area—Marine Planes in Action—Planning for the New Operation—The Jump-Off on 21 February—Stiffening of Chinese Resistance—General Smith in Command of IX Corps—The Advance to Phase Line ARIZONA—JOC Air Control System Criticised	59
V	Operation RIPPER . . . . Light Resistance the First Day—Seoul Abandoned by Enemy—Second Phase of the Operation—Changes in 1st MAW Units—General MacArthur Visits Marine Battalion—1st KMC Returns to Division—38th Parallel Recrossed by Marines—Renewal of Division's CAS Problems	79
VI	The CCF Spring Offensive . . . . Prisoners Reveal Date of Offensive—Hwachon Occupied by KMC Regiment—CCF Breakthrough Exposes Marine Flank—	99

Page

Marine Air in Support Everywhere—Plugging the Gap on the Marine Left—Repulse of Communist Attacks—Withdrawal to the KANSAS Line—Enemy Stopped in IX Corps Sector—1st Marine Division Returns to X Corps

VII Advance to the Punchbowl 127
Plan to Cut Off Communists—Initial Marine Objectives Secured—MAG–12 Moves to K–46 at Hoengsong—Fight of the 5th Marines for Hill 610—1st MAW in Operation STRANGLE—KMC Regiment Launches Night Attack—1st Marines Move up to BROWN Line—7th Marines Committed to Attack

VIII The Truce Talks at Kaesong 153
Communists Ask for Truce Talks—Patrol Bases on BADGER Line—Red Herrings at Kaesong—1st Marine Division in Reserve—Marine Helicopters Take the Lead—Marine Body Armor Tested in Korea—MAG–12 Moves to K–18—The Division Back in Action Again

IX Renewal of the Attack 173
Crossing the Soyang in Flood—Light Resistance at First—Supply Problems Cause Delay—Resumption of Division Attack—The Mounting Problem of CAS—First Helicopter Supply Operation of History—The Fight for Hill 749—5th Marines Attack Hill 812—The Struggle for the "Rock"

X The New Warfare of Position 199
Sectors of Major EUSAK Units—Statement by General Van Fleet—Hill 854 Secured by 3/1—Helicopter Troop Lift to Hill 884—Helicopter Operation BLACKBIRD—"To Organize, Construct, and Defend"—Marine Operations of November 1951—The Second Marine Christmas in Korea

XI Winter Operations in East Korea 227
Ambush Patrol on New Year's Eve—Marine Raid in Company Strength—Major General John T. Selden Assumes Command—Boot, Combat, Rubber, Insulated—500 Armored Vests Flown to Korea—Helicopter Operations MULE TRAIN and CHANGIE-CHANGIE—The Five Days of Operation CLAM-UP

XII The Move to West Korea 247
Truce Talks—Tactical Innovations—The Marines in Operation MIXMASTER—Operations of Fifteen Months in Retrospect

Appendixes

		Page
A	Glossary of Technical Terms and Abbreviations	263
B	Effective Strength of 1st Marine Division	267
C	Command and Staff List	269
D	Unit Citations	315
Bibliography		319
Index		325

Illustrations

Photographs

Sixteen-page sections of photographs following pages 86 and 214.

Maps and Sketches

		Page
1	Korea as a Battlefield	4
2	EUSAK Dispositions, December 1950	9
3	Combat Air Bases	16
4	EUSAK Front Lines, January 1951	23
5	Pohang Guerrilla Hunt	43
6	Operation KILLER	61
7	Area Gained, February–March 1951	77
8	Operation RIPPER Plan	82
9	Operation RIPPER Zone	87
10	CCF Offensive, April 1951	110
11	Actions of 1/1 and 3/1, April 1951	112
12	Night of 16–17 May 1951	124
13	Drive to Yangu, May 1951	129
14	X Corps Routes of Advance, May–June 1951	139
15	1st Marine Division Zone of Action, June–July 1951	140
16	1st MAW Operating Area, May–July 1951	145
17	1st Marine Division Area, September 1951	174
18	X Corps Zone of Action, September 1951	177
19	EUSAK Dispositions, September 1951	200
20	1st Marine Division Situation, September 1951	205
21	HMR–161 Operations 1951	209
22	EUSAK Dispositions, March 1952	252

CHAPTER I

Interlude at Masan

Return to the Bean Patch—1st Marine Division in EUSAK *Reserve—General Ridgway New* EUSAK *Commander—Ridgway's Declaration of Faith—Marine Personnel and Equipment Shortages—Marine Air Squadrons in Action—The Air Force System of Control—X Corps Conference at Kyongju*

A NEW CHAPTER in Korean operations began for the 1st Marine Division at 1800 on 16 December 1950 with the opening of the CP at Masan. By the following afternoon all units of the Division had arrived from Hungnam with the exception of VMO–6 and small groups of such specialists as the amphibian tractor troops left behind to assist with the redeployment of remaining X Corps elements to south Korea.

The 1st Marine Division and 1st Marine Aircraft Wing were separated for the first time since the Inchon landing. VMF–311, the new Panther jet squadron, was flying from K–9, an Air Force field near Pusan. Operating together as an all-Marine carrier group taking part in the Hungnam redeployment were the three Corsair squadrons: VMF–212 on the CVL (light carrier) *Bataan;* VMF–214 on the CVE *Sicily;* and VMF–323 on the CVE *Badoeng Strait.* The two Japan-based night fighter squadrons, VMF(N)–542 and VMF(N)–513, flying from Itazuke, patrolled the skies between Japan and Korea.

VMO–6, the observation squadron, consisting of helicopters and OY fixed-wing planes, was attached to various ships of the Seventh Fleet for rescue missions when pilots were forced into the sea. A detachment of Marine Ground Control Intercept Squadron–1 (MGCIS–1) and the entire Air Defense Section of Marine Tactical

Air Control Squadron–2 (MTACS–2) were also attached to the warships. They assisted in the control of hundreds of planes that flew over the Hungnam beachhead daily in support of the final stages of the X Corps evacuation.

The three Marine Corsair squadrons on the *Sicily*, *Badoeng Strait*, and *Bataan* represented the entire air strength of Escort Carrier Task Group (TG) 96.8, commanded by Rear Admiral Richard N. Ruble. Each squadron came directly under the operational command of the ship on which it had embarked. Supply, engineering, ordnance, billeting, and messing were of course provided through naval channels. The only relationship of the squadrons to their parent organization, MAG–33, derived from the administration of personnel and the storage of equipment at Itami.

Return to the Bean Patch

Masan, the new Division assembly area, was located about 27 air miles and 40 road miles west of Pusan on the Bay of Masan, which indents the southern coast of the peninsula (Map 1). In order to prepare for the arrival of the Division, Brigadier General Edward A. Craig, the assistant division commander (ADC), had flown from Hungnam with the advance party on 12 December to make necessary arrangements.

The small seaport, which skirts the bay for about two and a half miles, was untouched by the war as compared to the ravaged towns of northeast Korea. It had a protected anchorage, dock facilities, and good rail and road communications. There was an air strip at Chinhae, a few miles to the southeast.

Some sort of cycle seemed to have been completed by veterans of the 5th Marines when they found themselves back again in the familiar surroundings of the Bean Patch on the northern outskirts of Masan. This large, cultivated field is entitled to capital letters because of its historical distinction as bivouac area of the 1st Provisional Marine Brigade after the battle of the Naktong in August 1950. Barely four months had passed since that hard fight, but a great deal more history had been made during the combats of the Inchon-Seoul and Chosin Reservoir operations.

There was room enough in the Bean Patch for all three infantry regiments. Headquarters, the 11th Marines, the 1st Signal, 1st Tank, 1st Amtrac, 1st Ordnance, and 1st Motor Transport Battalions were located on the southern outskirts of town along with the 41 Independent Commando, Royal Marines. The 1st Combat Service Group, the MP Company, and the 1st Service, 1st Shore Party, and 1st Engineer Battalions occupied the dock area of Masan proper. A large building in the center of town housed the Division hospital, and the 7th Motor Transport Battalion was assigned to the Changwon area, four miles to the northeast.[1]

Peaceful as the surroundings may have seemed to troops who had just completed the 13-day running fight of the Chosin Reservoir Breakout, the Chidi San mountain mass some 50 miles northwest of Masan had been for many years the hideout of Korean bandits and outlaws. The Japanese had never been able to clear them out, and the Republic of Korea had met with no better success. After the outbreak of civil war, they made some pretense of aiding the Communist cause but were actually preying upon the ROK army and police for arms, food, clothing, and other loot. Operating in prowling bands as large as 50 or 60 men, the guerrillas were well armed with rifles, machine guns, and at times even mortars.

In order to assure the safety both of its own bivouac areas and the vital port of Masan, Division promptly initiated measures to maintain surveillance over a broad belt of countryside which described an arc from Chinju, some 40 miles west of Masan, around to Changwon (Map 1). The infantry and artillery regiments and the Division Reconnaissance Company were all assigned subsectors of this security belt. Daily motor patrols of not less than platoon strength were to be conducted in each subsector for the purpose of gaining information about the roads and the guerrillas as well as discouraging their activities.[2] As it proved, however, no hostile contacts were made by the Marines during the entire Masan interlude. The guerrillas preferred to restrict their attention to the local police and civilian population.

---

[1] This section is based on 1st Marine Division (1stMarDiv) Historical Diary (*HD*), Dec 50, 1–12; MajGen O. P. Smith, *Notes on the Operations of the 1st Marine Division During the First Nine Months of the Korean War* (hereafter Smith, *Notes*), 1239–1242; and BrigGen E. A. Craig, Comments, 4 Jun 57.

[2] CG 1stMarDiv FragO, 1515, 18 Dec 50.

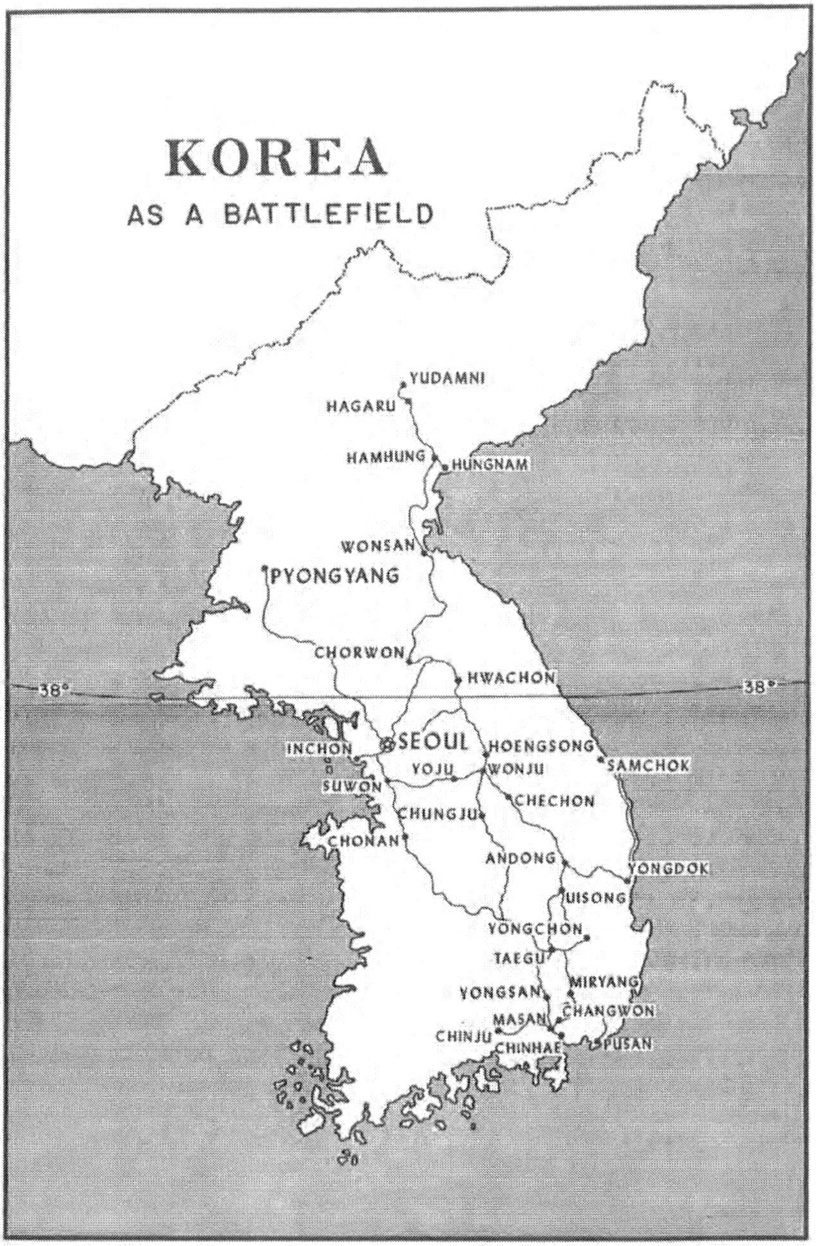
KOREA
AS A BATTLEFIELD
YUDAMNI
HAGARU
HAMHUNG
HUNGNAM
WONSAN
PYONGYANG
CHORWON
HWACHON
38°
38°
SEOUL
INCHON
HOENGSONG
SAMCHOK
YOJU
WONJU
SUWON
CHECHON
CHUNGJU
CHONAN
ANDONG
YONGDOK
UISONG
YONGCHON
TAEGU
MIRYANG
YONGSAN
MASAN
CHANGWON
CHINJU
CHINHAE
PUSAN

1st Marine Division in EUSAK *Reserve*

At 2240 on the 18th a dispatch from Major General Edward M. Almond, USA, commanding general of X Corps, informed the 1st Marine Division that it had passed to the operational control of the Eighth Army.[3]

Major General Oliver P. Smith reported in one of his first dispatches to EUSAK that the Marines had received fresh rations on only three days since landing in Korea. The Division commander invited attention to the importance of building up the physical condition of men who had lost weight during the Chosin Reservoir operation. An information copy went to Commander Naval Forces, Far East, (ComNavFE), who reacted promptly by ordering a refrigeration ship to Masan with 50,000 rations of turkey. The G–4 of EUSAK also responded with fresh rations from time to time until the Marines, in the words of General Smith, "had turkey coming out of their ears." [4]

Games of softball and touch football became popular in the crisp, invigorating weather as the men rapidly recuperated from fatigue and nervous tension. A series of shows was put on by troupes of U.S. Army and Korean entertainers, and the U.S. Navy sent Christmas trees and decorations.

The first Christmas in Korea was observed with a memorable display of holiday spirit by men who had cause to be thankful. A choir from the 5th Marines serenaded Division Headquarters with carols on Christmas Eve, and all the next day the commanding general and ADC held open house for staff officers and unit commanders.[5]

The United States as a whole rejoiced over the news that the last of 105,000 X Corps troops had embarked from Hungnam on 24 December without a single life being lost as a result of enemy action. President Truman spoke for the Nation when he sent this message to General MacArthur:

> Wish to express my personal thanks to you, Admiral Joy, General Almond, and all your brave men for the effective operations at Hungnam. This saving of our men in this isolated beachhead is the best Christmas present I have ever had.

---

[3] CG XCorps msg X 15292, 18 Dec 50; EUSAK msg GX–35290–KG00, 19 Dec 50.
[4] Smith, *Notes*, 1244–1245.
[5] The remainder of this section is based upon Smith, *Notes*, 1264–1274.

Photographers and press correspondents flocked to Masan during the holiday season for pictures and interviews about various aspects of the Chosin Reservoir campaign. Among them was Captain John Ford, USNR, a successful motion picture director who had been recalled to active duty to make a documentary film depicting the role of the Navy and Marine Corps in Korea. He used scenes in the Masan area for background material.

General Smith was informed that a motion picture company intended to produce a feature film entitled "Retreat, Hell," based on a remark attributed to him, "Retreat, Hell, we are just attacking in a different direction!" When asked if these actually were his words, the Division commander had a diplomatic answer. He said that he had pointed out to correspondents at Hagaru that the drive to Hamhung was not a typical withdrawal or retreat, and thus "the statement attributed to me described my thinking, that of my staff and unit commanders, and my situation."

During the Masan interlude Colonel S. L. A. Marshall, USAR, arrived as a representative of the Operations Research Office of Johns Hopkins University, which had been employed on military research projects by the Far East Command. Marshall, a well-known military analyst who had written several books about World War II operations, based his studies on personal interviews with scores of participants.

The researcher was given a free hand at Masan. Aided by a stenographer, he interviewed officers and men from privates to commanding general. The resulting thousands of words went into a classified report entitled, "CCF in the Attack (Part II), A Study Based on the Operations of the 1stMarDiv in the Koto-ri, Hagaru-ri, Yudam-ni area, 20 November–10 December 1950."

General Ridgway New EUSAK *Commander*

Shortly after arrival at Masan, General Smith called a conference of unit commanders and emphasized that their task was to re-equip, resupply, repair and rehabilitate. Officers and men of replacement drafts were to be integrated and given unit training as soon as possible. Both veterans and newcomers were soon training in regimental areas assigned by Colonel Alpha L. Bowser, the Division G–3, who arranged for a 200-yard rifle range and a mortar range.

On 23 December came the news that Lieutenant General Walton H. Walker, the Eighth Army commander, had been killed in a jeep accident. His successor, Lieutenant General Matthew B. Ridgway, USA, had commanded the U.S. XVIII Airborne Corps in Europe during the final operations of World War II. Commencing his flight from Washington on the 24th, he landed at Tokyo just before midnight on Christmas day.[6]

The new commander's task was made more difficult by the fact that the Korean conflict, at the end of its first six months, had become probably the most unpopular military venture of American history, both at the front and in the United States. From a mere "police action" at first, the struggle soon developed into a major effort in which the national pride suffered humiliations as a consequence of military unpreparedness. Far from building up the morale of the troops, letters and newspapers from home too often contributed to the doubts of men who asked themselves these questions:

"Why are we here? And what are we fighting for?"

Some of the answers were scarcely reassuring. It was insinuated, for instance, that Americans were fighting "to make South Korean real estate safe for South Koreans."

"I must say in all frankness," commented General Ridgway in his memoirs, "that the spirit of the Eighth Army as I found it on my arrival gave me deep concern. There was a definite air of nervousness, of gloomy foreboding, of uncertainty, a spirit of apprehension as to what the future held. There was much 'looking over the shoulder' as the soldiers say."[7]

These criticisms were not applicable to the 1st Marine Division. "Our men were in high spirits and busily engaged in getting ready to fight again," commented Brigadier General Edward A. Craig, ADC. "In my travels around the various units of the Division, and in talking to the men, I never even once noticed any air of nervousness or apprehension. . . . When General Ridgway visited the Division at Masan he made a tour of the entire camp area and observed training and general arrangements. He stated that he was quite satisfied with the

---

[6] Gen M. B. Ridgway as told to H. M. Martin, *Soldier, The Memoirs of Matthew B. Ridgway* (New York, 1956), 196-211, hereafter Ridgway, *Memoirs*.
[7] *Ibid.*, 204–205.

634040 O–62–2

1st Marine Division and its quick comeback after the Chosin fighting."[8]

General Ridgway learned soon after his arrival that the Eighth Army staff had prepared a plan for a phased withdrawal to Pusan in case of necessity. He called immediately for a plan of attack. Prospects of putting it into effect were not bright at the moment, but at least it served to announce his intentions.

Rumors were rife at this time that a general withdrawal from Korea, in virtual acknowledgment of defeat, was contemplated. In a letter of 1957, General Douglas MacArthur wrote an emphatic denial: "I have no means of knowing whether such action may have been seriously considered in Washington; but, for my own part, I never contemplated such a withdrawal and made no plans to that effect."[9]

The front hugged the 38th Parallel during the last week of December as the Eighth Army held a defensive line along the Munsan-Chunchon-Yangyang axis (Map 2). Three U.S. divisions were in a combat zone occupied largely by ROK units. The 24th and 25th Divisions both reduced a third in strength by casualties, remained in contact with the enemy in west Korea while the 1st Cavalry Division, also depleted in numbers, occupied blocking positions to the rear. Personnel and equipment losses suffered by the 2d Division during the CCF counteroffensive of late November had rendered it noneffective as a tactical unit until it could be reinforced and re-equipped, and the 3d and 7th Infantry Divisions had just landed in the Pusan-Ulsan area after the Hungnam redeployment.[10]

On 27 December 1950 the commanding general began a three-day tour of Eighth Army units at the front. He talked to hundreds of soldiers ranging from privates to unit commanders. There was nothing the matter with the Eighth Army, he assured them, that confidence wouldn't cure. "I told them their soldier forbears would turn over in their graves if they heard some of the stories I had heard about the behavior of some of our troop leaders in combat. The job of a commander was to be up where the crisis of action was taking place. In time of battle, I wanted division commanders to be up with their forward battalions, and I wanted corps commanders up with the

[8] LtGen E. A. Craig, USMC (Retd), ltr of 4 Jun 57. All letters, typed interviews, and other documentary sources cited in footnotes are on file in the archives of the Historical Branch, G–3, Headquarters Marine Corps.

[9] Gen Douglas MacArthur, ltr of 6 Jun 57 to MajGen E. W. Snedeker.

[10] EUSAK Command Report (*Cmd Rpt*), Dec 50.

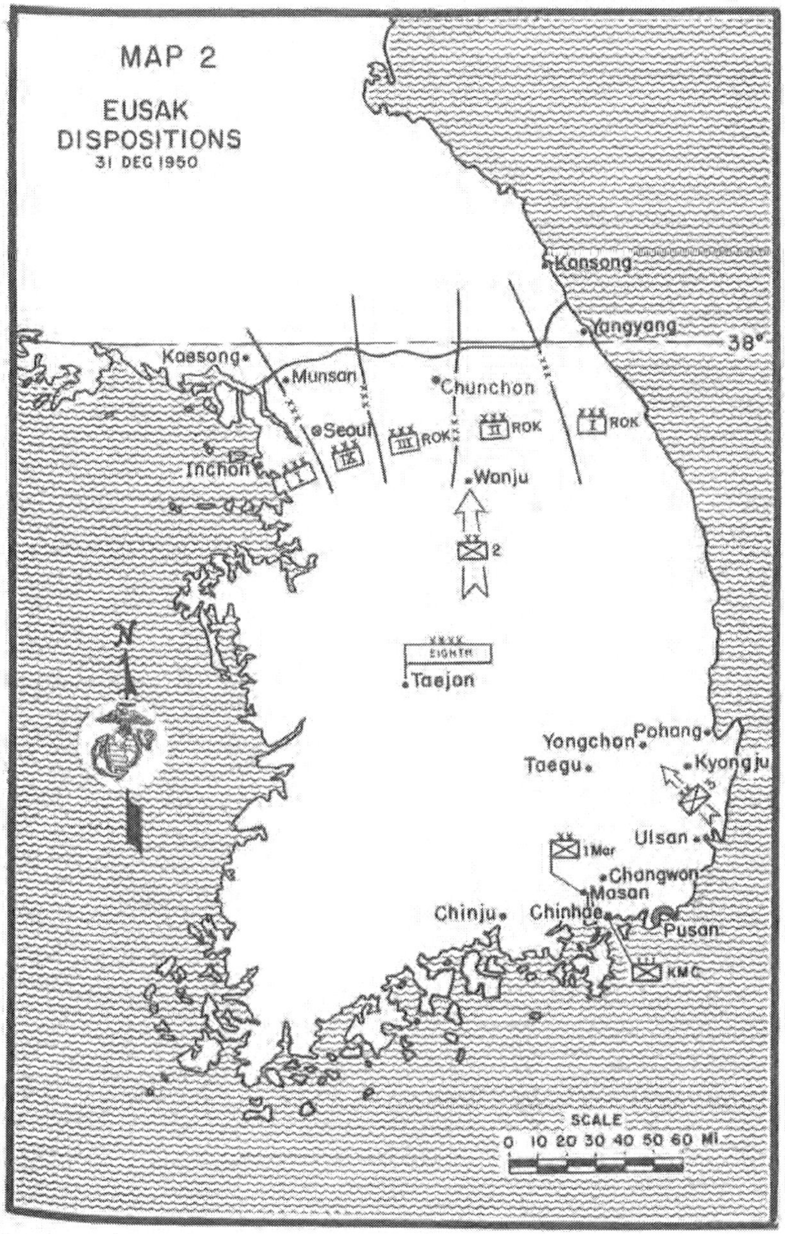

MAP 2
EUSAK
DISPOSITIONS
31 DEC 1950
Konsong
Yangyang
38°
Kaesong
Munsan
Chunchon
Seoul
ROK
ROK
ROK
Inchon
Wonju
2
EIGHTH
Taejon
Yongchon
Pohang
Taegu
Kyongju
5
Ulsan
1Mar
Changwon
Masan
Chinju
Chinhae
Pusan
KMC
SCALE
0 10 20 30 40 50 60 MI.

regiment that was in the hottest action. If they had paper work to do, they could do it at night. By day their place was up there where the shooting was going on."

It could never have been said that this professional soldier, the son of a Regular Army colonel, had failed to set an example in his own career. As the commander of an airborne division, he had jumped along with his men in Normandy.

Seldom seen in Korea without a grenade attached to his harness, Ridgway insisted that it was not a gesture of showmanship. In mobile warfare a man might be surprised by the enemy when he least expected it, he said, and a grenade was useful for blasting one's way out of a tight spot.

Ridgway's Declaration of Faith

After completing his tour of the combat area, the commanding general concluded that one thing was still lacking. Soldiers of the Eighth Army hadn't as yet been given an adequate answer to the questions, "Why are we here?" and "What are we fighting for?" In the belief that the men were entitled to an answer from their commanding general, he sat down in his room and wrote this declaration of faith:

> To me the issues are clear. It is not a question of this or that Korean town or village. Real estate is here, incidental. . . .
>
> The real issues are whether the power of Western civilization, as God has permitted it to flower in our own beloved lands, shall defy and defeat Communism; whether the rule of men who shoot their prisoners, enslave their citizens and deride the dignity of man, shall displace the rule of those to whom the individual and individual rights are sacred; whether we are to survive with God's hand to guide and lead us, or to perish in the dead existence of a Godless world.
>
> If these be true, and to me they are, beyond any possibility of challenge, then this has long since ceased to be a fight for freedom for our Korean allies alone and for their national survival. It has become, and it continues to be, a fight for our own freedom, for our own survival, in an honorable, independent national existence. . . .[11]

The deep conviction of this declaration could not be doubted. But Ridgway did not confine himself to moral leadership; he also insisted on a return to sound tactical principles. Upon learning that some of

[11] Memorandum from commanding general to all troops of Eighth Army, 21 Jan 51.

the infantry commanders in combat sectors had no knowledge of the enemy's strength or whereabouts, he ordered that aggressive patrolling be resumed at once. He directed further that every unit make a resolute effort to provide a hot reception for the Red Chinese patrols which had met too little opposition while prodding every night for soft spots along the thinly held 135-mile United Nations line.[12]

In his talks with officers and men, the new commander told them that too many weapons and vehicles had fallen into the hands of the enemy during the withdrawals in west Korea. He made it plain that in the future any man abandoning equipment without good cause would be court-martialed.

Not only did Ridgway stress the increased use of firepower; he requested in one of his first messages to the Pentagon that 10 additional battalions of artillery be sent to Korea. These guns were to provide the tactical punch when he found an opportunity to take the offensive.

Meanwhile, he had the problem of putting up a defense against a Chinese Communist offensive expected within a week. On his first day as Eighth Army commander he sent a request to President Syngman Rhee, of the Republic of Korea, for 30,000 native laborers to dig field fortifications. The energetic, 71-year-old Korean patriot provided the first 10,000 at dawn the following morning and the others during the next two days. Armed with picks and shovels, this army of toilers created two broad belts of defense, one to the north and one south of the river Han. The purpose of the first was to stop the enemy if American firepower could compensate for lack of numbers, and the second was a final line to be held resolutely.

Marine Personnel and Equipment Shortages

Although the Marine ground forces found themselves in the unusual situation of being 200 miles behind the front, they could be sure that this respite wouldn't last. Every effort was being pushed to restore the Division to combat efficiency by a command and staff acutely aware of shortages of men and equipment. The effective strength on 29 December 1950 was 1,304 officers and 20,696 men, including 182 attached U.S. Army troops and 143 Royal Marine Commandos. This

[12] The source for the remainder of this section is Ridgway, *Memoirs*, 205–207.

total also included 28 officers and 1,615 men who had arrived in a replacement draft of 17 December, and 4 officers and 365 men in a draft of three days later.[13]

Authorized Division strength was 1,438 officers and 24,504 men, indicating a shortage of 134 officers and 3,808 men. Most of the deficiencies were in the infantry and artillery units—29 officers and 2,951 men in the three infantry regiments, and 38 officers and 538 men in the artillery.

Division G–1 had been informed by the FMFPac representative in Japan that about 5,000 casualties were hospitalized there, and an unknown number had been evacuated to the United States because of overcrowding of hospitals in Japan. Such factors made it difficult to predict how many would return to the Division, but G–1 estimated from 500 to 1,000 in January.

The situation in regard to Division equipment might be summed up by saying that on 23 December there was a serious shortage of practically all essential items with the single exception of M–1 rifles. Upon arrival at Masan, units had been required to submit stock status reports. These lists were forwarded on 23 December to the Commanding General, Eighth Army, with a notification that requisitions had been submitted to the 2d Logistical Command, USA, in Pusan. It was requested that deliveries of supplies and equipment be speeded up, so that the Division could soon be restored to its former combat efficiency. A comparison of the totals of selected items on 23 and 31 December as listed on the following page shows that considerable progress was made during those eight days.

The 2d Logistical Command in Pusan, commanded by Brigadier General Crump Garvin, USA, deserved much of the credit for the week's restoration of Marine equipment. Progress passed all expectations, considering that General Garvin was supplying other Eighth Army units which had lost equipment during their withdrawal.[14]

There still existed on 29 December a requirement for clothing and individual equipment, and the spare parts problem remained acute. Ironically, the fact that the 1st Marine Division had brought most of its motor transport out from the Chosin Reservoir was a handicap at Masan. Eighth Army units which had lost their vehicles were given

[13] Sources for this section, unless otherwise indicated, are Smith, *Notes*, 1280–1292, 1294, 1295, 1303; 1stMarDiv Periodic Logistics Reports (*PLR*) 2, 11.

[14] LtGen E. A. Craig, ltr of 4 Jun 57.

Items of equipment	*T/E allowance*	*Shortages 23 Dec 50*	*Shortages 31 Dec 50*
Bags, sleeping	23,000	3,585	0
Machine gun, Browning, Cal. 30, M1919A4	1,398	338	0
BAR, 30 cal.	904	441	0
Carbine, 30 cal., M2	11,084	2,075	0
Launcher, rocket, 3.5", M20	396	105	0
Howitzer, 105mm	54	8	0
Howitzer, 155mm	18	9	0
Glasses, field, 7x50	1,740	1,305	1,006
Tank, Med., M4A3, dozer, 105 mm	12	7	7
Tank, med., M-26, 90mm	85	16	12
Truck, ¼ T., 4x4	641	105	58
Truck, 1½ T., 6x6, cargo	54	3	0
Truck, 2½ T., 6x6, cargo	737	124	33
Radio set, SCR 536	474	211	211
Radio set, SCR 619	137	74	49
Telephone, EE8	1,162	58	58

priority for receiving new ones. This meant that the Marines must make the best of war-worn trucks.

Marine Air Squadrons in Action

While the ground forces trained in the Masan area, the Corsair squadrons and the jet squadron flew combat missions. Support of the Hungnam redeployment had top priority until 24 December, when the last of the 105,000 troops were evacuated by Rear Admiral James H. Doyle's Task Force 90. Such totals as 91,000 Korean refugees, 17,500 vehicles, and 350,000 measurement tons of cargo were also recorded by the U.S. Navy's largest operation of the Korean conflict.[15]

No serious trouble was experienced from enemy action during the two weeks of the redeployment, although G-2 reports warned that several Chinese divisions were believed to be in the general area. Air strikes and naval gunfire shared the credit for this result. Nearly 34,000 shells and 12,800 rockets were fired by the support ships, and UN planes were on station or carrying out missions every moment that weather permitted. Marine fighters of VMF-212, VMF-214,

[15] ComPhibGruOne Action Rpt, Hungnam, 5-10, 25.

and VMF–323, flying from carriers after the closing of Yonpo Airfield, made a noteworthy contribution to the success of the Hungnam redeployment.[16]

VMF–212, commanded by Lieutenant Colonel Richard W. Wyczawski, was assigned the task of gathering the helicopters of VMO–6 from various ships of the Seventh Fleet and returning them to the operational control of the 1st Marine Division at Masan. There the OYs of the observation squadron were waiting after an overland flight, and Major Vincent J. Gottschalk's unit was complete.

With the Hungnam redeployment ended, the Navy offered to make its primary carrier-borne air effort in support of the Eighth Army. There was no single over-all commander of Navy and Air Force aviation in Korea (other than General MacArthur himself) and the two services were working under a system of mutual agreement and coordination.[17]

The Far East Air Forces (FEAF), under Lieutenant General George E. Stratemeyer, was the senior Air Force command in the Far East, on the same level as ComNavFE, Vice Admiral C. Turner Joy. The largest FEAF subordinate command was the Fifth Air Force, commanded by Major General Earle E. Partridge, with headquarters at Taegu, alongside that of the Eighth Army.

Strictly speaking, land-based Marine air had been under Fifth Air Force operational control throughout the Chosin Reservoir operation. Actually a verbal agreement between General Partridge and Major General Field Harris, commanding the 1st Marine Aircraft Wing (MAW), had given the Marines a good deal of latitude in making decisions relative to close air support. This was often the salvation of Marine units during the breakout, when every minute counted. Later, during the Hungnam redeployment, control of Marine aircraft became the responsibility of Admiral Doyle. His control agency was Tactical Air Control Squadron–1 (TacRon–1) in his flagship, the *Mount McKinley*. TacRon–1 kept in close touch not only with the 3d Infantry Division, USA, defending the shrinking perimeter, but also with the Eighth Army and Fifth Air Force.[18]

During the last days of 1950 the four Marine air squadrons were kept busy. VMF–212 on the *Bataan* was attached to TF–77. The

[16] For a detailed account of the "amphibious operation in reverse," see the last chapter of Volume III of this series.
[17] CinCPacFlt *Interim Evaluation Report* (*PacFlt Interim Rpt*) No. 2, II, 621–758.
[18] TacRon-1 War Diary (*WD*), Dec 50.

coastline of east Korea was its hunting grounds for such missions as knocking out warehouses, bridges, and railway tunnels between the 38th and 39th parallels.

Along the west coast, VMF–214 on the *Sicily* and VMF–323 on the *Badoeng Strait* were commanded respectively by Major William M. Lundin and Major Arnold A. Lund. These squadrons were part of Task Group–95.1 under Vice Admiral Sir William G. Andrews, RN. The Marine aviators found themselves in an organization made up of Royal Commonwealth naval forces and of French, Thai, and ROK units. TG–95.1 had the responsibility for patrolling the western coastline to prohibit enemy movement by water in military junks and by vehicle along the littoral.[10]

VMF–311, the jet squadron commanded by Lieutenant Colonel Neil R. MacIntyre, remained the only land-based Marine air unit in Korea. The Fifth Air Force had made space for it on crowded K–9, seven air miles northeast of Pusan, when General Harris expressed a desire to keep his jets in Korea for possible defense against Red air attacks (Map 3).

MacIntyre exercised his prerogative as squadron commander to fly the unit's first combat mission on 17 December. He was not, however, the first Marine aviator to pilot a jet in combat. That distinction went to Captain Leslie E. Brown on 9 September 1950. Assigned to the Fifth Air Force's 8th Fighter-Bomber Squadron as an exchange pilot, he made the first of several routine flights with an F–80 Shooting Star.

On 20 December, 17 officers and 51 enlisted men arrived at K–9 to boost VMF–311's total to 27 officer pilots and 95 enlisted men. Under Fifth Air Force control, they were employed to attack suspected CCF troop shelters, entrenchments, and gun positions on the eve of the expected enemy offensive. Missions of the jet planes averaged 12 a day at the end of the month.

The Air Force System of Control

It was seldom realized in the middle of the twentieth century that for the first time since the Middle Ages, a single human being represented in his person a decisive tactical unit. Just as the mailed knight

[10] *HDs* and *WDs* of VMF–323, VMF–214, and VMF–212; *PacFlt Interim Rpt* No. 2, II, 1939–1960.

MAP 3
COMBAT AIR BASES
KOREA – JAPAN
100
0
100
MILES
SEOUL
K-14
K-13
SUWON
KOREA
TAEJON
K-5
POHANG
K-3
TAEGU
K-2
CHINHAE
K-1
K-9
K-4
K-10
PUSAN
SACHON
MIHO
HONSHU
KOBE
ITAMI
OSAKA
OZUKI
BOFU
IWAKUNI
SHIKOKU
FUKUOKA
TSUIKI
ITAZUKI
SASEBO
KYUSHU

on his barded charger had ruled the battlefields of the medieval world, so did the pilot of a modern aircraft have the power to put an enemy battalion to flight with napalm, or to knock out an enemy stronghold with a 500-pound bomb.

A great deal depended, of course, on how the lightning of this human thunderbolt was controlled. The Marine Corps and the Air Force had different ideas on the subject. At the foundation of the Marine system was the concept that the needs of the ground forces came first, and control of air support should be exercised by the troops being supported. In each Marine infantry battalion a tactical air control party (TACP) included two aviators—one to be employed as a forward air controller (FAC) at the front, and the other as an air liaison officer in the battalion supporting arms center (SAC).[20]

In an emergency both could quickly be assigned to companies or even platoons to "talk" air strikes down on the enemy. The normal chain of command was bypassed in favor of direct radio from the TACP to the cognizant air control agency that had the authority to cross-check the request for possible conflict with other operations and to channel fighter-bombers to the attack.

Intermediate commands kept themselves informed of the over-all air picture and controlled the employment of aviation by their own subordinates as they listened in on these requests. They indicated approval by remaining silent, and disapproval by transmitting a countermand.

The hub of the Air Force system was the Tactical Air Control Center (TACC) of the Fifth Air Force-EUSAK Joint Operations Center (JOC), known by the code name MELLOW. An aviator coming on duty called up MELLOW and received his instructions from JOC.

FACs were assigned to U.S. Army and British units down to corps, division, and regimental levels, and to ROK corps and divisions. Further assignment to smaller front line units was possible but entailed a good deal of time and advance planning. And even the most urgent requests had to be channeled through division and regimental levels to JOC for approval.

If a Marine FAC wasn't able to control an air strike visually because of terrain conditions, he called for a "tactical air coordinator, airborne"

[20] The material in this section is derived from the following sources: *PacFlt Interim Rpt* No. 2, II, 463–620; Chief, Army Field Forces Headquarters, Tactical Air Command, Joint Training Directive for Air-Ground Operations; and CMC, ltr to Dist List re "Analysis of CAS Systems," 19 Aug 52.

(TACA) to locate the target from the air and direct planes to the attack. The Fifth Air Force also used special airborne coordinators. Known as "Mosquitoes," they flew low-winged, two-seater North American training planes, designated T–6s by the Air Force and SNJs by the Navy.

This plan was capable under favorable circumstances of providing the Fifth Air Force-EUSAK tactical air control system with a mobile and flexible means of directing air power at the front. Its chief weakness, according to Marine doctrine, lay in the separation of air power from ground force control. The Air Force claimed the advantage of projecting tactical air power deep into enemy territory; but as the Marines saw it, this was deep or interdictory support, and not to be compared to genuine close air support.

X Corps Conference at Kyongju

The command and staff of the 1st Marine Division could only speculate during this interim period as to what the near future might hold for them. Rumors had been circulated, during the first week at Masan, that the Division would be employed as rearguard to cover an Eighth Army withdrawal from Korea, with Pusan serving as the port of debarkation. And while plans cannot be made on a basis of rumor, General Smith and Colonel Bowser went so far as to discuss the possibility seriously. At last, on 24 December, a more definite prospect loomed when the EUSAK staff requested the Division to furnish logistical data for a move by rail and truck to Wonju, some 130 miles north of Masan.

It was not known whether an actual move was contemplated or the intention was merely to have available a plan for future use if the occasion warranted. General Smith sent the data but added a strong recommendation to the effect that any commitment of the Division be postponed until it was re-equipped and strengthened by replacements.[21]

At this time the Marine general received a copy of a map prepared by the Eighth Army staff which showed the phase lines of a 200-mile withdrawal from the combat zone to the Pusan port of debarkation. No enlightenment as to the employment of the Division was forthcoming until 27 December 1950, however, when a EUSAK dispatch

[21] 1stMarDiv *HD* Dec 50, 12–13; Smith, *Notes*, 1258–1259.

directed that the Marines be detached from Eighth Army reserve and reassigned to the operational control of X Corps.[22]

A message of the 28th requested General Smith to attend a conference at the X Corps CP at Kyongju (about 60 air miles northeast of Masan) on the 30th. He was directed to bring several members of his staff with him and to assign a liaison officer to X Corps.[23]

Two VMO-6 helicopters flew him to Kyongju along with his G-3, Colonel Bowser, and his aide, Captain Martin J. Sexton. Tossed by high winds, they landed just in time to meet General Ridgway, who gave a talk emphasizing the necessity for reconnaissance and maintaining contact with the enemy.

The new plan for X Corps employment, as modified after discussion with the Eighth Army commander, called for the recently reorganized 2d Infantry Division to be placed under operational control of General Almond. It was to move out at once to the Wonju front, followed by the 3d and 7th Infantry Divisions. The 1st Marine Division was to stage to Pohangdong (Map 3) on the east coast, some 65 miles north of Pusan, with a view to being eventually employed on this same front.[24]

"Certainly no one could accuse General Almond, the X Corps commander, of defeatism," was a tribute paid by General Smith. On the contrary, the Marine general had sometimes differed with him on the grounds that he was aggressive to the point of giving too little weight to logistical considerations and time and space factors.

It was realized at the conference that administrative decisions must depend to a large extent on the outcome of the impending enemy offensive. G-2 officers of the Eighth Army, forewarned by prisoner interrogations, were not surprised when the blow fell shortly before midnight on the last night of the year.

In spite of Air Force bombings of roads and suspected supply dumps, the Chinese Reds had been able to mount a great new offensive only three weeks after the old one ended. Attacking in the bitter cold of New Year's Eve, they made penetrations during the first few hours in ROK-held sectors of the central and eastern fronts. By daybreak it became evident that Seoul was a major objective, with the UN situation deteriorating rapidly.

---

[22] EUSAK msg GX-20179-K600, 27 Dec 50.
[23] CG X Corps msg X 16070, 28 Dec 50.
[24] This account of the Kyongju conference is derived from Smith, *Notes*, 1269-1271.

CHAPTER II

The CCF January Offensive

UN Forces Give Ground—Further Eighth Army Withdrawals—Marine Aircraft in the Battle—1st Marine Division Assigned Mission—Replacements by Air and Sea—The Move to Japanese Airfields—Red China's "Hate America" Campaign—A Tactical Formula for Victory

ON THE LAST DAY of 1950 the 1st Marine Division was alerted for two missions within an hour. At 1425 it was detached from X Corps, after only four days, and once more assigned to the operational control of the Eighth Army. The Marines were directed to resume their former mission of training, reorganizing, and replacing equipment so that they could be employed either to block enemy penetrations along the Ulchin-Yongju-Yechon axis (Map 4), or to take over a sector along the main line of resistance (MLR).

Forty minutes later another EUSAK dispatch alerted the Division to move to the Pohang-Andong area, where it would be in position to block any CCF penetration. This warning order came as no surprise, since X Corps had already contemplated such employment for General Smith's troops. In fact, General Craig and Deputy Chief of Staff Colonel Edward W. Snedeker had left Masan that very morning to select assembly areas and command posts.[1]

At a conference of G–3 and G–4 officers held at Masan on New Year's Day, it was recommended that the administrative headquarters remain in its present location when the rest of the Division moved up to Pohang. Although this headquarters had accompanied the Division CP in the past, it was believed that gains in mobility would result if

[1] EUSAK msgs GX 20332–KG00 and GX 20335 KG00 31 Dec 50; 1stMarDiv *HD*, Jan 51, 4.

the large number of clerical personnel and their increasing bulk of documents were left behind.[2] In view of the changing situation at the front, there was less danger of losing valuable records if the headquarters continued to function at Masan, maintaining contact with the forward CP by means of daily courier planes. The plan was approved by the Division commander and worked out to general satisfaction.

UN Forces Give Ground

Decisions were made during the first few days of 1951 in an atmosphere of suspense and strain as adverse reports came from the firing line. General Ridgway had assumed correctly, on the basis of prisoner interrogations, that the main Chinese effort would be channeled down the historical invasion corridor north of Seoul. He made his dispositions accordingly, and the Eighth Army order of battle on 31 December 1950 (Map 2) was as follows:

U.S. I Corps—Turkish Brigade, U.S. 25th Division, ROK 1st Division, from left to right northwest of Seoul. In Corps reserve, British 29th Brigade.

U.S. IX Corps—ROK 6th Division, U.S. 24th Division, from left to right north of Seoul. In Corps reserve, British Commonwealth 27th Brigade, U.S. 1st Cavalry Division.

ROK III Corps—ROK 2d, 5th, and 8th Divisions, from left to right on central front. In Corps reserve, ROK 7th Division.

ROK II Corps—ROK 3d Division, on east-central front.

ROK I Corps—ROK 9th and Capital Divisions, from left to right on eastern front.

The U.S. X Corps, comprising the newly reorganized U.S. 2d Infantry Division at Wonju and the 7th Infantry Division in the Chungju area, had been given a mission of bolstering the ROK-held line in central and east Korea and blocking enemy penetrations to the rear.

In Eighth Army reserve was the 187th Airborne RCT, with Thailand Battalion attached, in the Suwon area. Also under EUSAK operational control in rear areas were the 1st Marine Division (Masan), the 3d Infantry Division (Kyongju), the Canadian Battalion (Miryang), and the New Zealand Field Artillery Battalion (Pusan).

[2] Smith, *Notes*, 1315.

EUSAK FRONT LINES 3 Jan,51
MAP 4
3 Jan.51
38°
Chunchon
Seoul
24 Jan.51
Samchok
Suwon
Yoju
Wonju
Osan
Chechon
Pyongtaek
Line "D" (modified)
Chungju
Tanyang
Yongju
Ulchin
Yechon
Andong
Yongdok
Pohang
Yongchon
Kyongju
Miryang
1Mar
Masan
Chinae
Pusan
N
ROK
SCALE
0 20 40 60 MI.
Area of 1MarDiv Guerrilla Hunt

Altogether, the United Nations forces in Korea numbered 444,336 men as of January 1951. The cosmopolitan character of the fight against Communism is indicated by the aid given to the U.S. and ROK forces by contingents of combat troops from 13 other nations—Australia, Belgium, Canada, Ethiopia, France, Greece, Netherlands, New Zealand, Philippines, South Africa, Thailand, Turkey, and the United Kingdom.[8]

Enemy numbers at this time were estimated at a total of 740,000 men in Korea and near-by Manchuria. Seven CCF armies, the 37th, 38th, 39th, 40th, 42d, 50th, and 66th were identified among the troops attacking on New Year's Eve. The NKPA I and V Corps also participated. Estimated strength of the assaulting forces was 174,000 Chinese and 60,000 North Koreans.

Previously identified but not reported in contact with U.S. forces on 31 December were the 24th, 48th, 49th, and 65th CCF armies and the NKPA 1st, 3d, and 15th Divisions.

As another possibility which could not be overlooked, the five CCF armies which had opposed X Corps in northeast Korea might also take part in the new offensive. Elements of the 20th, 26th, 27th, 30th, and 32d Armies identified in that area early in December, had more than two weeks in which to reorganize and make their way to the Eighth Army front. If they got into the fight, it would mean a formidable addition to the enemy's forces.

With only five days at his disposal, after arrival in Korea, General Ridgway's preparations were limited. His dispositions could not be blamed, but it was the old story of the chain and its weakest link as the enemy scored a major breakthrough at the expense of the 1st ROK Division on the west-central front. Unfortunately, this unit represented the tactical joint between I Corps and IX Corps. The enemy widened the gap before dawn and drove on toward Seoul.

Early in the morning the EUSAK commanding general was on the road, waving his arms in an attempt to stop ROK soldiers streaming rearward in their vehicles after abandoning crew-served weapons. The short training period for these troops, their tactical inexperience, and the language barrier were the dissonant notes tolling the ominous chords of defeat. The whole front was endangered as the enemy poured through an ever widening gap, and Ridgway ordered that roadblocks be set up where MPs could halt the fugitives, rearm them,

[8] Sources for this section are EUSAK *Cmd Rpt,* Jan 51, 4–5, 7, 9, 27, 62–65.

and send them back to the front. At his request, President Syngman Rhee appealed to ROK soldiers over the radio and exhorted them to make a stand. By that time it was too late to save Seoul, and the commanding general gave orders for its evacuation.

"The withdrawal was initiated in mid-afternoon on the 3d," he commented in retrospect. "I stayed on the bridge site on the north bank until dark to watch the passage of the most critical loads. These were the 8-inch howitzers and the British Centurion tanks, both of which exceeded the safety limits of the bridge under the conditions existing at the time." [4]

It was a scene of terror and despair that Ridgway never forgot. Thousands of Korean civilian refugees were making their way over the thin ice of the river Han, many of them carrying children or old people on their backs. What impressed the observer most was the uncanny silence of this mass flight in the freezing winter dusk, broken only by the sound of a multitude of feet shuffling over the ice—a sound strangely like a vast whispering. It was as if these derelicts of war were trying incoherently to confide their misery to someone.

From a strategic viewpoint, the only course left to the Eighth Army was a continued retirement south of Seoul. "We came back fast," Ridgway admitted, "but as a fighting army, not as a running mob. We brought our dead and wounded with us, and our guns, and our will to fight." [5]

Further Eighth Army Withdrawals

EUSAK Fragmentary Operations Plan 20, issued as an order on 4 January, called for a further withdrawal to Line D (Map 4). In preparation, X Corps had moved up to the front on the 2d, after assuming operational control of the U.S. 2d and 7th Infantry Divisions and the ROK 2d, 5th, and 8th Divisions and occupied a sector between U.S. IX Corps and ROK III Corps.[6]

The U.S. 3d Division was attached to I Corps and the 187th Airborne RCT passed temporarily under operational control of IX Corps.

By 7 January the UN forces had pulled back to a modified Line D

[4] Gen M. B. Ridgway, ltr of 5 Jun 57.
[5] Ridgway, *Memoirs*, 215.
[6] EUSAK *Cmd Rpt*, Jan 51, 9, 62, 64, 82, 92.

extending from Pyongtaek on the west coast to Samchok on the east and taking in Yoju and Chechon. General Ridgway sent telegrams to all corps commanders expressing dissatisfaction with the personnel and material losses inflicted on the enemy during the withdrawal. "I shall expect," each message concluded, "utmost exploitation of every opportunity in accordance with my basic directive."

That evening, foreshadowing the offensive operations he was contemplating, the commanding general ordered a reconnaissance-in-force by a reinforced infantry regiment north to Osan to search out the enemy and inflict maximum punishment. No contacts were made, nor did strong patrols sent out by U.S. IX Corps flush out any sizeable groups of Chinese. But the Eighth Army had served notice that it intended to regain the initiative at the first opportunity.

One more blow remained to be absorbed. On the 8th the Communists struck in the Wonju area with an attack of four divisions. Elements of the newly reorganized 2d Infantry Division were forced to give up that important highway and rail center after counterattacks failed. The enemy now directed his main effort along the Chunchon-Wonju-Chechon corridor, and North Korean guerrilla forces infiltrated through the gap between the U.S. X Corps and ROK III Corps.

The salient created by this CCF attack caused Line D to be modified again so that in the center it dipped sharply downward to Chungju before curving northeast to Samchok (Map 4).

Marine Aircraft in the Battle

The pilots and aircrewmen of the three carrier squadrons and the land-based jet squadron were the only Marines in a position to take an active part in the battle. With but one TACP per division, close air support was out of the question for the ROKs on New Year's Day.

Control facilities were severely strained when scores of UN fliers made use of the frequencies which the Mosquitoes employed for tactical air direction. The voices were all in the English language, but with more than one person doing the sending, shrill side noises sliced in to garble the whole into a cacophony of jungle sounds. A Mosquito trying to coach a fighter-bomber attack at the crossings of the Imjin might be drowned out by a distant pilot calling up a controller in the Hwachon Reservoir Area.[7]

As a consequence, there was no coordinated air-ground attack in direct support of the man in the foxhole. Most of the JOC effort was directed to the enemy's rear in an effort to block supporting arms, reinforcements, and supplies.

The two Marine squadrons attached to Admiral Ruble's carriers were at sea, some 80 miles south of Inchon when news of the Chinese offensive filtered through the tedious communication channels from JOC and EUSAK. Major Lund, CO of VMF–323, led an eight-plane attack which destroyed enemy trucks and some 40 huts believed to be occupied by CCF troops in a village south of the Imjin.

Another Marine air mission of New Year's Day was the flight commanded by Major Kenneth L. Reusser for the purpose of wiping out a reported CCF concentration on the central front. Unfortunately, he could not get verification that the target consisted of enemy troops. Before a decision could be made, Reusser heard a Mosquito of the 2d ROK Division calling urgently for any flier in the area to hit another CCF concentration (this time verified) in a village to the enemy's rear of the Chorwon-Hwachon area. Under the Mosquito's direction the Corsairs bombed and napalmed the village, then strafed survivors trying to escape.

VMF–212, flying with Navy (Task Force) TF–77 on the eastern side of the peninsula, had a busy New Year's Day. Two eight-plane interdiction strikes were flown in the morning against rear area targets along the coastal highways. The afternoon brought an emergency call from JOC, and the squadron "scrambled" 14 planes which hit the east flank of an extensive enemy push south of the Hwachon Reservoir.[8]

More than 300 UN fighter-bombers were sent out under JOC, or MELLOW, control on the embattled first day of 1951. On the west coast TacRon–3 received more calls for air support than TG–96.8 could fill. Rear Admiral Lyman A. Thackrey sent a request to Admiral Struble in the *Missouri* for additional carrier planes, and within a few hours the Marines of VMF-212 were detached and on their way to the west coast to join the other two Corsair squadrons of TG–96.8.

[7] USAF Hist Study No. 72, *U. S. Air Force Operations in the Korean Conflict*, 1 Nov 50–30 Jun 52, (hereafter cited as AHS–72) pp. 188–196. The term "net" denotes a network of radio stations, all on a single frequency for a specific purpose.

[8] The term "scramble" in military aviation parlance refers to an expeditious takeoff in response to an emergency call or an alert. With highly trained personnel, it is a pattern rather than a panic.

All four Marine fighter-bomber squadrons took part daily in air operations as the Chinese Reds continued their advance south of Seoul. VMF-311 was badly handicapped, however, by mechanical difficulties. Engine or radio trouble accounted for five "aborts" of the 15 sorties launched on 4 January. The remaining pilots could not make radio contacts with their assigned Mosquito controller, and had little choice other than to attack targets of opportunity.

The jets continued in action, but it was realized that they were not giving the maximum of their capabilities. By mid-January the squadron had become almost ineffective through no fault of its own. Technical representatives from the companies that had manufactured both the engine and plane were flown to K-9 (Map 3), and on the 16th all jets were grounded. These inspectors did not work on the planes; they were empowered only to report the nature of the trouble to the airplane companies concerned. The companies in turn reported to BuAir in Washington, which sent instructions and if necessary mechanics to Itami, where major aircraft maintenance was done.

Meanwhile, the fall of Seoul meant that the Air Force was evicted by enemy action from such major fields as Kimpo and K-16 on an island in the river Han. The Sabre jets and Mosquitoes had to be pulled back, and soon the F-51s were no longer secure at Suwon from an advancing enemy.

Admiral Thackrey's Western Deployment Group completed the evacuation from Inchon of 70,000 tons of supplies, 2,000 vehicles, and about 5,000 troops.[9] As the Navy closed out activities on the west coast, TG-96.8 sent out its last combat air missions on 7 January. VMF-214 made its final reconnaissance patrols; VMF-212 flew 25 sorties in support of UN troops in central Korea; and VMF-323 took part in a series of Air Force raids on enemy troop assembly areas in the Hoengsong area.[9]

Until the last, the carrier Marines alternated their Eighth Army support missions with routine CAPs, coastal searches, and airfield bombings. Admiral Thackrey's Redeployment Group, including TacRon-3, completed its task in the Inchon area and departed on the 7th. On that same day HMS *Theseus*, flying the flag of Admiral Andrewes, was back in west coast waters as the British pilots resumed their coastal patrols and naval air support on that side of the peninsula. Within a week VMF-212 and the *Bataan* returned to fly alternate

[9] AHS-72, 47, 48; PhibGru-3 *WD*, Jan 51.

tours of duty with the pilots of the *Theseus.* The other two carrier squadrons found themselves unemployed for the time being. Not only were they out of a job, they were also homeless, since the United Nations had been forced to give up airfields at Yonpo, Wonsan, Seoul, Kimpo, and Suwon. Only K–1, K–2, K–4, K–9, K–10, and two small fields near Taegu remained (Map 2), and they would scarcely serve the needs of FEAF. Thus it was that VMF–214 and VMF–323 found a temporary haven at Itami, along with VMF–311 and most of the administrative and service units of the 1st MAW. There was nothing to do but wait until a new home could be found for the fighter-bomber squadrons.

1st Marine Division Assigned Mission

The Marine aviators might have found some consolation in the fact that their comrades of the ground forces were also groping in a fog of uncertainty. At the most critical period of the CCF thrust in the Wonju area, General Smith was summoned to Taegu on 8 January for a conference with General Ridgway. The Eighth Army commander proposed to attach one of the Marine RCTs to X Corps in the Andong area, about 95 air miles north of Masan. The remainder of the Division would then move to the Pohang-Kyongju-Yongchon area, some 60 air miles northeast of Masan (Map 4).

Ridgway asked the Marine general to discuss the prospect with his staff. He realized, he said, that no commander liked to have his division split up, and he assured Smith that as soon as the X Corps zone became stabilized, the RCT would be sent back to him.

They parted with this understanding, but a few hours after his return by air to Masan the following message was received from Ridgway:

> Subsequent your departure, alternate plan occurred to me on which I would like your views soonest. It follows: 1st Mar Div, under Army control, move without delay to general area outlined to you personally today, to take over responsibility at date and hour to be announced later for protection of MSR between Andong and Kyongju, both inclusive, and prevent hostile penetration in force south of Andong-Yongdok road.[10]

[10] 1stMarDiv *HD*, Jan 51, 4; EUSAK msg G–1–628–KGG, 8 Jan 51.

At 1115 on the 9th the plan was made official. An Eighth Army dispatch ordered the 1st Marine Division to move without delay to the Pohang area (Map 4), remaining under EUSAK control, with the following missions:

(a) Prevent enemy penetrations in force south of the Andong-Yongdok road;

(b) Protect the MSR connecting Pohang, Kyongju, Yongchon, Uihung, and Uisong.[11]

Based on these directives, Division OpnO 1–51 was issued at 1600 on the 9th. RCT–1 was directed to move by motor to Yongchon and to protect the MSR, Yongchon-Uisong inclusive, from positions in the vicinity of Yongchon and Uihung. The 1st and 7th Motor Transport Battalions, plus other Division elements, were ordered to provide the required trucks.[12]

General Ridgway arrived at Masan by plane on the morning of 9 January. He was met by General Smith and driven to Headquarters, where the Division staff officers and regimental commanders were presented to him. In a brief talk he reiterated the necessity for reconnaissance and for regaining and maintaining contact with the enemy. The Marine officers were told that limited offensive actions by Eighth Army units would be put into effect soon.[13]

Division OpnO 2–51, issued at 1300 on the 10th, provided for the completion of the Division movement by road and water from Masan to the objective area.[14]

Shortages both of personnel and equipment were much reduced during the first two weeks of January. Returns to duty of battle and nonbattle casualties added 945 to the Division strength. Corresponding improvements had been made in the material readiness of the Division. Early in January a large resupply shipment arrived from Kobe, and a Navy cargo ship brought supplies and equipment which had been left behind at Inchon in October. Thus the situation was generally satisfactory except for nearly 1,900 gaps in the ranks that remained to be filled.[15]

---

[11] EUSAK msg GX–1–661–KG00, 9 Jan 51.
[12] 1stMarDiv *OpnO 1–51*, 9 Jan 51.
[13] Smith, *Notes*, 1279.
[14] 1stMarDiv *OpnO 2–51*, 10 Jan 51.
[15] Smith, *Notes*, 1285-1286, 1307-1308.

Replacements by Air and Sea

Facilities for air transport across the Pacific were limited, since the Army was also moving replacements to the Far East. A piecemeal process of shuttling Marines in plane-load increments could not be completed before 30 January. Lieutenant General Lemuel C. Shepherd, Jr., commanding FMFPac, took a dim view of this delay. It would be better for the Division, he maintained, to receive even a part of its replacements before it went back into action. As a compromise, he proposed a combined air-sea lift which met the approval of Rear Admiral Arthur H. Radford, commanding Pacific Fleet.

Three replacement drafts were already on the way, with the 3d in Japan and the 4th and 5th at Camp Pendleton. General Shepherd scraped the bottom of the manpower barrel so closely that he dug up an additional 700 men from Marine security detachments in Japan, the Philippines, and other Pacific Ocean bases.

Seven trainloads of Marines from Camp Lejeune arrived at San Francisco on 10 January to join those from Camp Pendleton. On the same day 230 of these replacements were flown to Hawaii by the Military Air Transportation Service (MATS), by the R5D's of Marine VMR-352 and of Navy VR-5, and by the "Mars" flying boats of Navy VR-9. The next day 799 Marines sailed on the fast transport USNS *General W. O. Darby.* The remainder were transported at the rate of one plane load a day by MATS and at the rate of three or four plane loads a day by the Navy and Marine transport planes of Fleet Logistics Air Wing, Pacific (FLogAirWingPac).[16]

Five days later, on the 16th, the airlift had cleared the last Marine out of Treasure Island. On 21 January, 1,000 men of the special draft were already with the 1st Division at Pohang and the 799 on board *General Darby* were due to dock at Pusan.

It had been a fast job of coordination by the Navy, Army, Air Force, and virtually all major units of the Marine Corps. Much of the special airlift was flown by the R5Ds of VMR-352 and of VMR-152. The former, commanded by Colonel William B. Steiner, had been flying the El Toro-Tokyo flights since October, but most of its effort had been in shuttling between the mainland and Hawaii. VMR-152 had concentrated on the Hawaii-Japan leg of the long trip. During

---

[16] Material relative to Marine replacements has been derived from FMFPac *HD*, Dec 50 and Jan 51.

the Chosin campaign, the squadron commander, Colonel Deane C. Roberts, had maintained his headquarters and 10 planes at Itami to support the shuttle to Korea. He had barely returned to Hawaii from that job when his squadron was alerted not only for the special lift of Marine replacements but also for a return to the Far East.[17]

Hawaii had been the bottleneck in this special troop lift. Land and seaplanes were discharging their human cargo at Barbers Point, Hickam Air Force Base, and Keehi Lagoon. From there FLogAirWingPac had to space the planes over the long stretches of sea at approximately four-hour intervals. The guiding factor was other air traffic over the same route and the servicing, messing, and rescue capabilities of Guam and other points along the way, such as tiny Johnston Island. The latter was barely big enough for its single 6,100-foot runway.

VMR–152 and the Navy's VR–21 were assigned the mission of flying the long Hawaii-Japan portion of the big lift. Itami became another collection center for the airborne replacements and five of the VMR–152 planes were retained there to shuttle the troops the last 300 miles to K–3, near Pohang (Map 2). On 21 January the troop lift reached virtual completion, but Admiral Radford authorized the 1st MAW to retain a couple of R5D's at Itami a little longer. Thus the Marines were able to avoid highway and rail traffic jams in Korea by flying men and materials from troop and supply centers in Japan to K–1, K–3, or K–9.

Looking back at the troop lift from a historical distance, the observer is most impressed by its demonstration of teamwork on a gigantic scale. The Marine Corps had functioned as a single great unit, even though a continent and an ocean separated the vanguard in Korea from the rear echelons in North Carolina.

The Move to Japanese Airfields

The seven remaining UN airfields in Korea were of course not enough to accommodate the 25 FEAF and Marine tactical squadrons. Logistics and lack of space proved to be knotty problems. Thirty tank cars of gasoline a day were needed for normal flight operations of K–2 alone.

[17] The balance of this section is derived from the Dec 50 and Jan 51 historical diaries of VMF–352, VMR–152, 1st MAW, and FMFPac.

Yet it took these cars eight days to make the 120-mile Pusan-Taegu round trip, such was the strain put on the railway system by the CCF offensive.

FEAF had standby plans to evacuate Korea entirely in an emergency. Some of the secondary airfields of the Itazuke complex in Japan had been reevaluated for this purpose. Originally built by the Japanese for World War II, they were obsolescent by 1951 and because of weather, neglect, and misuse badly deteriorated.[18]

The most promising of these secondary airfields were Tsuika, Ozuki, and Bofu (Map 3), ranging from 30 to 65 miles east of Itazuke and facing one another around Japan's Inland Sea. Nearest to Itazuke and on the same island of Kyushu was Tsuika. Across the narrow Shimonoseki Strait, on the shore line of Honshu, were Ozuki and Bofu.

General Stratemeyer, the FEAF commander, informed General MacArthur that it was necessary to start air operations from Ozuki and Bofu as soon as possible. A good deal of work had already been done on Tsuika, even to moving a major Japanese highway in order to lengthen the runway to 7,000 feet. The Air Force general wanted to repair Ozuki for his F–51 squadrons, and Bofu was to be reserved for the 1st MAW.

This decision meant a revision of plans for the Marines. MAG–12 had recently been lifting a hundred men a day to K–1 (Pusan west) with a view to making it into a major base. These preparations came to an abrupt halt, pending the final decision on Bofu.

A Marine survey of that World War II airfield showed it to be in serious disrepair. The Air Force had already rejected it as a base for night-harassing B–26s. Although the runway was only 7 feet above sea level, a 720-foot hill complicated the traffic pattern. Nevertheless, Bofu was considered suitable for the time being, and the Air Force assured the 1st MAW that its use would be but temporary.

FEAF proposed that the Marines start flying out of Bofu immediately, operating under field conditions. There were, however, essential repairs to be made. The 5,300-foot runway remained in fair condition, but much of the taxiway was not surfaced and couldn't stand heavy use by the Corsairs. Three of the four hangars needed extensive

---

[18] This section is based on the 1st MAW and MAG–33 *HDs* of Jan 51; Fifth Air Force History (hereafter to be designated FAF), Dec 50 and Jan 51; AHS–72, 35–37; EUSAK *Cmd Rpt*, Jan 51, Sec II, 12, 63, 64–65; *PacFlt Interim Rpt* No. 2, II, 969, 1062; Mobile Construction Battalion Two (hereafter MCB–2) Report of Activities for Jan 51; Col T. J. Noon, interv of 5 Jun 58.

repairs, as did the barracks and mess hall. Fuel would have to be stored in drums.

The Wing had the capability for minor construction but lacked the equipment, men, and fiscal authority to handle major work on the runways and taxiways. The Air Force offered to furnish the labor and materials, provided that the Navy pay for them. The Navy in its turn was too limited in funds to restore an Air Force field for only temporary use by Marines.

Finally, a compromise solved the problem. The Navy agreed to have the engineering work done by a detachment of its Mobile Construction Battalion 2 (Seabees) and furnish the concrete for patching the runways and rebuilding the warm-up aprons. The Air Force was to provide the pierced steel planking for the runways.

On 15 January MAG–33 sent an advance detachment of 125 officers and men to Bofu to do some of the preliminary work, and on the following day the Seabees initiated the heavy construction. The restoration of K–1 was meanwhile resumed by MAG–12.

Until these two fields were made ready, VMF–212 on the *Bataan* would be the only Marine squadron in combat.

Red China's "Hate America" Campaign

The middle of January was also a transition period for the 1st Marine Division. In accordance with Division Orders 1–51 and 2–51, the movement from Masan commenced at 0545 on 10 January when the first serial of RCT–1 departed by motor for the Pohang-Andong area. LSTs 898 and 914 sailed the next day with elements of the Tank, Ordnance, Engineer, and Service Battalions. The new Division CP opened at Sinhung, about five miles southeast of Pohang, at 1600 on 16 January; and by the 17th all designated motor and water lifts were completed. Thus the 1st Marine Division and 1st Marine Aircraft Wing were poised to begin new operations which will be described in the following chapter.[19]

By 15 January relative quiet prevailed along the entire front; the Chinese Reds had shot their bolt. In terms of territorial gains (Map 4) the Communists could claim a victory, for they had inflicted heavy

[19] 1stMarDiv *HD*, Jan 51.

losses both in troop casualties and equipment on the UN forces. Yet the CCF January offensive could not compare with the November–December attacks either in moral or material damage done to the Eighth Army. This time the UN divisions had withdrawn for the most part in good order after the rout of ROK units at the outset. Nor were Ridgway's troops always driven from their positions by enemy action. Whenever he had an option between sacrificing men or Korean real estate, it was the latter he chose. And by his insistence on good combat discipline, he made the enemy pay an exorbitant price.

Nevertheless, the blunt fact remains that the United Nations forces had been beaten in spite of an overwhelming superiority in aircraft, artillery, armor, and transport as well as command of the sea. Stateside Americans can scarcely be blamed for asking themselves why their well-equipped divisions had been defeated twice within six weeks by an Asiatic peasant army using semiguerrilla tactics and depending largely on small arms, mortars, and light artillery.

The answer cannot be given in simplified terms. Although the Chinese Reds were represented by a peasant army, it was also a first-rate army when judged by its own tactical and strategic standards. Military poverty might be blamed for some of its deficiencies in arms and equipment, but its semiguerrilla tactics were based on a mobility which could not be burdened with heavy weapons and transport. The Chinese coolie in the padded cotton uniform could do one thing better than any other soldier on earth; he could infiltrate around an enemy position in the darkness with unbelievable stealth. Only Americans who have had such an experience can realize what a shock it is to be surprised at midnight with the grenades and submachine gun slugs of gnomelike attackers who seemed to rise out of the very earth.

Press correspondents were fond of referring to "the human sea tactics of the Asiatic hordes." Nothing could be further from the truth. In reality the Chinese seldom attacked in units larger than a regiment. Even these efforts were usually reduced to a seemingly endless succession of platoon infiltrations. It was not mass but deception and surprise which made the Chinese Red formidable.

They also had an advantage over Western soldiers in their ability to withstand hunger and cold while making long night marches. After all, the rigors of a winter campaign in Korea were not much worse than the hardships the Chinese peasant had endured all his life. Usually he was a veteran of at least five years' combat experience, for

China had known little but war since the Japanese invasion of 1935. Many of Mao Tse-tung's troops, in fact, were former Nationalists who had fought for Chiang Kai-shek.

The Chinese Reds held another advantage in Korean terrain well suited to their tactical system. This factor has been ably summarized by U.S. Military Academy historians:

> The mountains are high, and the deep gorges between them are a bar to traffic even when the streams are dry or frozen. Roads are few, and those that do exist are not suited for heavy traffic. Transportation then becomes a problem for the pack mule and the human back rather than the self-propelled vehicle. Telephone wires are difficult to lay and, with guerrillas on every hand, are doubly hard to maintain. Even radio is limited by such terrain, with a considerable reduction in range. In all, most observers have agreed that American forces have seldom fought in terrain to which modern means of war are less adaptable.[20]

The fanaticism and political indoctrination of the CCF soldier must also be taken into account. His introduction to Communism began when he was persuaded that China's small farms would be taken away from the hated landlords and divided among the people. This is the first stage of every Communist upheaval. Next comes a reign of terror calculated to liquidate the entire class of landlords and small shopkeepers. Communist China, almost literally wading in blood, had reached this second phase in 1951, the "year of violence." Mass trials were held in which the People's Tribunals, keyed up to a frenzy of fury, sentenced group after group of "Capitalist oppressors" to death without bothering about the evidence. The executions were public spectacles. An estimated million and a half of them took place in 1951 alone as loudspeakers on street corners blared out first-hand descriptions.

Drives were organized for everything in Red China. So rapidly did they multiply that humorless Communist leaders saw no absurdity in announcing a new drive to reduce the number of drives. And when the Youth League tried too zealously to please, a drive was launched "to Correct the Undesirable Habit of Filing False Reports." [21] Under these circumstances it is understandable that great emphasis was placed on Red China's "Hate America" drive early in 1951. The illiterate masses were made to believe that Americans practiced all manner of bestialities, including even cannibalism. This was the

[20] U.S. Military Academy, *Operations in Korea*, 28–29.
[21] Richard L. Walker, *China Under Communism* (New Haven, 1955), 119, 307.

indoctrination of the CCF soldier in Korea, and political commissars with a captain's authority were attached to each company to see that no backsliding occurred. In case of doubt, it was a simple matter to compel the suspected political deviate to kneel at the roadside and await a bullet from behind.

A Tactical Formula for Victory

It might well be inquired where Red China raised the funds, for even wars waged with human cannon fodder do not come cheaply. Much of the money was donated by new farm owners as "voluntary" contributions exceeding by far the rent and taxes of pre-Communist years. The slave labor of millions of Chinese sent to concentration camps also helped to foot the bill. In the long run, however, the Communist lords found perhaps their most effective means in the extortion of ransom from Chinese living outside the country on pain of torturing or killing relatives dwelling within its borders. Enormous sums were collected in spite of the efforts of foreign governments to put an end to this form of secret terrorism.[22]

Altogether, the army of Red China may be appraised as a formidable instrument on terrain suited to its tactics. Several of America's foremost military thinkers were convinced, nevertheless, that Eighth Army reverses of the first few months in Korea were the penalty paid for a national preoccupation with airborne atomic weapons at the expense of preparations for limited wars.

It was only natural that the American public and its political and military leaders in Washington should have been much concerned about a weapon with the capability of wiping out a medium-size city in a minute.[23] Their anxiety was heightened by President Truman's announcement on 23 September 1949 that Soviet Russia had exploded an atomic bomb. A great many Americans, probably a majority, sincerely believed that it was hardly worthwhile to prepare for an old-fashioned limited war when the Armageddon of the future would be fought to an awesome finish with thermonuclear weapons. National policy was shaped by this line of reasoning; and though we had every

[22] *Ibid.*, 13.

[23] The day had not yet dawned when the hydrogen bomb would have a much greater potentiality for frightfulness.

opportunity to study Chinese tactics prior to 1950, few if any preparations were made to cope with them. The outbreak of Korean hostilities found the four U.S. skeleton divisions in Japan woefully unready, both morally and materially.

At a later date three high-placed U.S. Army generals, Matthew B. Ridgway, James M. Gavin, and Maxwell D. Taylor, would retire because they could not reconcile their views with a national policy which they interpreted as placing all our strategic eggs in the basket of intercontinental bombers and guided missiles. Afterwards, as advocates of preparedness for limited as well as atomic warfare, they published books presenting their side of the case.[24]

On 15 January 1951 these developments were still in the future, of course. But even at the time it had already been made evident that the armed forces of Red China were not an exception to the age-old rule that there is no such thing as an invincible army. When they came up against well trained and led U.S. Army outfits in both of their offensives, they always had a fight on their hands and frequently a repulse.

The Marines had proved beyond doubt in their Chosin Reservoir campaign that the Chinese Reds could be beaten by ground and air firepower engendered by sound training, discipline, and combat leadership. Five Chinese armies, of three or four divisions each, were identified in northeast Korea during the November–December operations. Three of them were directly or indirectly opposed to the 1st Marine Division, with a U.S. Army battalion and smaller Army units attached. Yet the beleaguered American forces seized the initiative and fought their way for 13 days and 35 miles through enveloping CCF units which had cut the mountain MSR in five places.

Throughout the CCF January offensive, EUSAK G–2 officers anxiously sought every scrap of evidence as to the whereabouts of the five CCF armies identified in northeast Korea as late as 10 December. Even if reduced by casualties, they would have been a formidable and perhaps even decisive reinforcement to the seven CCF armies engaged. But they did not appear. Nor were they encountered again until the middle of March 1951, when similarly numbered units filled with replacements reached the front.

---

[24] Gen M. B. Ridgway, *Memoirs* (New York, 1956); Gen J. M. Gavin, *War and Peace in the Space Age* (New York, 1958); Gen M. D. Taylor, *The Uncertain Trumpet* (New York, 1959).

The full story may never be known, since the Chinese Reds are not fond of acknowledging their disasters. But it is a likely conjecture that the fatal combination of Marine firepower and General Winter created terrible havoc among Communists who had been so certain of an immediate victory that they were neither armed, clothed, nor supplied for a 13-day campaign in subzero weather.

634040 O–62–4

CHAPTER III

The Pohang Guerrilla Hunt

The New Marine Zone of Operations—1st MAW Moves to Bofu—Marine Rice Paddy Patrols—Operations THUNDERBOLT and ROUNDUP—*Action in the Pohang-Andong Zone—KMC Regiment Joins 1st Marine Division—10th* NPKA *Division Scattered—New Mission for the Marines*

ON 15 JANUARY 1951 a reinforced regiment of the U.S. 25th Infantry Division drove northward from Line D (Map 4) to a point about half a mile from Suwon in the I Corps sector. VMF–212, flying from the CVE *Bataan,* supported the movement along with land-based Air Force planes. No CCF troops were encountered during a two-day thrust dignified with the name Operation WOLFHOUND. Its only importance lay in its distinction as the first Eighth Army counterstroke in reply to the enemy's January offensive. Other EUSAK advances were soon to follow, each more ambitious than the last and bearing a more bristling code name.

General Ridgway proposed by this means to exert continual and increasing pressure on an enemy paying for victory with extended supply lines. Meanwhile, he hoped to build up the morale of his own troops without asking too much of them at first.

In less than seven weeks, from 1 December 1950 to 15 January 1951, the Eighth Army had been pushed back an average distance of 200 miles. Never before in the Nation's history had an American army given up so much ground and equipment in so short a time, and damage to morale was inevitable. Yet the commanding general was confident that a cure would be effected by better combat leadership and discipline. He planned to emphasize the need for these remedies until he restored the Eighth Army to tactical health.

The New Marine Zone of Operations

Ridgway agreed with Marine generals that the 1st Marine Division had come out of its 13-day battle in the Chosin Reservoir area with its fighting spirit undulled. Minor respiratory ills seemed to be the only consequences felt by the survivors. "A hacking cough," recalled a Marine staff officer long afterwards, "was the symbol of the Bean Patch." [1]

Such ills soon responded to rest and medical care, and it was a physically fit division that made the move to the new zone of operations. About one man out of three in the infantry and artillery battalions was a newcomer to Korea. These replacements were shaping up nicely, and the new operation promised to be ideal combat training.

The move took nearly a week. While the other troops proceeded by motor, LSTs 898 and 914 sailed with elements of the Tank, Ordnance, Engineer, and Service Battalions. The Division CP opened at Sinhung (Map 5), about 5 miles southeast of Pohang, on 16 January. By the following day all designated motor and water lifts were completed.

On the 18th the Marines were assigned a three-fold mission by Division OpnO 3–51: (1) the protection of the Pohang-Kyongju-Andong MSR (main supply route); (2) the securing of Andong and the two airstrips in the vicinity; and (3) the prevention of hostile penetrations in force to the south of the Andong-Yongdok road. The following zones of patrol responsibility were assigned to Marine units:

Zone A—RCT–1: an area about 10 miles east and west of the Uisong-Andong road, including both Uisong and Andong.

Zone B—RCT–5: an area some 15 to 20 miles wide astride the Kyongju-Yongchon-Uisong road, including Kyongju but excluding Uisong.

Zone C—RCT–7: an area 20 to 25 miles wide from east to west and extending north from the latitude of Pohang to the Andong-Yongdok road.

Zone D—11th Marines: a strip seven miles wide along the coast astride the road from Pohang to a point about 10 miles north of Yongdok.

Zone E—1st Tank Battalion: the area bounded by the road from

---

[1] MajGen F. M. McAlister ltr, 17 Jun 57.

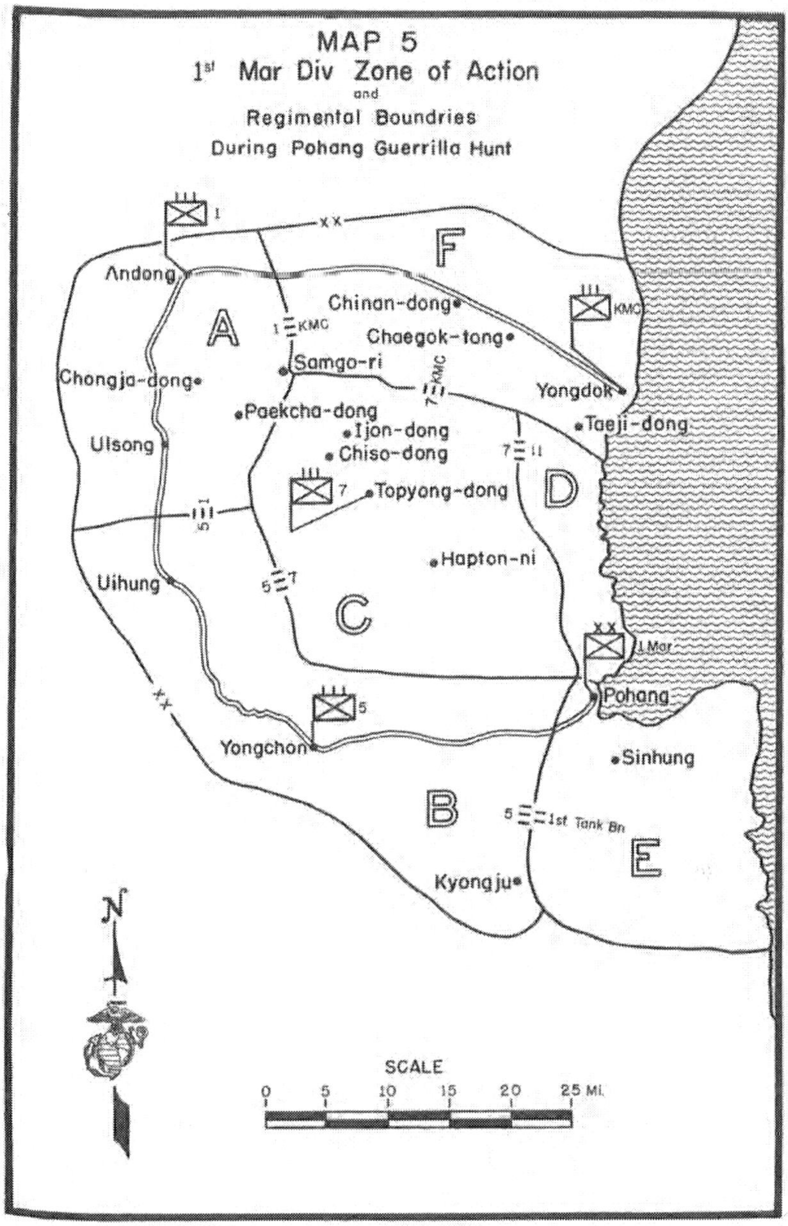

MAP 5
1st Mar Div Zone of Action
and
Regimental Boundries
During Pohang Guerrilla Hunt
Andong
Chinan-dong
Chaegok-tong
Chongja-dong
Samgo-ri
Yongdok
Paekcha-dong
Ijon-dong
Taeji-dong
Ulsong
Chiso-dong
Topyong-dong
Hapton-ni
Uihung
Pohang
Yongchon
Sinhung
1st Tank Bn
Kyongju
KMC
1 Mar
A
B
C
D
E
F
N
SCALE
0 5 10 15 20 25 MI.

Pohang to Kyongju and thence to the east coast at a point about 19 miles southeast of Pohang.

Keeping open the 75-mile stretch of MSR from Pohang to Andong was considered the principal mission of the Division. Strong points were set up at Pohang, Yongchon, Uisong, and Andong.

Captured documents indicated that enemy forces in unknown numbers had already infiltrated through gaps in the eastern sectors of the Eighth Army's Line D. Guerrilla activity was reported as far west as Tanyang, on the MSR of IX Corps, and as far south as Taejon, threatening the supply line of I Corps. Train ambushes occurred on 13 January in the Namchang area and to the south of Wonju. Other attacks took place on the rail line about 60 miles north of Taegu. In expectation of further attempts, trains were provided with a sandbagged car, pushed ahead of the engine, to absorb the shock of landmine explosions. Another car was occupied by guards who had the duty of dealing with direct guerrilla attacks.[2]

The tactical problem of the Marines was quite simple—on paper. About 1,600 square miles, most of them standing on end in mountainous terrain, were included in the new zone of operations. The experience of World War II had demonstrated how effective guerrilla warfare could be as an adjunct to large-scale military operations. Officers of the 1st Marine Division had no illusions about their mission, therefore, when they received unconfirmed reports of NKPA guerrilla infiltrations behind the EUSAK lines toward Andong.

All uncertainty vanished on 18 January, shortly after the issuing of OpnO 3–51, when a patrol of the 3d Battalion, 1st Marines, flushed out an undetermined number of North Korean troops east of Andong. They took to their heels so earnestly that the Marines barely managed to catch three of them after a long chase.

The prisoners identified their unit as the 27th Infantry of the NKPA 10th Infantry Division. The other two regiments, the 25th and 29th, were also in the general area. All three were supported more in theory than fact by artillery, mortar, medical, and engineer units organic to the division. In reality, however, the estimated total of 6,000 troops consisted largely of infantry. A few mortars, according to the prisoners, were the largest weapons.

Following the Inchon-Seoul operation, the remnants of the badly mauled NKPA 10th Infantry Division had straggled back across the

---

[2] EUSAK *Cmd Rpt*, Jan 51.

38th Parallel to the Hwachon area. There they were reorganized by the Chinese for guerrilla operations and placed under the command of NKPA Major General Lee Ban Nam.[3]

Late in December the rebuilt division, still short of arms and equipment, departed Hwachon with a mission of infiltrating through the UN lines to cut communications and harass rear installations of the Andong-Taegu area. Shots were exchanged with United Nations troops near Wonju, but General Lee Ban Nam and his troops contrived to slip to the east through the mountains. Stealthily moving southward, marching by night and hiding by day, they were soon in a position to heckle the rear of the X Corps sector. This advantage did not last long. Before they could strike a blow, the element of surprise was lost along with the three prisoners taken by the Marines.

As the Marine units moved into their assigned zones, General Ridgway flew to Pohang to confer with General Smith. Not only did he express confidence that the Marines would soon have the situation well under control; he also suggested the possibility of small amphibious landings along the east coast. The purpose was to block a possible southward advance of the three CCF armies that had operated in Northeast Korea during the Chosin Reservoir campaign.[4]

The east coast littoral was considered the most likely route of approach. Smith was of the opinion, however, that an amphibious landing should be made in strength, if at all. And there the matter rested.[5]

1st MAW Moves to Bofu

During the operations of the first few days the Marine ground forces had to depend for air support on FEAF planes sent by JOC. The 1st Marine Aircraft Wing had its hands full at this time with housekeeping activities. Work began at Bofu (Map 3) on 20 January as a Seabee detachment arrived with its graders and bulldozers. They were assisted by details of Marines from MAG–33.[6]

[3] 1stMarDiv Periodic Intelligence Report (*PIR*) 87–94, 17–22 Jan 51.

[4] Units of these armies, it may be recalled, were not identified again in the CCF order of battle before the middle of March 1951. Until that time, the possibility of these enemy troops being used for a surprise stroke had to be taken into consideration.

[5] Smith, *Notes*, 1339.

[6] This section, unless otherwise stated, is based on the January historical diaries of the following organizations: 1st MAW; MAG–33; MAG–12; VMF–312; VMF–214; VMF–323; VMF(N)–513; VMF(N)–542; MWSS–1; FMFPac. Another source is *PacFlt Interim Rpt* No. 2, II, 969.

The job went ahead with typical Seabee efficiency. While specialists installed plumbing for the galleys and barracks, other crews graded taxiways, laid pierced steel planking, and poured concrete to patch up runways, parking ramps, and warmup aprons.

MAG–12 kept busy at the task of moving men and equipment from Itami and other Japanese fields to Korea. Aircraft of VMR–152, commanded by Colonel Deane C. Roberts, provided transportation. Since safety measures precluded the use of the K–1 runway during construction activity, K–9 substituted temporarily. As fast as the planes unloaded, passengers and gear were trucked 15 miles through Pusan to K–1.

It was a transition period in more ways than one for the 1st MAW. Following are the changes of commanders that took place during the last 2 weeks of January:

Colonel Radford C. West, relieved by Lieutenant Colonel Paul J. Fontana as commanding officer of MAG–33;

Lieutenant Colonel Frank J. Cole, joined MAG–33 staff as personnel officer after being relieved of VMF–312 command by Major Donald P. Frame;

Major Arnold A. Lund of VMF–323, relieved by Major Stanley S. Nicolay and assigned to General Harris' staff as assistant operations officer;

Major William M. Lundin, relieved of VMF–214 command by Major James A. Feeney, Jr., and transferred to the command of Service and Maintenance Squadron–33 (SMS–33).

This left only Lieutenant Colonel Richard W. Wyczawski of VMF–212 and Lieutenant Colonel Max J. Volcansek, Jr., of VMF(N)–542 still in command of the tactical squadrons they brought to Korea; and the latter was to be relieved by Lieutenant Colonel James R. Anderson in February.

The only combat operations of the 1st MAW during the week of housekeeping from 16 to 23 January were carried out by VMF–212 from the deck of the *Bataan*.[7] This CVL carrier alternated with the British light fleet carrier HMS *Theseus* on the Korean west coast blockade. Their activities were coordinated by Vice Admiral Andrewes, RN, commanding the group blockading the Korean west coast.

VMF-212 sent out a morning and afternoon reconnaissance flight

[7] VMF–212 *HD*, Jan 51; USS *Bataan* (CVL–29) Action Rpt, "Operations off the West Coast of Korea, 15 Jan–7 Apr 51;" USS *Bataan WD*, Jan 51; Col R. W. Wyczawski, interv of 2 Jun 58.

each day up the coastline as far as the 39th parallel. On the trip north the pilots scanned the coastal waters for small enemy shipping which might indicate reinforcement from Chinese ports on the Yellow Sea. The return trip along the highways and railroads of the littoral was made to detect signs of any new enemy activity on land. Four aircraft flew each of the two coastal sweeps; eight maintained a defensive patrol over the carrier itself; and any remaining flights were under control of JOC, with FEAF Mosquitoes providing liaison between fighter-bombers and ground forces.

To insure sea room beyond the islands and mudbanks of the west coast, the *Bataan* had to stay outside the 100-fathom curve. This meant that the pilots must fly across 65 to 80 miles of open sea in order to reach the coast. The winter weather varied from unbelievable to unbearable, and bulky, uncomfortable survival suits were a necessity. They could be a death trap, however, if a leak developed or if they were not adjusted tightly at the throat and wrists. Captain Alfred H. Agan, for instance, was shot down southeast of Inchon and had to choose between landing in enemy territory and ditching in the sea. He tried for a small island offshore but crash-landed into the surf. Before a helicopter from the *Bataan* could fly 65 miles to the rescue, he died from the shock of icy water which partially filled his survival suit.

The pilots of VMF–212 reported an increase in enemy antiaircraft fire, particularly in CCF rear areas. They were amazed to find troops dug in along the coast as far back as 50 or 60 miles from the battle lines. These precautions were the enemy's tribute to Marine capabilities for amphibious warfare. The fear of another Inchon caused the Chinese to immobilize thousands of men on both coasts to guard against another such decisive landing far behind the front.

On the squadron's third day of sea operations, three planes were hit by rifle and machine gun fire on reconnaissance missions. One of them, flown by Captain Russell G. Patterson, Jr., was shot down behind the enemy lines but a FEAF helicopter rescued the pilot. First Lieutenant Alfred J. Ward was not so fortunate. His plane was riddled the following day by enemy fire and he crashed to his death in the midst of CCF soldiers.

Not until 22 January did the reconditioning of Bofu reach such an advanced stage that Lieutenant Colonel Fontana could set up his MAG–33 command post. VMF–312 moved in the next day and the first combat missions were launched to the vicinity of Seoul, 300 miles away. On the 24th General Harris established his headquarters. A few hours later VMF–214 and VMF–323 arrived from Itami, where they had put in an idle week, with no place to go, after their carrier duty. On the 26th, when they flew their first missions as land-based squadrons, MAG–33 was back in business and Bofu was a going concern.

No such claim could have been made for MAG–12 and K–1. Although Colonel Boeker C. Batterton set up his command post on 27 January 1951, two more weeks were to pass before the K–1 runway was fit for the flights to tactical aircraft. Meanwhile, the MAG–12 squadrons had to make out as best they could at K–9.

Marine Rice Paddy Patrols

Operations of the first few days demonstrated to 1st Marine Division ground forces that locating the enemy was more of a problem than defeating him. Obviously, the NKPA 10th Division had few if any of the advantages which make for effective guerrilla warfare. Far from receiving any voluntary support from the inhabitants, the Korean Reds had their own movements promptly reported to the Marines. Retaliations on civilians, such as burning mountain villages, were not calculated to improve relations. Nor did the enemy possess any of the other requisites for successful operations in an opponent's rear—a base, a source of supply, good communications, and a reliable intelligence system.

If it came to a fight, there could be little doubt about the outcome. But Marine staff officers must have been reminded of the old recipe for rabbit pie which begins, "First, catch your rabbit."

Such a situation called for systematic patrolling in all Marine zones of action. Secondary roads and mountain trails were covered by "rice paddy patrols." Numbering from four men to a squad, these groups ranged far and wide on foot in an area that was more often vertical than horizontal. On a single day the 5th Marines alone had 29 of these rice paddy patrols in action.[8] No better training for replacements could have been devised. Sometimes the men were on their own for several days, depending for supplies on helicopter drops. And while

---

[8] 5thMar *WD*, Jan-Feb 51.

casualties were light, there was just enough danger from sniping and potential ambushes to keep the replacements on the alert.

Roads fit for vehicles—especially the 75-mile stretch of MSR from Pohang to Andong—were under the constant surveillance of motorized patrols, each supported by at least one tank or 105mm howitzer. The farthest distance was 15 miles between the main Marine strong points at Pohang, Yongchon, Uisong, and Andong.[9]

Close air support was seldom needed against such an elusive enemy as the Marines faced. General Craig put in a request, however, for an air squadron to be based at Pohang or Pusan (Map 2). The two Marine all-weather squadrons, VMF(N)–513 and VMF(N)–542, were General Harris' first and second choices. They had been flying under Air Force (314th Air Division) control in the defense of Japan, a mission of dull routine and waiting for something to break the monotony of patrolling.

The twin-engined F7F–3N Tigercats of VMF(N)–542 were well equipped with electronics equipment for night interceptor work. VMF(N)–513 flew F4U–5Ns, the night-fighter modification of the latest Corsair.[10]

General Harris' plan for VMF(N)–542 to take over the duties of VMF(N)–513 at Itazuke had the approval of General Partridge. This made it possible to send the latter squadron to K–9 at Pusan to replace the VMF–311 jets, which in turn left for Itami to await corrections of engineering defects.

VMF(N)–513 flew its first combat missions from K–9 on 22 January. These consisted of routine armed reconnaissance flights and an occasional deep support mission for the Eighth Army. Not until the 25th did the squadron respond to a request from Marine ground forces. And out of 49 combat missions (110 sorties) during the remaining 6 days of the month, only three (10 sorties) were in support of the 1st Marine Division.

For routine operations the Marine ground forces found the support of VMO–6 sufficient. The nimble little OY observation planes were ideal for seeking out an enemy who had to be caught before he could be fought. And the helicopters did their part by dropping supplies, evacuating casualties, and laying wire.

---

[9] 1stMarDiv *HD*, Jan 51.

[10] The remainder of this section, except when otherwise noted, is based on 1stMAW *HD*, Jan 51 and VMF(N)–513 *HD*, Jan 51. See Glossary in Appendix A for explanations of aircraft designations.

Meanwhile, the 1st Marine Aircraft Wing strengthened its administrative ties with the 1st Marine Division. Although the two organizations had no common operational commander other than General MacArthur, they maintained a close liaison. Harris attached two TBM Avengers to VMO–6 for use as radio relays when ground-to-ground communications failed in the mountainous Pohang-Andong area. He also set up daily courier flights, at General Smith's request, to provide fast administrative liaison between widely dispersed Marine air and ground units in Korea and Japan.

Operations THUNDERBOLT *and* ROUNDUP

On 25 January two corps of the Eighth Army jumped off in Operation THUNDERBOLT. Advancing side by side, I Corps and IX Corps had orders to launch limited objective attacks and regain solid contact with the enemy, who was obviously preparing for a new offensive.

The EUSAK commander moved his CP from Taegu to Chonan (Map 1), the I Corps headquarters, in order to maintain personal control of the operation. He requested the Navy to step up offshore patrolling on the west coast as left-flank protection. Emphasis was also placed on aerial reconnaissance, both visual and photographic, as well as deep support directed by the Mosquitoes.

Even VMF(N)–542 at Itazuke had orders to conduct long flights to Seoul and maintain continuous patrols to report any attempt of the enemy to retire across the frozen Han River. The F7F–3N pilots shot up camp areas, convoys, and other lucrative targets but found no indications of large-scale crossings over the ice.[11] So varied were the missions of the squadron that it came as no surprise to be assigned to naval gunfire spotting for the USS *St. Paul* and the other British and American cruisers shelling Inchon.

All Marine tactical squadrons were in action on 28 January for the first time since December. Nearly two-thirds of the flights from Bofu and K–9 were diverted from armed reconnaissance to troop support. A typical operation was carried out by four VMF–312 planes on their second day of duty at Bofu. After reporting to MELLOW they were directed to Mosquito Cobalt, which had received a message that enemy troops were hiding in a village just north of Suwon, occupied that day

[11] VMF(N)–542 *HD*, Jan 51.

by the U.S. 35th Infantry. Under the Mosquito's direction they bombed, strafed, and napalmed some 40 buildings containing CCF soldiers.[12]

The fall of Suwon opened the way to Inchon and Seoul as Chinese resistance stiffened. Eighth Army progress was anything but reckless, but Ridgway had served notice on the enemy that he held the initiative and intended to keep it. Operation ROUNDUP followed on the heels of THUNDERBOLT. Merely a change in name was involved, for the advance continued at the same prudent pace without any important amendments to the original mission.

Action in the Pohang-Andong Zone

The Marines in the Pohang-Andong zone had their first brush with the elusive enemy on 22 January. A patrol of the 1st Battalion, 1st Marines, flushed out a guerrilla force near Mukkye-dong, several miles southeast of Andong (Map 5). Captain Robert P. Wray's Charlie Company deployed for action at sunset and shots were exchanged. The Marines had no casualties and the enemy could not have suffered many losses before he disappeared into the winter dusk.

Even at this early date the Korean Reds seemed to have lost confidence in their guerrilla operations. In a message dated 23 January taken from a prisoner, the commanding general of the II NKPA Corps directed General Lee Ban Nam to withdraw if possible. It read as follows:

> Get all of your troops out of the enemy encirclement and withdraw to north of Pyongchang without delay. Liaison team sent with radio. If you will inform us of your escape route we will assist by clearing your advance. If you cannot escape, stay in the rear of enemy as guerrillas.[13]

By the 24th an enemy drift southeast from the zones of the 1st and 5th Marines to 7th Marines territory was apparent. The 1/7 command post and Company A received scattered mortar fire late that afternoon. Action picked up the next morning when dawn brought an attack by an estimated 100 guerrillas on the regimental command post. After a brisk 90-minute fire fight the Korean Reds withdrew to the east, leaving seven dead behind and taking with them an unknown number of wounded.

---

[12] VMF-312 *HD*, Jan 51.
[13] 1stMarDiv *PIR* 116, Encl 1.

Later that morning the 7th Marines teamed up with the National Police against the Chiso-dong area. Nine bodies were counted as the 3d battalion seized its objective, but 1/7 was slowed by an entrenched enemy who offered an unyielding defense. The Marine battalion ground to a halt just one mile short of Chiso-dong and dug in for the night as artillery continued to pound the enemy. The air strikes on the 25th were flown by VMF(N)–513 and VMF–323, both based at K–9, but the pilots could not contact the FAC and had to make dummy runs over the enemy.

Marine planes and artillery cleared the way on 26 January as 1/7 advanced against scattered opposition. Nearly 400 guerrillas put up a ragged and futile resistance, but by 1530 Marine firepower prevailed and Chiso-dong was taken. The 2d Battalion had meanwhile occupied Hapton-ni, eight miles southeast of Topyong-dong (Map 5). A light enemy counterattack was repulsed with ease.[14] Altogether, enemy casualties for the day amounted to 161 KIA or POW.

The VMF–323 flight led by Captain Don H. Fisher and Captain Floyd K. Fulton's VMF(N)–513 flight merit recognition as the first successful instance of Marine air-ground cooperation since the Chosin Reservoir campaign.

While the 7th Marines served eviction notices on the enemy in its area, action elsewhere was light. Task Force Puller[15] hastened on the 26th to Chongja-dong, seven miles northeast of Uisong, to investigate a police report that 300 enemy had seized the town. A Marine attack, following an artillery preparation, was planned for 1500. Captain Thomas J. Bohannon led Able Company in but discovered that the shells had fallen on empty huts.[16]

During the next few days the rice paddy patrols continued to range over the countryside, searching out the enemy. Combat units were sent to areas where the G–2 red arrows indicated an NKPA build-up. On the morning of the 29th, the 5th Marines tried to organize an attack on a large enemy force reported near Chachon-dong, 12 miles west of Topyong-dong. Captain Jack R. Jones' Charlie Company, moving out at night in small foot patrols to maintain secrecy, scoured the area in an attempt to pin down the enemy.

---

[14] *HDs* of 1stMarDiv, 7thMar, VMF(N)–513, and VMF–323 for Jan 51.

[15] Organized from units of RCT–1 on 25 January when Colonel Puller was promoted to the rank of brigadier general. On that date Colonel McAlister assumed command of RCT–1.

[16] 1stMarDiv *HD*, Jan 50, 11.

Marine intelligence reports had warned of a dawn raid on the town for the purpose of plundering food from the inhabitants and arms from the Korean police station. First Lieutenant Richard J. Schening, executive officer, led a scouting force ahead of the main body to reconnoiter the area. He urged that a trap be set for the enemy, and the company commander has left a description of one of the most elaborate ambushes ever attempted by the Marines during the war:

> Well before daylight, a cordon was stealthily braided around Chachon-dong and we settled down to await the raiders. A later daylight inspection of the deployment showed that the men had done a splendid job of locating themselves so as to avoid detection. They were concealed under porches, beneath the brambles, and in the heaviest foliage and trees. But no guerrilla attack materialized, probably due to a "grapevine" warning of our movement and intent. . . . During the remaining days in the village we conducted extensive patrolling in an attempt to catch at least one guerrilla for our effort. Patrols were kept small to maintain secrecy. We even dressed Marines in clothing worn by the "locals" and sent them out in the hills with wood-gathering details. Larger patrols up to a platoon in size were sent on combat missions at night. One thing was certain: it was easier to talk about capturing guerrillas than it was to lay a hand on them.[17]

The elusiveness of the enemy could not always be credited to effective guerrilla tactics. Often it was due to distaste for combat. As evidence of low NKPA morale, Major Yu Dung Nam, a battalion commander, was condemned to death and shot late in January because he planned to surrender, according to POW testimony. Rations were at a bare subsistence level and typhus had claimed many victims.[18]

Unrelenting Marine pressure throughout the first week of February wore the guerrillas down until groups larger than 50 men were seldom encountered. On the 3d an NKPA second lieutenant surrendered voluntarily to a RCT–7 patrol and brought three of his men with him. NKPA morale had sunk so low, he divulged, that all ranks were striving only for survival. The division commander, Major General Lee Ban Nam, had apparently become a victim of acute melancholia. He spent nearly all his time, according to the prisoner, in the solitude of foxholes dug into the slopes of hills for added protection. There he brooded constantly over his predicament, but without arriving at any better solution than alternate hiding and flight.[19]

Certainly the military situation didn't offer much to gladden this

[17] Maj J. R. Jones ltr, 24 Jun 57.
[18] 1stMarDiv *PIR* 105, 5 Feb 51; 1stMarDiv *HD*, Jan 51, 13.
[19] *Ibid.*

Hamlet of the rice paddies, and the Marines continued to give him fresh causes for pessimism. His footsore remnants eluded RCT–5 only to stumble into the zone of RCT–1, northeast of Uisong. Neither rest nor sanctuary awaited them, for the 1st and 2d battalions penetrated into the mountains near Sangyong to surprise and rout a force estimated at 400 men.[20]

KMC Regiment Joins 1st Marine Division

Late in January the 1st KMC Regiment got into the fight after being attached once more to the 1st Marine Division by a EUSAK dispatch of the 21st. Lieutenant Colonel Charles W. Harrison headed a new group of Division liaison and advisory officers as the four KMC battalions moved out from Chinhae by LST and truck convoy to the Pohang area. Division OpnO 4–51 (26 January) assigned the regiment Sector F, astride the Yongdok-Andong road, which had been carved out of Sectors C and D, held by the 7th and 11th Marines respectively. The KMCs were ordered to conduct daily patrolling from positions near Yongdok, Chaegok-tong, and Chinandong and prevent enemy concentrations in their sector.[21]

Although the ROK Army and Eighth Army had the responsibility for supplying the KMCs, it proved necessary for the 1st Marine Division to cope with some of the gaps in equipment and rations. Contrary to a prevalent Western belief, Koreans did not subsist on a diet of rice alone. They were accustomed to having "side dishes" with their rice, such as eggs, meat, fish, or vegetables. Colonel Kim Sung Eun, the regimental commander, had an allotment of money for these purchases, but the sum was insufficient to meet inflation prices even if there had been enough food left in a district eaten bare. As a consequence, the KMCs had to get along on a monotonous and vitamin-poor diet until the ROK Army belatedly came to the rescue with issues of food for side dishes.

On 29 January the KMC Regiment opened its CP at Yongdok. Regimental OpnO 1 of that date divided Sector F into three parts, assigning the western, central, and eastern subsectors to the 3d, 1st,

[20] 1stMar *WD*, Feb 51.

[21] References to the 1st KMC Regt in this chapter are based on Smith, *Notes*, 1450–1458, and Col C. W. Harrison, *Narrative*, n.d., 1–15. Previous periods of KMC attachment to the 1stMarDiv are described in Vols I, II, and III of this series.

and 2d Battalions respectively. The 5th Battalion was attached to the 1st Marines and assigned to patrolling operations in the Andong area.[22]

The first few days of February saw a brief flurry of activity before NKPA guerrilla resistance breathed its last gasps. Reports that the remnants of the NKPA 25th and 27th Regiments were in flight toward the zone of the 5th Marines led to a concentration for a knockout blow, but the enemy stole away to the north in the vicinity of Topyong-dong. There he discovered that he had jumped from the frying pan into the fire. The 2d and 3d Battalions of the 1st Marines closed in from one side while the 1st and 3d Battalions of the KMC Regiment blocked roads in the vicinity of Samgo-ri and Paekcha-dong. Only a wild flight in small groups saved the guerrillas from annihilation.

The nearest approach to effective NKPA resistance was encountered on 5 February after the 1st and 2d KMC Battalions had established blocking positions in zone at the request of the 7th Marines, which was driving the enemy northward. A platoon-size patrol of the 2d KMC Battalion came up against Korean Reds dug in with 81mm mortars and heavy and light machine guns a few miles southwest of Yongdok. The KMCs were scattered with losses of 1 KIA, 8 WIA, and 24 MIA in addition to all arms and equipment, though the missing men returned later.

It was the single NKPA success of the entire campaign.

An assault was launched the following morning on this enemy stronghold by a composite KMC battalion, supported by four VMF(N)–513 aircraft which attacked with rockets and bombs. The largest combat of the guerrilla hunt appeared to be in the making, but again the enemy vanished after putting up an ineffectual resistance with small arms and mortars.[23]

An unusual air tactic was tested on 4 February in the 7th Marines zone when an interpreter in an R4D plane hailed the guerrillas by loud speaker in their own language with a demand that they surrender or suffer the consequences. Marine fighter-bombers were on station to back the threat, and about 150 supposed NKPA soldiers came in with uplifted hands while VMF–323 planes delivered the consequences to the holdouts in the form of bombs, rockets, and napalm. Unfortu-

[22] What would normally have been the 4th KMC Battalion was designated the 5th because the Korean word for 4th is the same as the word for death and is considered unlucky.

[23] Col C. W. Harrison, *Narrative*, 8–9; VMF(N)–513 *HD*, Feb 51.

nately, it developed that practically all of the prisoners were terrified civilians seeking an escape from the slave labor imposed upon them by the guerrillas.[24]

10th NKPA Division Scattered

Reports of enemy activity were received daily from Korean civilians and police, and seldom was a smaller number than "about two thousand" mentioned. In reality, Marine patrols had difficulty in tracking down as many as ten of the skulking, half-starved fugitives split up into small bands hiding in the hills. On 5 February the situation was summed up by General Smith in reply to a EUSAK request for an estimate of the time required to complete the Marine mission:

> The original 10th NKPA Div forces in the 1st Marine Division area have been dispersed into many groups, reduced to an effective strength of 40 per cent, and are no longer capable of a major effort while dispersed. . . . It is considered that the situation in the Division area is sufficiently in hand to permit the withdrawal of the Division and the assignment of another mission at any time a new force to be assigned the responsibility for the area assumes such responsibility and the 1st Marine Division can be reassembled.[25]

Patrolling continued as usual in all Marine regimental zones during the second week in February. Some units, such as the 11th Marines and the Division Reconnaissance Company, had made few enemy contacts throughout the operation. But at least the cannoneers had found good pheasant hunting and enjoyed a change in the bill of fare.

It was just as well that the tactical situation seldom made it necessary to call for air support at this stage, since the 1st MAW was once again in the throes of moves which will be described in the following chapter. Bofu had been only a temporary base for MAG–33 squadrons which were making another transfer to K–9 while MAG–12 completed its shift to K–1.

VMO–6 took care of the reduced air requirements of the Division adequately. Another helicopter "first" was scored when First Lieutenant John L. Scott received credit for the first night casualty evacuation by a HTL (Bell), which then had no instruments for night

---

[24] VMF–323 *HD*, Feb 51.
[25] Smith, *Notes*, 1378.

flying. For a harrowing moment, however, it would be hard to beat the experience of Captain Clarence W. Parkins and Corpsman R. E. Krisky. While they were flying a casualty to the hospital ship *Consolation,* the patient became wildly delirious. It took the combined efforts of pilot and corpsman to subdue him and make a safe landing.[26]

Any excitement would have been welcomed by the troops in general. For the area was as tranquil as if the guerrillas had never troubled its snowbound heights. Recently arrived Marines might have been pardoned for concluding that the NKPA 10th Division and its gloomy commander were but creatures of the imagination—phantoms to be compared to the crew of the *Flying Dutchman,* that legendary ship condemned to sail on endlessly until the Day of Judgment. The NKPA 10th Division also seemed doomed to perpetual flight as its ghostly survivors made their way from crag to crag of the remote ridgelines.

Thanks to the rice paddy patrols, the replacements were ready for combat and the Division was organizing a rotation draft for return to the States. Five officers and 600 men had already been selected on a basis of combat time, wounds received, and length of service. Major General Edward A. Craig, who commanded the first Marines to land in Korea, was given a farewell dinner and congratulated on his second star. Two new brigadier generals were named, with Lewis B. "Chesty" Puller relieving Craig as ADC and Gregon A. Williams accompanying him on the voyage back to the States. Captain Eugene R. "Bud" Hering, (MC) USN, was also returning with the gratitude of all Marines for his care of casualties in the "frozen Chosin" campaign.[27]

All Marine missions in the guerrilla hunt had been successfully accomplished, so that the Division could be relieved at any time by the 2d ROK Division. There were 120 counted enemy dead and 184 prisoners. Only estimates are available for the wounded, but there is no doubt that the total NKPA casualties were crippling. At any rate, the NKPA 10th Division was destroyed as a fighting force without accomplishing any of its objectives. Marine casualties from 18 January to 15 February were 19 KIA, 7 DOW, 10 MIA, 148 WIA, and 1,751 of a nonbattle classification, largely frostbite cases soon restored to duty.[28]

---

[26] VMO–6 *HD,* Feb 51.
[27] Smith, *Notes,* 1369.
[28] 1stMarDiv Periodic Operations Report (*POR*) 18 Jan–15 Feb 51, 71–159.

New Mission for the Marines

On 11 February, General Smith flew to Taegu to discuss the next Marine mission with General Ridgway. The EUSAK commander spoke favorably of employing the 1st Marine Division to relieve the 24th Infantry Division in the critical Han River corridor, where recent UN advances had been made. He also recognized the advantages of committing the Marines to the east coast, so that they could be held in readiness for an amphibious operation. A third possibility was the Yoju corridor of the IX Corps zone (Map 1). As "the most powerful division in Korea," said Ridgway, "the Marines would be astride what he considered the logical route for an expected enemy counterthrust." [29]

No decision was reached that day. At midnight the CCF attack materialized; and the central front was the area of decision, as Ridgway had predicted.

Naturally, the next mission for the Marines had to be reconsidered in the light of this development. On 12 February EUSAK warning orders alerted the 1st Marine Division to be prepared to move to Chungju, in the rear area of the IX Corps front where the heaviest CCF attacks were taking place. The Division was further directed to make an immediate reconnaissance of the Chungju area while the 1st KMC Regiment prepared for a move to Samchok on the east coast and attachment to the ROK Capitol Division. The following day brought orders from the Eighth Army to initiate these movements on 15 February 1951.[30] Thus the Pohang-Andong guerrilla hunt came to an end with the Marines on their way to new employment in the battle line of the Eighth Army.

---

[29] Smith, *Notes*, 1441–1445.
[30] 1stMarDiv *HD*, Feb 51.

CHAPTER IV

Operation Killer

The Move to the Chungju Area—Marine Planes in Action—Planning for the New Operation—The Jump-Off on 21 February—Stiffening of Chinese Resistance—General Smith in Command of IX Corps—The Advance to Phase Line ARIZONA*—JOC Air Control System Criticized*

THE CCF COUNTERATTACK which began northeast of Wonju on 11 February 1951 came in reaction to the unremitting pressure exerted during the previous month by the Eighth Army. Twice beaten during a recent six-week period and pushed back some 200 miles, EUSAK had shown amazing powers of recuperation.

"It is hard for me to put into words the magnificent competence, the fierce, combative, aggressive spirit of that force once it picked itself off the ground and waded back into the fight," commented General Ridgway in retrospect.[1]

During Operations THUNDERBOLT and ROUNDUP he had kept a tight rein on the Eighth Army by insisting on vigorous artillery preparations and close lateral contacts between units. On 10 February, however, caution was relaxed as CCF resistance suddenly collapsed west and south of Seoul. That day the U.S. 24th Infantry Division forged ahead 11,000 yards to occupy the port of Inchon and Kimpo Airfield, both so wrecked that weeks of repair would be necessary to make them operational. Seoul was within sight of the U.S. forces on the left bank of the Han when an aroused enemy struck back on the subzero night of the 11th.

Apparently the CCF drive on the central front had as its objective the relieving of UN pressure on the Seoul area to the west. The CCF

[1] Ridgway, *Memoirs*, 216.

40th and 66th Armies and NKPA V Corps struck in the IX Corps sector north of Hoengsong (Map 6). Two ROK divisions being dislodged by the initial blows, their retreat made necessary the withdrawal of other IX Corps units. As a consequence, Hoengsong had to be abandoned on 12 February to the Communists hammering out a salient northeast of Wonju.[2]

The UN forces were not bound by any unrealistic concept of holding ground to the last ditch. General Ridgway deemed it more important to inflict maximum punishment on the enemy at a minimum cost in casualties. While fighting on the defensive, he had already made up his mind to launch an offensive of his own to catch the Chinese off balance the moment their counterattack ground to a halt. His new limited objective operation emphasized the destruction of the enemy's fighting strength as the major objective rather than the acquisition of territory. A high attrition rate would preclude the Communists' capacity to hold and enable EUSAK commander to recover the critical hill mass north of Wonju. It was for this purpose, he informed Major General Bryant E. Moore, IX Corps commanding general, that the 1st Marine Division would be employed.

"The force which holds Wonju," he said, "has the situation in hand."[3]

The Move to the Chungju Area

The 1st Marine Division had instructions to report its order of march to the Eighth Army, and to keep the Taegu headquarters informed of progress. Meanwhile, the Marines were to remain under EUSAK operational control but would pass to IX Corps control at a date and hour to be announced.

General Puller flew to Chungju with a reconnaissance party on 13 February to look over the road and select CP sites. On the following morning Major Walter Gall's Division Reconnaissance Company arrived at Chungju for patrol duty, and movement by rail and road commenced on the 15th in accordance with Division OpnO 5-51, issued the day before.

The 1st Marines, with the 7th Motor Transport Battalion attached,

---

[2] IX Corps *Cmd Rpt*, Feb 51; 1stMarDiv *HD*, Feb 51; Smith, *Notes*, 1462–1465.

[3] EUSAK *Cmd Rpt*, Feb 51, Sec 1, 52. Comments by Gen O. P. Smith, USMC (Ret), 13Oct 57, and BrigGen A. L. Bowser, 14 Feb 58.

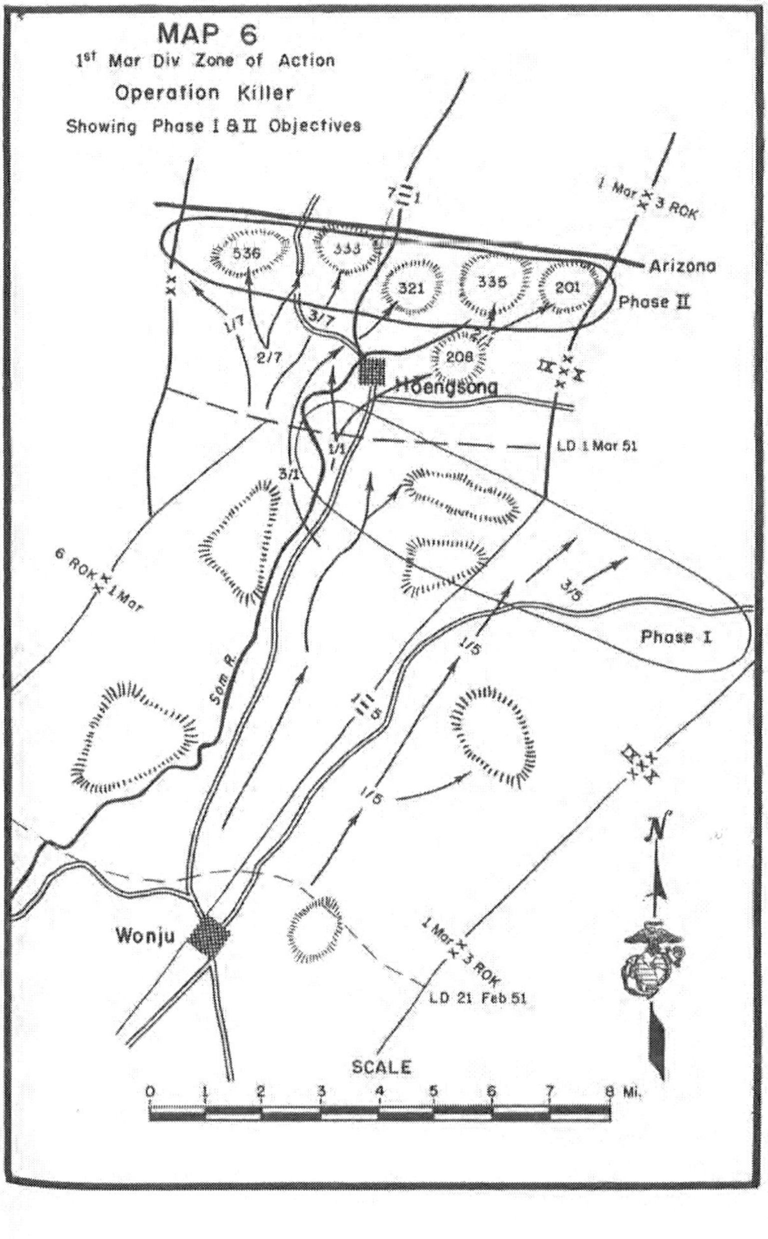
MAP 6
1st Mar Div Zone of Action
Operation Killer
Showing Phase I & II Objectives
7
1
1 Mar
3 ROK
536
333
321
335
201
Arizona
Phase II
1/7
2/7
3/7
2/1
208
Hoengsong
1/1
LD 1 Mar 51
3/1
6 ROK
1 Mar
3/5
Phase I
Som R.
1/5
5
1/5
N
Wonju
1 Mar
3 ROK
LD 21 Feb 51
SCALE
0
1
2
3
4
5
6
7
8 Mi.

led the motor march, and the 5th and 7th Marines followed in that order. Tracked vehicles were outloaded by rail from Andong and Pohang in a total of 67 flat cars. Owing to a shortage of cars, Company B and H&S Company of the 1st Tank Battalion made the move of 120 miles by road. These tankers claimed the all-time Marine Corps distance record for armor.[4]

While the Marine move was in progress, the CCF counterattack went on full blast along the central front. Driving southeast from the IX Corps area to the X Corps front, the Chinese cut off and surrounded the 23d Infantry of the 2d Infantry Division, USA. Colonel Paul Freemen and his men put up a fight that is one of the classics of the war. Supported by Marine and Air Force planes, they gave more fire than they received and held out until rescued by a tank column.[5]

February was also a transition period for Marine fighter squadrons which had been more or less on the move since the middle of January. Even before the transfer to Bofu, it had been decided that K–3, four miles south of Pohang, was to be the ultimate home of MAG–33. While awaiting completion of this field, VMFs–214, –312, and –323 would find temporary lodging at K–1, near Pusan, recently assigned to MAG–12.[6]

On 6 February, Brigadier General Thomas J. Cushman, assistant commanding general of the 1st MAW, radioed General Harris that K–1 would be ready to receive a squadron a day, starting on the 8th. Harris ordered Squadrons 323, 214, and 312 to make their moves on 8, 9, and 10 February respectively. Transport aircraft were to lift ground crews, extra pilots, and light equipment directly to K–1. Pilots had orders to fly combat missions en route.

By the 13th most of the vehicles, heavy equipment, and general supplies had been loaded on a train for Kobe, there to be transshipped on LSTs to Pohang. That same day Lieutenant Colonel Fontana set up his MAG–33 command post at K–3 and directed the three fighter squadrons to report from K–1.

The new field occupied a bench overlooking a wide, sandy beach. Built originally by the Japanese, the strip had 5,200 feet of concrete

---

[4] LtCol H. T. Milne, ltr of 3 Dec 57; 1stMarDiv *HD*, Feb 51.

[5] EUSAK *Cmd Rpt*, Feb 51; Comments by Gen M. B. Ridgway, USA (Ret), 4 Oct 57, and BrigGen A. L. Bowser, 14 Feb 58.

[6] The balance of this section, unless otherwise specified, is derived from the *HDs* for Feb 51 of the 1st MAW, MAG–33, VMF–214, VMF–312, VMF–323, VMF(N)–513, and VMF–311.

runway. The Air Force had extended it to 5,700 feet with pierced steel planking. This addition brought the end of the runway to the brink of a 60-foot drop-off—a hazard in the event of a "hot" landing to the northwest or too low an approach from the southeast.

Next to arrive at K–3 were the F9F–2Bs of VMF–311. Four weeks of adjustments at Itami had restored the jets to operative condition. An advance echelon went ahead to establish squadron living and operating areas, and the pilots ferried the 19 aircraft. Ground crews and equipment followed on transport planes.

Plans were made for VMF(N)–513 to move from Itami to K–3 before the end of the month. The other all-weather squadron, VMF(N)–542, now commanded by Lieutenant Colonel James R. Anderson, completed the transfer from Itami and Itazuke to K–1.

This field was also the destination of the photo pilots of Headquarters Squadron, 1st MAW, who flew their F7F–3P and F4U–5P fighters from Itami. Major Donald S. Bush commanded a unit, formerly a squadron, which had been one of the first aviation organizations to see action in Korea. Among its accomplishments were the preliminary beach studies for the Inchon and Wonsan landings.

With the completion of the moves of February 1951, the 1st MAW was again based on Korean soil. Fifteen types of Marine aircraft were being flown. For the heavy hauling, the R4D and R5D transports shifted troops and supplies. Included among the fighters were F9F Panthers, F4U Corsairs, and two models of F7F Tigercats—a stripped-down photo plane, and a radar-armed night fighter. Stinson OY Grasshoppers, TBM Avengers, and Beechcraft SNBs rounded out the list of conventional planes. Three types of rotary-wing aircraft were represented: the Sikorsky HO3S–1, and two models of the Bell HTL.[7]

Marine Planes in Action

By 15 February the brief CCF counterstroke had spent its force. Hoengsong had fallen to Communists who hammered out a salient on a 20-mile front extending as far southward as the outskirts of Wonju (Map 6). But the enemy's main purpose had failed of accomplishment, for the grip of the Eighth Army on Inchon and

[7] *Naval Aviation News*, Apr 51, 8.

Kimpo Airfield was not shaken. Nor did the Chinese gain a breathing spell in their preparations for a third great offensive as a followup to the December and January drives.

More by coincidence than design, the Fifth Air Force launched a new system of air tactics a few days after the beginning of the CCF counterstroke. Called "Reconnaissance Plan Fighter," it was based on a division of enemy-held Korea into 22 sections. Squadrons were given the mission of making hourly surveys of the same areas, day after day, until pilots became so familiar with them that any change hinting at CCF activity would be noticed at once.[8]

If these surveys revealed any sign of any enemy concentration, either of men or supplies, JOC scrambled special bombing strikes against them.

Although Marine fliers could readily see the advantages of covering the same ground daily, it made for monotony on reconnaissance missions. Only a highly unusual spectacle would startle a pilot, but First Lieutenant Weldon R. Mitchell blinked when he saw a camel in his gunsights.[9] Shaggy little Mongolian horses were no novelty as ammunition bearers, and after recovering from his first astonishment the VMF–311 pilot cut loose with .50 caliber machine gun slugs. As he suspected, the camel's pack contained ammunition and the animal was all but vaporized in the explosion.

Major Bush's photographic unit had an important part in keeping the enemy under constant surveillance. The Fifth Air Force directed on 16 February that all photo requests were to be screened by the Fifth Air Force's 543d Tactical Support Group at Taegu. Under the tactical coordination of this Group, the Marine unit was to fill all Navy and Marine requests. When not on such missions, it would be fitted into the Fifth Air Force photographic reconnaissance program.[10]

Pinpoint photos of suspected troop areas and such terrain features as defiles, junctions, detours, and bridges were in demand. The fact had to be faced that the enemy was almost unbelievably clever at camouflage and concealment. In one instance it was found that the Chinese had constructed bridge sections which they hid by day and put to use at night.[11] On another occasion they sank a bridge by means of weights so that it remained far enough beneath the surface

[8] VMF–323 *HD*, 15 Feb 51.
[9] *Naval Aviation News*, Apr 51, 8.
[10] 1st MAW and MAG–12 *HDs*, Feb 51.
[11] MajGen H. L. Litzenberg, ltr of 14 Jun 57.

of the water in the daytime to avoid detection by reconnaissance aircraft.

When the photo planes carried out missions as far north as MIG Alley [12] they flew in pairs. A fighter circled overhead to protect the photo pilot from an enemy air attack while he paid full attention to the task of "shooting" the terrain with his camera.

Planning for the New Operation

Adaptability to changing circumstances had already become perhaps the outstanding quality of the revitalized Eighth Army. No better example could be found than the evolution of Operation KILLER, which completed the cycle from concept to plan and execution in just three days.

On 18 February 1951, General Ridgway learned that the enemy was apparently withdrawing. IX Corps and X Corps units had probed forward that morning without meeting any opposition. Before nightfall the commanding general decided to launch a limited objectives offensive by the entire Eighth Army. He called a planning conference for the 19th and set the 21st as D-Day for the new operation.

The 1st Marine Division found itself detached from X Corps on the 19th and placed under the operational control of General Moore of IX Corps. This was not the first time in Marine Corps history, of course, when "soldiers of the sea" have fought alongside U.S. Army units in conventional land warfare. One of the best-known occasions was in World War I, when two Marine regiments distinguished themselves in France as a brigade of the U.S. 2d Infantry Division.

The Marines had been a part of X Corps in 1950, but always under tactical circumstances which permitted more or less independent operations with the support of organic aircraft. Now the Division was to be closely integrated with the other major IX Corps units, the 24th Infantry Division, the 1st Cavalry Division, the 6th ROK Division, and the 27th British Commonwealth Brigade. Marine calls for air strikes would continue to be made through JOC, as they had been since the Hungnam redeployment.

General Ridgway was on hand for the planning conference held

[12] MIG Alley was the name the American airmen gave the area along the Yalu River where Communist jets were active.

on 19 February in General Moore's CP at Yoju and attended by officers from IX and X Corps. General Smith, Colonel McAlister, and Colonel Bowser represented the 1st Marine Division.

The scheme of maneuver called for the Marines to relieve elements of X Corps and attack in a northeasterly direction from a line of departure north of Wonju (Map 6) through the Wonju basin. The object was to cut off enemy forces which had penetrated south and east of Hoengsong, and to recover control of the roads running eastward by seizing the high ground just south of the town.[13]

In the X Corps zone to the east, on the right flank of the Marines, the 7th Infantry Division was to attack to the north along the Yongwol-Pyongchang road. On the other Marine flank would be elements of the 6th ROK Division.

Simultaneous advances were planned for I Corps to the west, where patrols had found evidence that Seoul was lightly held.

Two U.S. Army units were designated at the 19 February conference to support the 1st Marine Division—the 74th Truck Company and the 92d Armored Field Artillery, then en route to the Chungju area.[14] These cannoneers and their commanding officer, Lieutenant Colonel Leon F. Lavoie, USA, were well and favorably known to the Marines, having given effective support during the Chosin Reservoir operations.

First Marine Division OpnO 6–51, issued on 20 February, directed the two assault regiments, the 1st and 5th Marines, to jump off at 0800 on the 21st and seize the first objective, the ridgeline about three and a half miles south of the high ground dominating Hoengsong (Map 6). RCT–1, with Division Recon Company and C/Engineers attached, was to pass through elements of the 2d Infantry Division in zone while RCT–5, with A/Engineers attached, passed through elements of the 187th Airborne Infantry, USA. RCT–7 had been designated the reserve regiment; but since it could not arrive from the Pohang-Andong area in time, a battalion of the 5th Marines was assigned this mission.[15]

The objective area was believed to be defended by the 196th Infantry Division of the 66th CCF Army and unknown elements of the 39th and 40th CCF Armies. Ahead of the Marines and other IX Corps units lay some uninviting terrain. Rocky heights and narrow valleys were laced by swift streams, the largest being the river Som, running

[13] 1stMarDiv *HD*, Feb 51, 1–2, 20.
[14] *Ibid.*
[15] *Ibid.*, 2, 22, and 1stMarDiv OpnO 6–51.

from northeast to southwest through a defile cutting across the western part of the Division sector. Bordering this twisting stream was the Wonju-Hoengsong "highway"—a poor dirt road even by Korean standards. Through the right half of the Division zone an even more primitive road, scarcely fit for vehicular traffic, wound northeast from Wonju.[16]

All Eighth Army forces were to be tightly buttoned up and to keep in close physical contact while maintaining integrity of units. Patrol observation and reconnaissance were stressed by the EUSAK commanding general, and even lack of opposition would not justify a unit in advancing ahead of schedule. Again, as in previous operations, real estate was to be secondary to the inflicting of maximum personnel and materiel damage.

On the eve of Operation KILLER, a message from IX Corps emphasized to all units the necessity for making sure "that no hostile force of sufficient strength to jeopardize the safety of your forces has been bypassed. Maintenance of lateral contact between all units is of prime importance." [17]

Marine ground force and aviation officers alike realized that the forthcoming offensive would be the first real test of the operational control of the 1st MAW by the Fifth Air Force and the Eighth Army. General Smith was uneasy about the outlook. On 13 February 1951, the day he was alerted for the move to Chungju, he had requested in a message to EUSAK that the 1st MAW be assigned to the support of his division. Both Marine ground and air officers, he said, believed that this change would fit into the JOC overall air control system without any disruption.[18] But no approval of General Smith's proposal had been received before D-Day.

The Jumpoff on 21 February

From the outset the transport and supply situation was a G-4 officer's nightmare. Heavy traffic broke the back of the MSR before the jumpoff, so that mud delayed the 5th Marines in reaching the line of departure (LD).

General Puller, the ADC, telephoned the Division commander

[16] *Ibid.*, 22–24.
[17] IX Corps msg in 1stMarDiv *In&Out#9*.
[18] CG 1stMarDiv, msg of 12 Feb 51 to CG EUSAK.

for a decision in the event that all elements of the regiment were unable to arrive in time. This question was already under discussion between General Moore and General Smith in the new 1st Marine Division CP, just opened at Wonju. After later reports of troop arrivals reached him, Smith decided with few minutes to spare that he would attack with only the troops able to reach the LD in time—three battalions of the 1st Marines, a battalion of the 5th Marines, two battalions of the 11th Marines, and a company of tanks. Moore then confirmed 1000 as H-hour and notified Puller of the decision.

The last-minute arrival of the 1st Battalion, 5th Marines, reminded Smith of the occasion in France, 32 years before, when the 5th Marines of World War I had to double-time across the wheat fields in order to attack on schedule at Soissons on 18 July 1918. For at Wonju the lone battalion scrambled out of trucks on the double and advanced without taking time for reorganization.[19]

Snarled traffic conditions were complicated by the arrival of high-ranking officers for the jumpoff. General MacArthur visited the zone of the 187th Airborne RCT, recently attached to X Corps. General Ridgway and General Moore were on hand when the Marines attacked. The EUSAK commander, surveying the scene from a snow-covered embankment, was disturbed to see a Marine corporal stumbling over an untied shoe lace while carrying a heavy radio.

"I hesitated just a moment," commented Ridgway, "knowing that what I wanted to do might be misconstrued as showmanship. Then I slid down the bank on my tail, landed right at his feet, knelt down and tied his shoe. Later, when this incident was reported in the States, there were some who did report it as a theatrical gesture. This was not true. It was purely an impulse to help a fighting soldier, a man in trouble." [20]

The Eighth Army commander was not the only one to see the advantages of tobogganing in terrain consisting of mud on the sunny slope of hills and snow on the shady side. When Captain Jack R. Jones' Charlie Company of 1/5 reached its first steep decline, the Marine leading the 2d Platoon slipped and fell in the snow, sliding about a hundred feet down the embankment. The man behind him profited from his example to make a purposeful slide, as did the rest of First Lieutenant William E. Kerrigan's men.[21]

[19] LtGen O. P. Smith, ltr of 28 Jul 53.
[20] Ridgway, *Memoirs*, 218–219.
[21] Maj W. E. Kerrigan, ltr of 25 May 57.

This was but one of the unwarlike incidents which enlivened the jumpoff of Operation KILLER. Seldom if ever have Marines taken part in an offensive which began so inoffensively, for 21 February was distinguished for lack of enemy resistance in the Marine zone. Only a few rounds of scattered rifle fire were encountered until late afternoon. Then the 1st Battalion, 5th Marines, leading the column of attack, had two long-distance fire fights before digging in for the night. Three Marines were slightly wounded and the enemy withdrew with such casualties as he may have suffered.[22]

The word "light" could never have been applied to the resistance put up by the weather and terrain. Lieutenant Colonel Joseph L. Stewart, commanding 3/5, described it as "a mixture of thawing snow, rain, mud, and slush." His men spent the night in foxholes half filled with water. Every one of them was "wet to the bones, including his clothes, parka, weapons, and ammo." [23]

The 1st Marines led the attacking column of battalions on 22 February, with 1/1 in the lead. More long-distance small-arms fire was encountered than on the first day, but again there were no close contacts with a retreating enemy.[24]

Stiffening of Chinese Resistance

Not until the 23d did either Marine regiment run into determined opposition. Then the 1st and 2d Battalions of the 1st Marines, advancing abreast, had a fight while going up against two hills of a ridge just south of the first phase objectives.

So far the Marines had found JOC air support satisfactory in quantity. The statistics show that the Fifth Air Force supported the Eighth Army during the first phase of Operation KILLER (21–24 February, inclusive) with an average of 600 sorties a day.[25] There was no room for complaint until the morning of the 23d, when an air strike the 5th Marines requested the preceding evening for 0800 failed to materialize on time. On this occasion the combination of an intense Marine artillery preparation and light enemy resistance compensated for lack of air support and the hill was taken with ease.

---

[22] 1stMarDiv *HD*, Feb 51, 4.
[23] Comment by Col J. L. Stewart, 25 Oct 57.
[24] 1stMarDiv *HD*, Feb 51, 4–5.
[25] Statistics are from EUSAK *Cmd Rpt*, Feb 51, G–3 Air Rpt.

That afternoon it took a brisk fight to evict an enemy in estimated battalion strength from log-covered bunkers on the second hill. This time JOC responded to Marine requests with two effective air strikes. Sixty Chinese dead were counted, and the Marines reported 1 KIA and 21 WIA.[26]

On the whole, however, the 5th Marines encountered only slight resistance. "About all we did was walk—walk—walk!" recalled Captain Franklin B. Mayer, commanding Easy Company of 2/5. "I don't think I've ever been so tired or footsore in my life—exception the retreat from Chosin, but not by much." [27]

On the 24th the 1st and 3d Battalions of the 5th Marines had little trouble in taking two hills designated as the main Phase 1 objectives. The 1st Marines on the left sent a tank and infantry patrol into Hoengsong after artillery preparation and an air strike. Captain Robert P. Wray, commanding Charlie Company of 1/1 and a platoon of tanks, entered the ruins of the town only to encounter machine gun and mortar fire from the hills to the west.[28]

When the antennae were shot off two tanks, Wray directed their 90mm fire by runner and knocked out the enemy positions. After proceeding further into the town, he was recalled by his battalion commander, Lieutenant Colonel Donald M. Schmuck, because an aerial observer had reported that Chinese were waiting to ambush the patrol.

An air strike was directed on them while Wray rescued several survivors of "Massacre Valley," northwest of Hoengsong, where a U.S. Army truck convoy had been ambushed during the recent CCF counterattack. The patrol returned before the ground had completely thawed. Only a few hours later a jeep passing over the same road was blown up by a land mine which killed the driver. This was one of the first object lessons illustrating the danger from enemy mines which were harmless until the midday sun thawed out the ground.

Chinese artillery fire from the hills north of Hoengsong accounted for one Marine KIA and four WIA late that afternoon before counter-battery fire by 2/11 silenced the enemy. This exchange ended the

---

[26] 2/1 *HD*, 16 Dec 51, 15–16.
[27] LtCol F. B. Mayer, ltr of 8 May 57.
[28] This account of the tank-infantry patrol is based on Maj R. P. Wray's ltr of 6 May 57.

first phase of Operation KILLER at dusk on 24 February with all preliminary objectives seized.[29]

Air support had been rendered, for the most part, by Fifth Air Force planes. This gave rise to grumbling by Marine ground forces, who felt that they had been unnecessarily deprived of their own close air support. The fact was, however, that U.S. Army and British Commonwealth troops also preferred Marine air and were outspoken about it. As a disgruntled Marine ground force officer put it, Marine air was "too good for our own good."

During the first phase of Operation KILLER most of the sorties by 1st MAW planes were in support of U.S. Army units. On 23 February the Marines flew 101 of the Fifth Air Force total of some 800 sorties for the day.[30] The experience of VMF–312 was fairly typical of the other Marine fighter-bomber squadrons. In the morning VMF–312 took part in a 16-plane strike behind the CCF lines. That afternoon two special flights of four planes each were scrambled in support of 2d and 7th Infantry Division units of X Corps. The following morning Major Daniel H. Davis, executive officer of the squadron, scrambled with four planes and reported to a FAC attached to the Canadian and Australian battalions of the British Commonwealth Division. These troops were engaged near Chipyong-ni in the hottest fight of the first phase of Operation KILLER. After the FAC marked the CCF strongholds with white phosphorus, the Corsairs came snarling in with napalm, rocket, and strafing runs just ahead of the infantry. The enemy was driven out of positions defended by 20mm antipersonnel fire, but Major Davis paid with his life on his eighth run when he lost a wing and crashed to his death.

General Smith in Command of IX Corps

On 24 February 1951 came the news that General Moore had suddenly died as the indirect result of a helicopter accident. The aircraft had plunged into the Han River, after hitting a telephone wire, and the IX Corps commander was rescued unhurt only to die of a heart attack half an hour afterwards.

[29] 1stMarDiv *HD*, Feb 51, 2, 5–6.

[30] EUSAK *Cmd Rpt*, Sec III, Bk 4, Pt 5, 23 and 24 Feb; 1st MAW *HD*, 22–24 Feb 51.

634040 O–62–6

Commander of the 8th Infantry Division in European operations of World War II, General Moore later became Superintendent of the U.S. Military Academy at West Point. As his successor, pending a permanent appointment, General Ridgway named General Smith to the command of IX Corps. When announcing this decision, the Eighth Army commander said, "General Smith is to be taken into their hearts in IX Corps, and, by definite action, made to feel that he belongs there." [31]

Marines with an interest in Corps history could recall only two similar occasions when Marines commanded major U.S. Army units. Major General John A. Lejeune had headed the 2d Infantry Division in World War I, and Major General Roy S. Geiger led the U.S. Tenth Army to victory during the closing days of the Okinawa operation after a Japanese shell killed Lieutenant General Simon Bolivar Buckner, Jr., USA.

On 24 February, with General Puller taking command of the 1st Marine Division, General Smith flew to Yoju by helicopter to begin his new duties. His military competence and complete lack of ostentation made him cordially accepted at the IX Corps CP. The following day General Ridgway arrived for a conference. Wishing to change the boundary between IX and X Corps, so as to orient the former more to the north, he directed the Marine general to reach an agreement with X Corps. He also asked for a recommendation as to future operations of the Marines, and General Smith replied that he knew of no better employment for his division than to continue attacking along the Hoengsong-Hongchon axis.[32]

The change in boundaries, as decided at a conference of corps commanders, meant that in the zone of the 1st Marine Division the 5th Marines on the right would be pinched out by the 3d ROK Division of X Corps. On the left, the zone was to be extended by bringing the 7th Marines into line to the left of the 1st Marines while the 5th Marines dropped back into reserve.[33]

Logistics became the better part of valor on 25 February as Ridgway called a halt in the fighting until enough ammunition, fuel, and other supplies could be brought up for a resumption of the attack toward the

---

[31] EUSAK *Cmd Rpt*, Feb 51, Sec 2, 23.

[32] MajGen Oliver P. Smith, *Chronicle of the Operations of the 1st Marine Division During the First Nine Months of the Korean War*, 1950–1951 (MS), (hereafter, Smith, *Chronicle*), 24–25 Feb 51.

[33] 5th and 7th Marines *HDs*, Feb 51.

final objective, Phase Line ARIZONA (Map 6). Napoleon's famous remark that mud should be recognized as a separate element was apt as violent rains turned all roads into swamps. Operations might have come to a standstill except for air drops. On the 25th the Combat Air Command flew 480.7 tons of freight and 1,004 passengers, followed by 604.9 tons and 1,193 passengers the following day.[34] Corps and Division engineers strove meanwhile with indigenous labor to repair the roads.

By a prodigious effort, enough progress in logistics was made so that the EUSAK commanding general could issue orders on 25 February for the second phase of Operation KILLER to commence on 1 March. He made it known that he was not satisfied with results so far. The assigned physical objectives had been taken, but the enemy's withdrawals had saved him from the full extent of the personnel and material losses Ridgway had hoped to inflict. He called on his staff officers, therefore, for plans aiming at a new operation "having the primary intent of destroying as many enemy and as much equipment as possible and, by continued pressure, allowing the enemy no time to mount a counteroffensive." [35] A secondary mission was that of outflanking Seoul and the area between Seoul and the Imjin River, "so that this territory may be taken either by attack from the east or by enemy default."

The name of the new drive was to be Operation RIPPER, and it was to jump off as soon as possible after the finish of KILLER.

The Advance to Phase Line ARIZONA

From newly won positions in the high ground south of Hoengsong, the Marines could look across the soggy plain to their Phase II objectives, the hills to the north of the battered town. Hoengsong occupied a valley at the confluence of two rain-swollen streams. Thus a triangular area of low, flat ground lay between the ruins and the hills which must be taken in the final phase of Operation KILLER. The 1st and 7th Marines were the combat units, with the 5th Marines in reserve. (The KMC Regiment, it may be recalled, had been temporarily detached for service with the ROK army.)

[34] FEAF *Operations History*, Vol II, 300–306; Comment by Col J. H. Partridge, 10 Dec 57.

[35] EUSAK *Cmd Rpt*, Mar 51, Sec 1, 53.

Before the 1st and 7th Marines could launch their combined attack, the latter had to fight its way up to the point of junction after relieving elements of the 6th ROK Division.[36] The scheme of maneuver then called for Lieutenant Colonel Virgil W. Banning's 3/1 to sideslip into the zone of Major Maurice E. Roach's 3/7, in order to be in position for the advance across the Hoengsong plain. This meant a crossing of the river Som for 3/1 and a combined assualt with 3/7 on the high ground along the west bank.

The problem of crossing the river, 200 feet wide and chest-deep at the most likely site, was turned over to Banning with the explanation that the engineer company supporting the regiment could not be diverted from road repairs. To meet this emergency Major Edwin H. Simmons, commanding Weapons Company of 3/1, produced a field manual with instructions for building a "Swiss bent bridge." [37] His Antitank Assault Platoon was given the task under the command of energetic Technical Sergeant Carmelo J. Randazzo, a veteran on his third enlistment.

There was no lack of trees for timbers, and rolls of telephone wire were sworn to be beyond salvaging by the battalion communications officer. The A-shaped bents, or trusses, were lashed together with wire and enthusiasm, then carried out into the ice-cold water to be attached to spars and stringers.

It was a great triumph for "war by the book." Before dark on 28 February two spans, one 120 feet long and another half that length, were linked by a sandbar in midstream. The improvised bridge stood up well next morning when the battalion crossed to the west bank. There 3/1 echeloned itself behind 3/7, which gained the first 1,000 yards under cover of a vigorous artillery preparation and belated air strikes.

On the left, Major James I. Glendinning's 2d Battalion of the 7th Marines ran into increasingly stubborn opposition from CCF mortar and small-arms fire. Before noon the attacks of both battalions of the 7th Marines were brought almost to a halt in difficult terrain which the Communists had booby-trapped. Neither artillery nor air strikes had a decisive effect against an enemy sheltered by log-covered

---

[36] Except when otherwise noted, this section is derived from the 1stMarDiv *HD*, Mar 51, 2–5; 7thMar *HD*, Mar 51, 2–6; IX Corps *Cmd Rpt* No. 4, Mar 51, 31–32; EUSAK *Cmd Rpt*, Sec 1, Mar 51, 53–59; LtCol Edwin H. Simmons narrative, n.d.; Comment by Col Wilbur F. Meyerhoff.

[37] FM 70–10, *Mountain Operations*, 41–46.

bunkers. So many delays were encountered that it was decided in mid-afternoon to postpone the advance until the following morning, 2 March.

Artillery and air strikes supported 2/7, 3/7, and 3/1 as they attacked at 0800 west of the river. Meanwhile, 1/7 patrolled on the division left flank while maintaining contact with the 6th ROK Division.

Apparently the enemy put up a hard fight only when he could not withdraw in time to avoid one. Resistance was light on the west bank, and east of the river Lieutenant Colonel Allen Sutter's 2/1, supported by tanks, had little trouble. His battalion linked up with 3/1 in the afternoon and dug in after taking its assigned objective, Hill 208, with casualties of three men wounded.

The only determined opposition of 2 March took place during the afternoon in the zone of 2/7. There the attackers could only inch forward over rocky terrain which the enemy defended, ridge by ridge, in spite of air strikes and 1,600 artillery rounds fired by the 11th Marines.

At daybreak on the 3d the men of the 1st and 7th Marines could look to the north and see their final objectives. Five hills lay along Phase Line ARIZONA from west to east—Hills 536 and 333 in the zone of the 7th Marines, and Hills 321, 335, and 201 in the zone of the 1st Marines.

The last two positions were in the path of 2/1, which seized them after several brisk fire fights. Casualties of three KIA and 28 WIA were incurred while inflicting losses of 70 counted CCF dead. The terrain gave 3/1 more trouble than the enemy in taking Hill 321, where the CCF troops had already begun their withdrawal.

It was in the zone of the 7th Marines that Communist resistance was hottest. The 1st battalion was summoned to cover the regimental left flank and aid in the attack of 2/7 on Hill 536 while 3/7 continued its struggle for Hill 333. Both battalions had their hardest fight of the entire operation that afternoon. They lost most of the 14 KIA and 104 WIA which the Division reported for 3 March, and the enemy still held the topographical crests.

The 1st Marines had reached the mopping-up stage on 4 March, while the 7th Marines prepared to go up against an expected last-ditch stand of the enemy on Hills 536 and 333. The parkas of the assault troops were powdered with snow as the men moved out to the attack at 0800, following an intensive artillery preparation. There was something ominous about the silence in the objective area, but no trap had been set for the attackers. The Communists actually had pulled out under cover of darkness, leaving behind only enough outpost troops for delaying operations.

Operation KILLER ended at nightfall on the 4th for the Marines, though mopping up continued throughout the following day. Total Marine casualties for the 8 days of fighting were 395—48 KIA, 2 MIA, and 345 WIA. Enemy losses amounted to 274 counted dead and 48 prisoners. It is certain, however, that the actual KIA and WIA figures were much higher, since the withdrawing Communists buried their dead and took their wounded with them.

Any evaluation of this limited objective operation must credit it with achieving its main purpose—keeping the Communists off balance while they were striving desperately to make ready for another great offensive (Map 7). This explains why the enemy as a whole put up a half-hearted resistance. He preferred to withdraw whenever possible and fight another day.

JOC *Air Control System Criticized*

Operation KILLER was the first real test of the JOC system as far as the Marines were concerned, and both the flying and ground-force Marines felt that it had shown grave shortcomings. Air support on 1 March proved so disappointing that General Puller, as temporary commander of the 1st Marine Division, reported the situation to General Shepherd, commanding FMFPac. His letter is quoted in part as follows:

> We are having very little success in obtaining Marine air for CAS [close air support] missions and practically no success in having Marine air on station for CAS missions. . . . Most of our CAS missions in the current operation have been Air Force or Navy Carrier planes. They do a good job and we are glad to have them, but our Marine air, with whom we have trained and operated, can do a better job. We have attempted to insure that Marine air would support us, and to cut down the delays in receiving such support, as evidenced by the attached dispatches. We have received no decision relative to our requests. Apparently, the answer is no by default.[38]

[38] Shepherd Papers, 27 Apr 51, Encl. 4. This is a file of documents in the Marine archives relating to problems of JOC control in the spring of 1951.

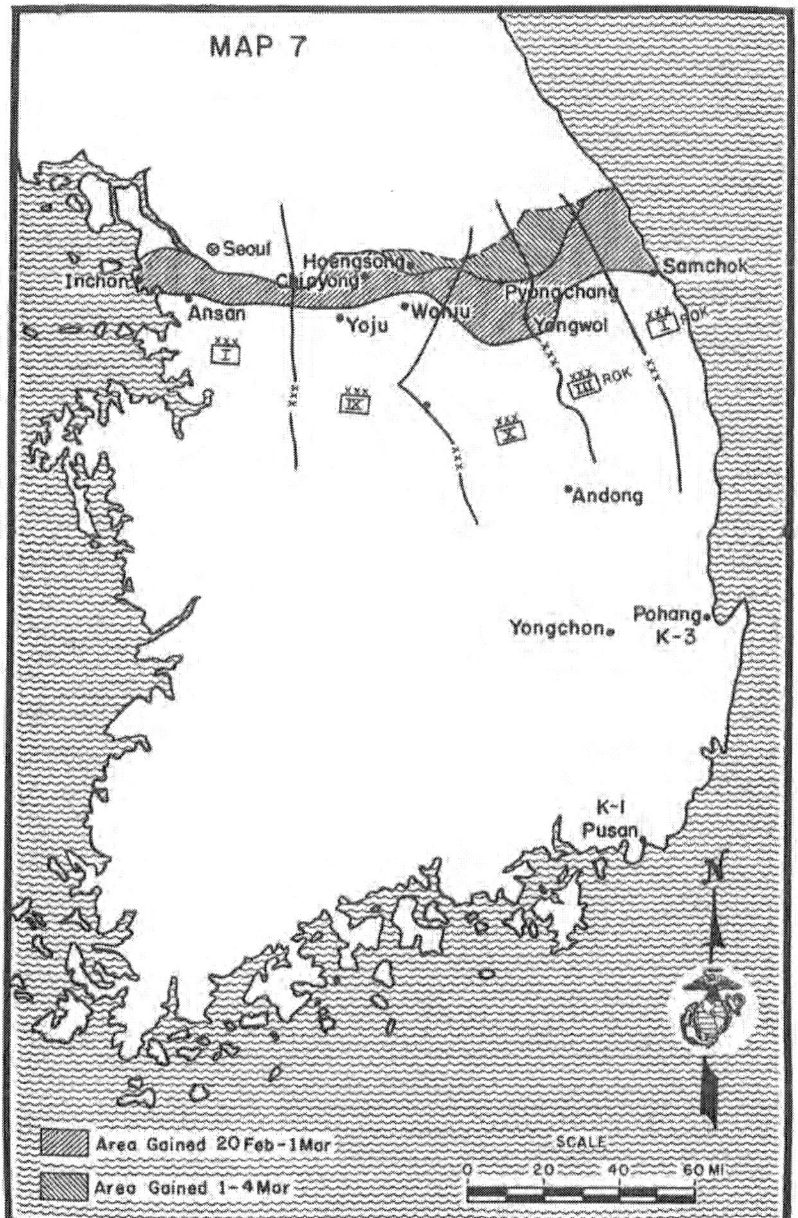

MAP 7
Seoul
Inchon
Ansan
Hoengsong
Chipyong
Yoju
Wonju
Pyongchang
Yongwol
Samchok
ROK
ROK
Andong
Yongchon
Pohang
K-3
K-1
Pusan
N
Area Gained 20 Feb-1 Mar
Area Gained 1-4 Mar
SCALE
0
20
40
60 MI

General Puller's report was obviously written for the record, since General Shepherd was present at the 1st Marine Division CP at the time. He witnessed personally the Marine attacks of 2 and 3 March and the air support they received. On the 3d, the day of heaviest fighting in the entire operation, there could be no complaint that few Marine aircraft supported Marine ground forces. The Corsairs flew 26 CAS sorties that day and cleared the way more than once for the 2d and 3d Battalions of the 7th Marines. The trouble was that air support as administered by JOC was so often late in arriving, even when requested the evening before. More than once the infantry had to go ahead with only artillery support. Such delays threw the whole plan of attack out of gear, for air and artillery had to be closely coordinated to be at their best.

General Shepherd had a series of talks with General Harris. Both then conferred with General Partridge, commander of the Fifth Air Force. They requested that he authorize the 1st MAW to keep two planes on station over the 1st Marine Division whenever it was engaged. General Partridge did not concur. He maintained that Marine aircraft should be available to him if needed elsewhere in an emergency. He did consent, however, to permit 1st MAW armed reconnaissance sorties to check in with DEVASTATE BAKER for any CAS requests.[39]

This conference did much to clear up the situation. On 5 March no less than 48 Marine sorties reported to DEVASTATE BAKER, though there was little need for them in mopping-up operations. And during the next two weeks an average of 40 sorties a day was maintained.

---

[39] Shepherd Papers, Encl 2, a ltr from Gen Shepherd to CMC, dtd 9 Mar 51. DEVASTATE BAKER was the call sign of Marine Tactical Air Control Squadron–2 (MTACS–2), which coordinated the assignment of aircraft to tactical air control parties (TACPS).

CHAPTER V

Operation Ripper

Light Resistance the First Day—Seoul Abandoned by Enemy—Second Phase of the Operation—Changes in 1st MAW Units—General MacArthur Visits Marine Battalion—1st KMC Regiment Returns to Division—38th Parallel Recrossed by Marines—Renewal of Division's CAS Problems

THE NEW IX CORPS COMMANDER, Major General William H. Hoge, USA, arrived at Yoju on 4 March 1951. He relieved General Smith the next day and a color guard turned out to render honors to the Marine commander when he returned by helicopter to his own Division CP. Upon Smith's arrival, General Puller resumed his former duties as ADC.

The jumpoff of the new operation was scheduled for 0800 on 7 March, so little time remained for last-minute preparations. The basic plan called for the drive of IX and X Corps toward the 38th Parallel on the central front. Protection was to be given on the left flank by I Corps in the area south and east of Seoul. On the right the ROK divisions had the mission of maintaining lateral security with a limited northward advance.

It was no secret that General Ridgway had been disappointed in the numbers of enemy soldiers put out of action during Operation KILLER. The primary purpose of RIPPER was to inflict as many Communist casualties as possible, and by means of constant pressure to keep the enemy off balance in his buildup for a new offensive. A secondary purpose was to outflank Seoul and the area between that city and the river Imjin, thus compelling the enemy to choose between default and a defense on unfavorable terms.[1]

[1] EUSAK *Cmd Rpt*, Mar 51.

CCF strategy in the early spring of 1951 was obviously conditioned by preparations for a third great offensive. The enemy's emphasis on caution is shown in a translation of a CCF training directive of this period:

> There must absolutely be no hasty or impatient attitude toward warfare. Consequently, even though we have a thorough knowledge of the enemy situation and the terrain, if one day is disadvantageous for us to engage in combat, it should be done the next day; if day fighting is disadvantageous, fighting should be conducted at night, and if engagements in a certain terrain are not to our advantage, another location should be selected for combat engagement. When the enemy is concentrated and a weak point is difficult to find, one must be created (by agitating or confusing them in some way), or wait until the enemy is deploying. Engagements must be conducted only when the situation is entirely to our advantage.[2]

Light Resistance the First Day

United Nations forces held a line extending across the peninsula from Inchon (Map 8) in the west by way of Hoengsong to the east coast in the vicinity of Chumunjin. The IX Corps order called for the 1st Marine Division to maintain lateral contact with the 1st Cavalry Division on the left and the 2d Infantry Division on the right. Hongchon and Chunchon, two of the main objectives of Operation RIPPER, lay directly in the path of the IX Corps advance. Both were important communications centers which could be utilized to advantage by the enemy for his forthcoming offensive.

The first phase line in the IX Corps zone was ALBANY. The Marines did not need a map to locate an objective just beyond Oum Mountain, a stark 2,900-foot peak about five and a half miles from the line of departure. Distance in this area was conditioned by terrain, and it was a natural fortress of wooded hills and swift streams that confronted the 1st Marine Division. Highways were conspicuous by their absence, and extensive maintenance would be required to utilize the Hoengsong-Hongchon road as a MSR. So few and poor were the secondary roads that it would sometimes prove necessary for vehicles to detour along the rocky stream beds.[3]

---

[2] IX Corps *PIR* #169; IX Corps *Cmd Rpt*, Mar 51, 21.

[3] This section, except when otherwise stated, is derived from the IX Corps *Cmd Rpt* and the 1stMarDiv *WD* for Mar 51.

The last offensive had not developed major or prolonged resistance at any point. Yet that possibility had to be anticipated by Marine planners. At least the enemy was an old acquaintance—the 66th CCF Army,[4] commanded by General Show Shiu Kwai. The 196th Division was on the left and the 197th on the right, with the 198th in reserve. These units were believed to comprise about 24,000 men.

Wednesday, 7 March, dawned cold and clear, with snow falling in the afternoon. The Hoengsong-Hongchon road, winding through Kunsamma Pass, paralleled the boundary between the two Marine assault regiments, the 7th Marines on the left and the 1st Marines on the right. They jumped off to attack in line abreast, employing all three battalions when the broken terrain permitted, while the 5th Marines continued its patrolling activities in the Hoengsong area as Division reserve.

The 11th Marines had to ration its artillery ammunition, owing to supply shortages. JOC came to the rescue nobly by ordering MAG–33 to place 11 flights of four planes each at the disposal of DEVASTATE BAKER on D-minus-one. These aircraft reported at hourly intervals to work over targets in the area of the next day's Marine operations. For the ground forces, it was an embarrassment of riches. They had more air support than they could use at times, and DEVASTATE BAKER sent the surplus to hit reserve concentrations and other targets of opportunity in the enemy's rear.[5]

The two Marine assault regiments met with light resistance on D-Day. Both took their objectives with little trouble except for scattered bursts of machine gun fire. Total casualties for the day were seven men wounded.

It was like old times to have Marine planes supporting Marine ground forces. MAG–12 aircraft were on the job the next day, when CCF resistance stiffened without ever becoming serious. Heavy CCF mortar and small-arms fire was received by 3/1, supported by Company A of the 1st Tank Battalion. Well placed rounds by the 11th Marines silenced the enemy in this quarter, and both battalions of the 1st Marines reached their assigned positions by nightfall.

The second day's advances gave added proof that the enemy was up to his old trick of putting up a limited defense while pulling back

---

[4] A CCF Army, composed of three or four infantry divisions supported by artillery, is comparable to a U.S. corps.

[5] Compilation from 1st MAW Sqdns' *HDs*, 7 and 8 Mar 51.

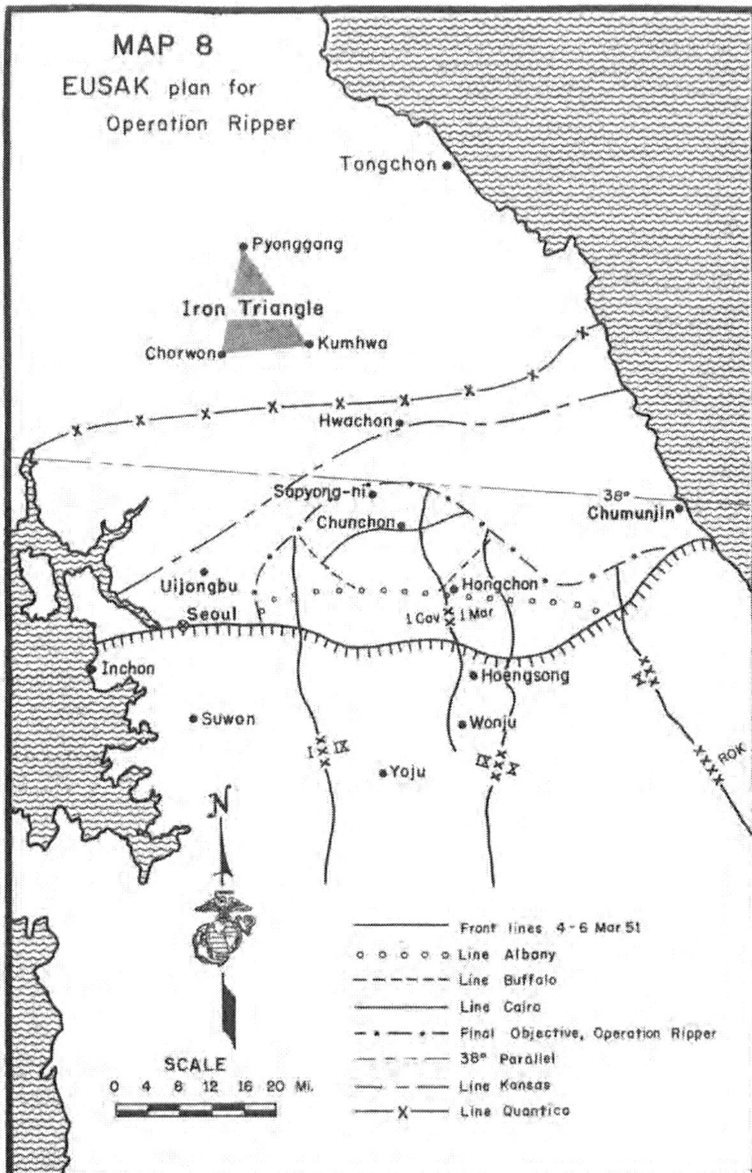

MAP 8
EUSAK plan for
Operation Ripper
Tongchon
Pyonggang
Iron Triangle
Chorwon
Kumhwa
Hwachon
38°
Sapyong-ni
Chunchon
Chumunjin
Uijongbu
Hongchon
Seoul
1 Cav
1 Mar
Inchon
Hoengsong
Suwon
Wonju
I
IX
Yoju
IX
X
ROK
N
SCALE
0 4 8 12 16 20 Mi.
Front lines 4-6 Mar 51
Line Albany
Line Buffalo
Line Cairo
Final Objective, Operation Ripper
38° Parallel
Line Kansas
Line Quantico

before the Marines could come to grips. Log bunkers were ideal for these CCF delaying tactics; each was a little fortress that might enable a squad to stand off a company while larger CCF units withdrew.

The Marine assault troops found that a preliminary treatment of napalm from MAG–12 aircraft, followed by well-aimed 90mm fire from the tanks, did much to soften up the bunkers for an infantry attack with hand grenades.

Company A of the 7th Marines had the hardest fight of all Marine units on 8 March. Second Lieutenant Clayton O. Bush and the 2d Platoon led the attack on the company objective, a hill mass to the left of Oum San. With 300 yards still to be covered, the Marines were pinned down by well aimed CCF small-arms and mortar fire, including white phosphorus. A high explosive shell scored a direct hit on the platoon, killing two men and wounding three. Bush was evacuated, with his right arm mangled. First Lieutenant Eugenous Hovatter, the company commander, ordered the 1st Platoon to pass through the 2d and continue the attack with air and tank support. The flat-trajectory fire of the 90mm rifles did much to help the company clear the enemy from the hill and the 7th Marines reached all assigned regimental objectives for the day.[8]

The Marine advance came to a halt on 9 March to wait for Army units to catch up on the right. While the 2d Battalion of the 1st Marines took blocking positions, the 1st and 7th Marines sent out patrols on both flanks in an effort to regain lateral contact. For the next two days, 1st Marine Division operations were limited to patrolling. A good deal of activity took place in the rear, however, as Marine service units moved up to Hoengsong.

Seoul Abandoned by Enemy

The advance was resumed on 11 March after the relief of 2/1 by Major Walter Gall's Division Reconnaissance Company, reinforced by a platoon of tanks. Although the enemy withdrew from most of his positions without putting up much resistance, a patrol of George Company, 3/1, had a hot fire fight on Hill 549. Opening fire at 50 yards from camouflaged, log-faced bunkers, the Chinese killed one

[8] 7thMar *HD*, Mar 51; VMF–323 *HD*, Mar 51; Capt Clayton O. Bush, ltr of 11 Aug 57.

man and wounded nine. Marine infantrymen, supported by flat-trajectory 90mm fire, approached within grenade-throwing range to destroy five bunkers and kill 16 of the defenders. As the patrol withdrew, it called on the 11th Marines to finish the job. The cannoneers were credited with several direct hits.[7]

Chinese resistance continued to be light as the two Marine regiments occupied rather than seized ground on 12 and 13 March. By the 14th all units were dug in along Phase Line ALBANY.

CCF withdrawals were also reported by other Eighth Army units. On 15 March a patrol from the 1st ROK Division of I Corps found Seoul abandoned by the enemy. The Chinese Reds had made their choice and UN forces took over a devastated city with some 200,000 civilians dragging out a miserable existence in the ruins. Dead power lines dangled over buildings pounded into rubble, and even such a famous landmark as the enormous red, brass-studded gates of the American Embassy Compound had been destroyed.

It was the fourth time that Seoul had changed hands in 9 months of war. Air reconnaissance having established that the enemy had withdrawn about 15 miles to entrenched positions in the Uijongbu area, General Ridgway enlarged the mission of I Corps by directing it to advance on the left of IX Corps.[8]

During the first phase of Operation RIPPER, from 7 to 13 March, counted casualties inflicted on the enemy by X Corps amounted to 6,543 KIA and 216 POW. IX Corps casualties during the same period were reported as 158 KIA, 965 WIA, and 35 MIA—a total of 1,158.[9]

The total strength of the Eighth Army (less the Marines) was 185,229 officers and men in March 1951. Adding the 25,642 of the 1st Marine Division, the 4,645 of the 1st Marine Aircraft Wing, plus 11,353 of the American Air Force and 355 attached from the U.S. Navy, 227,119 Americans were serving in Korea. This does not count 13,475 South Koreans serving in various U.S. Army divisions.[10]

The 1st Marine Aircraft Wing, with an authorized total of 728 officers and 4,216 enlisted men, had an actual strength of 626 and 4,019 respectively on 31 March 1951. Of an authorized 29 officers and

---

[7] 1stMarDiv *HD*, Mar 51, 8.
[8] EUSAK *Cmd Rpt*, Mar 51, 68–70.
[9] IX Corps *Cmd Rpt*, Mar 51, 35–37.
[10] *Ibid.*

93 enlisted men from the Navy, 22 and 83 in these categories were on duty.[11]

Troops to the number of 21,184 from the ground forces of other United Nations were represented as follows:

United Kingdom and Australia	10,136
Turkey	4,383
Philippines	1,277
Thailand	1,050
Canada	858
New Zealand	816
Greece	777
France	749
Belgium-Luxembourg	638
Netherlands	500
	[12] 21,184

The 249,815 officers and men of the ROK Army make a total UN combat strength of 493,503. There were an additional 671 in three noncombat units: the Danish hospital ship *Jutlandia,* 186; the 60th Indian Ambulance Group, 329; and the Swedish Evacuation Hospital Unit, 156.[13]

Chinese forces in Korea, including confirmed and probable, totaled 16 armies, each comparable to a U.S. corps. Eight others were reported. Assuming that these CCF units averaged a field strength of 24,000 officers and men, the total would have been 384,000 for the 16 armies. The reorganized forces of the North Korean People's Army (NKPA) were credited with five armies. Adding these 120,000 men to the 16 Chinese armies, the enemy had 504,000 troops in Korea plus whatever might have been the strength of the eight reported armies and the rear area service elements. In addition, large reserves stood just over the border in Manchuria.[14]

Second Phase of the Operation

With scarcely a pause on Phase Line ALBANY, the second phase of Operation RIPPER began on 14 March with a drive toward Phase Line

[11] FMF Status Sheet, 31 Mar 51.

[12] EUSAK *Cmd Rpt,* Mar 51, Plate 17. Figures do not include personnel in hospitals or clearing stations. UK and Australian statistics are not separated in available records.

[13] *Ibid.*

[14] EUSAK *Cmd Rpt,* Mar 51, 95.

BUFFALO (Map 8). Despite the difficulty of maneuver over muddy roads in mountainous terrain, an Eighth Army directive of that date called for a pincers movement to be initiated by means of a rapid advance of the 1st Marine Division on the right and the 1st Cavalry Division on the left. It was hoped that the Chinese forces south of Hongchon might be trapped and destroyed after the 187th Airborne Regiment cut off escape by landing north of the town. General Ridgway having urged his corps commanders to stress maneuver, IX Corps sent this message to division commanders:

> It is desired that more use be made of maneuver within and between division zones with a view toward trapping and annihilating the enemy through such maneuver. Movements should be less stereotyped; it is not desirable that units always advance toward the enemy abreast. Well planned and successfully executed maneuver using companies and battalions has previously been conducted; this should be extended to include regiments. This Headquarters is studying and will continue to study and order into execution the maneuver of divisions with the same intent and purpose.[15]

Both the 1st Marine Division and 1st Cavalry Division made rapid progress toward Phase Line BAKER (Map 9), established by IX Corps as an intermediate control. Unfortunately for the purposes of the envelopment maneuver, the Chinese withdrew from the Hongchon area before the pincers could close or the 187th Airborne make an air drop. CCF resistance was confined to machine gun fire covering hasty retirements. The 7th Marines on the left occupied its objective without once calling for air or artillery support, and the 1st Marines was virtually unopposed. Division casualties for the 14th were six men wounded.

Flash floods and roads churned into hub-deep mud were the greatest enemies of progress. Serious as the resulting supply problems were, they might have been worse but for the efforts of the recently organized Civil Transport Corps formed from members of the ROK National Guard who lacked the necessary training for military duties. There was no shortage of willing indigenous labor, for these auxiliaries received pay as well as rations and clothing. Formed into companies, they worked with the wooden "A-frames"—so-called because of their shape—used from time immemorial in Korea as a rack for carrying heavy burdens.

The Civil Transport Corps proved to be a boon for the Eighth

---

[15] IX Corps msg IXACT–1053 (122100) in 1stMarDiv *HD*, Mar 51.

USA Photo SC 356353

Flight From The Foe—A Korean carries his aged father across the icy Han River in the flight southward to escape the advancing Communist troops in their drive of January, 1951. A1

634040 O–62–7

USMC Photo A 158073

At the Critical Moment—Above, a Marine tank blasts an enemy emplacement while a rifleman stands by for the final assault; below, Marine riflemen hug the ground as they advance under fire during Operation RIPPER *in March, 1951.* A2

USMC Photo A 6862

USMC Photo A 8064

USMC Photo A 161016

Attack—Above, Marine machine gunner climbs a ridge while a flame-thrower operator burns straw to deny cover to the enemy; below, a young rifle-man hurriedly reloads after emptying a clip at Chinese Communist soldiers. A3

USMC Photo A 159023

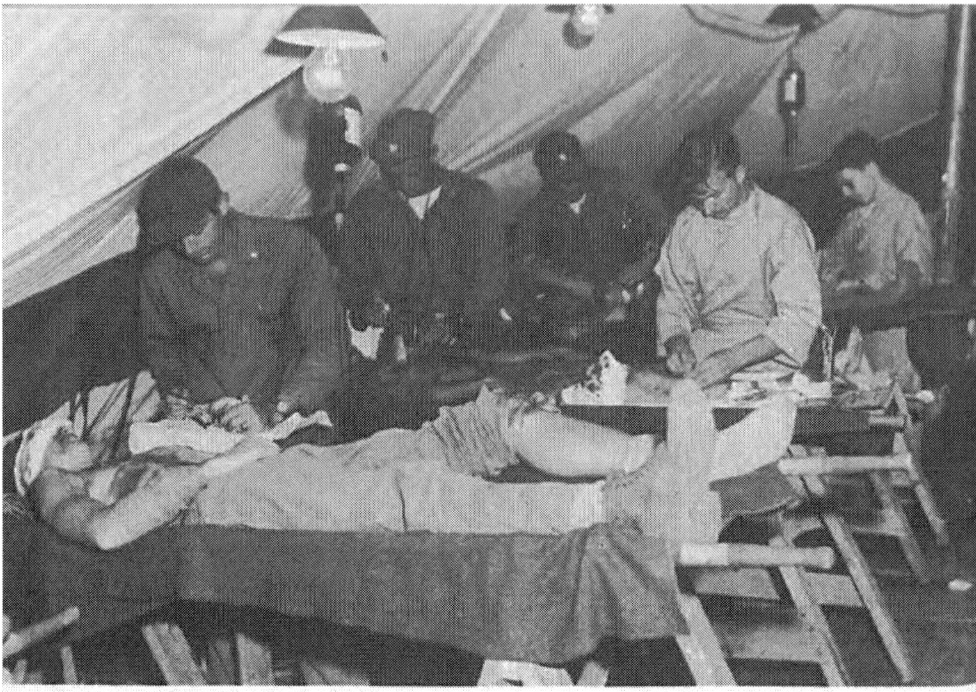

USMC Photo A 155354

Supporting The Fighting Man—Above, a group of surgeons are at work in a minor surgery ward; below, a cargador train carries ammunition and rations to the front-line companies. A4

USMC Photo A 9857

USMC Photo A 8756

Moving Out—Above, a Marine rifle company, C/1/1, moves to a rest area after almost two months of fighting; below, camouflaged Korean Marines on patrol pause to check the route.

USMC Photo A 161982

USMC Photo A 9458

USMC Photo A 9457

"Calling DEVASTATE BAKER!"—*Above, forward air controllers with attacking companies use the call sign of supporting air; below, a dependable Corsair responds to the infantry's call.* A6

USMC Photo A 133540

USMC Photo A 346714

"Launch and Attack"—Above, a Panther jet takes off on a close air support mission; below, the attack is pressed at close range on a stubborn enemy by a Tigercat (left) and a Corsair (right). A7

USMC Photo A 130948

USMC Photo A 7283

USMC Photo A 158968

Supporting The Troops—Above, a tank commander emerges from his steel shell in order to scan the hills for targets; below, a rocket battery harasses the enemy near Chunchon in May, 1951. A8

USMC Photo A 155643

USMC Photo A 159109

Power-Packed Punch—Above, a Marine tank topples Korean trees while moving into position to support an attack by fire; below, a rocket battery firing a ripple at night. **A9**

USMC Photo A 169791

USMC A 3597

Prisoners of War—Above, Marines guarding captured prisoners awaiting interrogation; and below, after hearing the familiar order to "saddle up" a Marine awaits the word to move out. A10

USMC A 159197

USMC A 9769

Random Scenes—Above, the result of what happens when a jeep runs over an anti-tank mine; and below, a Marine fire direction control center in operation.

USMC A 155903

USMC Photo A 159962

Aerial Workhorse—Above, after refueling at a mountainside fuel dump, a Marine helicopter loads men for evacuation; below, pre-fab bunkers are unloaded on a Korean hillside.

USMC Photo A 168493

USMC Photo A 159962

Have Chopper, Will Travel—Above, Marine helicopter demonstrating rescue technique; and below, staff officers board helicopters in preparation for front line inspection trip. A13

USMC Photo A 168493

USMC Photo A 160193

USMC Photo A-158811

Fighting Faces—Above, left, PFC H. W. Hodges pauses to drink from a "refrigerated" spring; above, right, PFC J. W. Harnsberger relaxes on the MLR; below, Sgt E. L. Whitlow and Capt W. F. Whitbeck scan the front for signs of enemy action. **A14**

USMC Photo A 167904

USMC Photo A 131819

Family Reunion—Above, Col W. S. Brown, CO 1st Marines, visits his son, a Marine corporal; below, left, Capt G. H. Parker directs an air strike from a tree top observation post; below, right, PFC K. L. Spriggs receives his Purple Heart. A15

USMC Photo A 132390

USMC Photo A 158743

USMC Photo A 158079

Necklace of Boots—Above, a South Korean cargador carries a necklace of boots to a front line company; below, new thermal boots are issued to combat troops. **A16**

USMC Photo A 157927

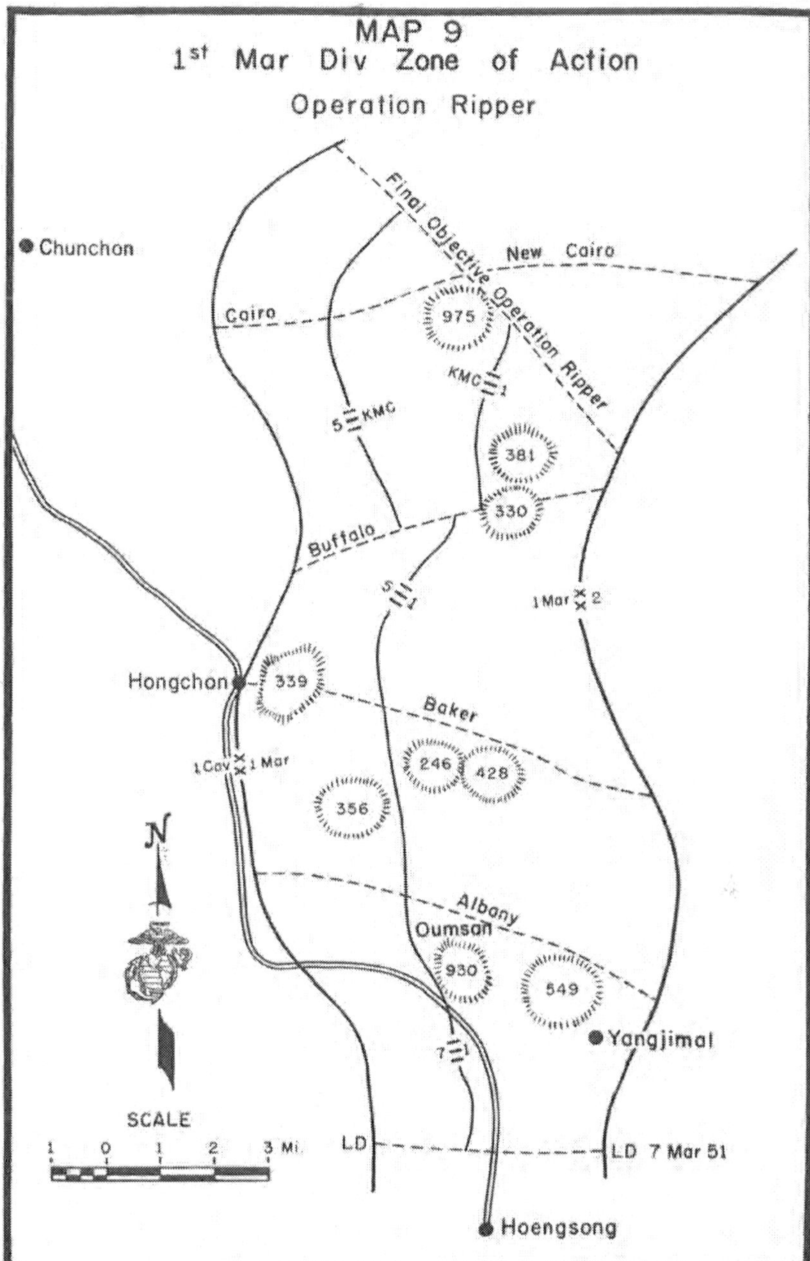
MAP 9
1st Mar Div Zone of Action
Operation Ripper
Chunchon
Final Objective Operation Ripper
New Cairo
Cairo
975
KMC
1
5
KMC
381
330
Buffalo
5
1
1 Mar
2
Hongchon
339
Baker
1 Cav
1 Mar
246
428
356
N
Albany
Oumsan
930
549
Yangjimal
7
1
SCALE
1
0
1
2
3 Mi.
LD
LD 7 Mar 51
Hoengsong

634040 O–62–8

Army. Veteran porters could manage a load of 100 to 125 pounds over ground too rugged for motor vehicles. Several hundred were attached to each regiment during Operation RIPPER.

Any lingering hope of rounding up Chinese prisoners in the Hongchon area was blasted on the 15th when evidence of Chinese withdrawal came in the form of an enemy radio message intercepted at 1230. "We cannot fight any longer," the translation read. "We must move back today. We will move back at 1400. Enemy troops will enter our positions at 1300 or 1400. Enemy troops approaching fast." [16]

Hongchon fell without a fight to the 1st Battalion of the 7th Marines on the afternoon of 15 March. Major Webb D. Sawyer, the commanding officer, sent a motor patrol through the ruins without flushing out any Chinese, but on the return trip a truck was damaged by a "butterfly bomb." This led to the discovery that the Hongchon area was covered with similar explosives that had been dropped by U.S. planes to slow up the CCF counterattacks in the middle of February.

Butterfly bombs, so-called because of the whirling vanes that controlled the drop and armed the 4-pound projectiles, could be set for air or ground bursts. Usually, however, they were dropped in clusters to remain on the ground until disturbed. Apparently the enemy had not troubled to clear them from the Hongchon area, and that three-day task was begun by Company D of the 1st Engineer Battalion while 1/7 seized the high ground northwest of the town.[17]

Changes in 1st MAW Units

Air support for the ground forces continued to be more than adequate in quantity. Since the agreement between Generals Partridge and Harris, 40 1st MAW sorties a day had been allotted to the 1st Marine Division. The timing was not all that could have been asked on occasion, but on the whole the Marine infantry had no complaint.

The 1st MAW had undergone an extensive reshuffling of units on the eve of Operation RIPPER. VMF(N)-542 was sent back to El Toro,

[16] CO 7thMar msg to CG 1stMarDiv; CO 1stMar, 1300 15 Mar 51.

[17] LtCol W. D. Sawyer, interv of 30 Aug 57; Field Manual 9-1980, AF 136-137. This was not the first nor the last time that M-83 fragmentation (butterfly) bombs became a deadly nuisance to friendly forces.

California, for conversion to F3D jet all weather fighters. The squadron's F7F-3N's and two F-82's were left with VMF(N)-513. The former commanding officer of 542, Lieutenant Colonel James R. Anderson, assumed command of 513. He relieved Lieutenant Colonel David C. Wolfe, who returned to the States.

The California-bound cadre of 542 included 45 officers and 145 enlisted men under Major Albert L. Clark. VMF(N)-513 was now a composite squadron, attacking from K-1 during the day with its F4U-5N's and at night with its F7F-3N's.

Another change took place when VMF-312 replaced VMF-212 on the CVE *Bataan*. The former squadron had been preparing for weeks to perform carrier duty, so that the change was made without a hitch. VMF-212, after nearly 3 months on the *Bataan*, established itself at K-3 under a new commanding officer, Lieutenant Colonel Claude H. Welch, who relieved Lieutenant Colonel Wyczawski.

The transportation jam in Korea made necessary the permanent assignment of a VMR-152 detachment to 1st MAW Headquarters. Transports had heretofore been sent to the Wing on a temporary basis and returned to Hawaii when missions were completed.

Mud and inadequate rail facilities doubled the demands on FEAF's aerial supply of combat forces. The Wing's courier service to Marine air and ground forces scattered over Korea reached the limit of its capabilities. As a solution General Harris requested a five-plane VMR-152 detachment on a long-term assignment, and Colonel Deane C. Roberts took command of this forward echelon at Itami.

It was now possible to handle cargo and troop transport at the cargo and passenger terminals of all Marine air bases. In one 4-day period, early in April, approximately 2,000 replacement troops were lifted from Masan to Hoengsong by the five R5Ds. About a thousand rotated veterans were flown back on the return trips.[18]

A further change involved the coordination of the Wing's air control organizations. As the enemy's air power increased, obviously the problems of UN air defense multiplied. At K-1 the Marine Ground Intercept Squadron-1 (MGCIS-1) and the Air Defense Section of Marine Tactical Air Control Squadron-2 (MTACS-2) were hard pressed to identify and control the hundreds of aircraft flying daily over Korea.

---

[18] Unless otherwise specified, references to changes in 1st MAW units are derived from the historical diaries of the units concerned for February, March, and April 1951.

There was no adequate system of alerting these air defense stations to the effect that planes were departing or incoming. Many of them failed to send out their standard identification friend or foe (IFF) signals; and those that did so were still suspect, since U.S. electronics equipment on UN planes had fallen into enemy hands. As a consequence MGCIS–1 was kept busy vectoring air defense fighters to verify that certain bogeys were friendly transports, B–29s, or enemy bombers.[19]

In an effort to cope with the situation, General Harris requested that another Marine ground control intercept squadron, MGCIS–3 be sent to Korea. He desired that Marine Air Control Group–2 (MACG–2) also be made available to coordinate the Wing's air control functions. These units sailed on 5 March from San Francisco.

Until March 1951 the Air Force's 606th Aircraft Control and Warning Squadron had participated in the air surveillance of the Pusan area from the top of 3,000-foot Chon-San—the encroaching mountain that made K–9's traffic pattern so hazardous. This Air Force unit displaced to Taejon early in March, and the MGCIS–1 commanding officer, Major H. E. Allen, moved his radio and radar vans to the mountain top to take over the job.[20]

General MacArthur Visits Marine Battalion

Following the occupation of Hongchon on the 15th, the Marine ground forces ran into stiffening enemy opposition during the next two days. The 2d and 3d Battalions of the 7th Marines were pinned down by intense CCF mortar and artillery fire when attacking Hill 356 (Map 9). Three out of six friendly 81mm mortars were knocked out on 15 March in the 3d Battalion area, and at dusk 2/7 and 3/7 had barely won a foothold on the hill.[21]

The 1st Marines also met opposition which indicated that the enemy planned to make a stand on the high ground east and north of Hongchon. An intricate maneuver was executed when Lieutenant Colonel Robert K. McClelland's 2/1 swung from the right flank, where no enemy was encountered, to the extreme left. As a pre-

[19] *PacFlt Interim Rpt* No. 2, II, 1038, 1039. A "bogey" is an unidentified plane.
[20] LtCol H. E. Allen, interv of 26 Feb 59.
[21] CO 7thMar msg to CG 1stMarDiv, 2130 15 Mar 51.

liminary, the battalion had to circle to the rear, then move by truck up the MSR and through the zone of the 7th Marines as far as the village of Yangjimal (Map 9). Dismounting, the men made a difficult march across broken country toward Hill 246. At 1230 on the 15th the column deployed to attack Hill 428 in conjunction with Lieutenant Colonel Virgil W. Banning's 3d Battalion.

Easy Company (Captain Jack A. Smith) and Item Company (First Lieutenant Joseph R. Fisher) engaged in a hot fire fight with the enemy. Both sides relied chiefly on mortars, but the Chinese had the advantage of firing from camouflaged bunkers. Smith called for an air strike and four planes from VMF–214 responded immediately. Fox Company (Captain Goodwin C. Groff) and Dog Company (Captain Welby D. Cronk) were committed in the attempt to carry Hill 428, but the enemy continued to resist stubbornly until dusk. McClelland then ordered a withdrawal to night defensive positions around Hill 246. The two assault battalions had suffered 7 KIA and 86 WIA casualties. Counted enemy dead were reported as 93.[22]

Lieutenant Colonel Donald R. Kennedy's 3/5 was attached to the 1st Marines to protect the right flank as the Marines prepared to resume the attack on the morning of the 16th. But the enemy had pulled out from Hill 428 during the night and patrols advanced more than 300 yards without making contact.

Another hard action awaited the 7th Marines on the 16th, when Major Sawyer's 1st Battalion moved up to Line BAKER (Map 9). The Chinese resisted so hard on Hill 399 that the Marines had to attack bunker after bunker with grenades.

The following morning was the occasion of a visit to the front by General MacArthur. Accompanied by Generals Ridgway and O. P. Smith, he drove in a jeep from Wonju over the mountain pass to Hongchon, where Marine engineers were still clearing mines. The jeep stalled after crossing the Hongchon-gang at a ford and a tow was necessary. This did not deter the commander in chief, who had asked to visit a Marine battalion in a combat area. He was taken to the CP of Major Sawyer, whose 1/7 was mopping up on Hill 399 after the hard fight of the day before.

Five hours of riding over miserable roads had not daunted the 71-year-old veteran of two World Wars. He seemed fresh and rested as he shook hands with 1/7 officers. "Although we had not

[22] 2/1 *HD* 16 Dec 50 to 30 Apr 51, 21–22; VMF–214 *HD* Mar 51, 25–26.

passed the word regarding General MacArthur's visit," commented General Smith, "there were dozens of cameras in evidence." [23]

IX Corps orders were received on the 17th for the 1st Marine Division to attack from Line BAKER to Line BUFFALO (Map 9). The Division plan of maneuver called for the 5th Marines to pass through and relieve the 7th Marines while the 1st Marines continued to advance on the right.

Again the enemy chose withdrawal to resistance, and five of the six Marine battalions reached Line BUFFALO on 20 March after encountering only sniper fire and a few scattered mortar rounds. Enemy opposition was reserved for 2/1 on the 19th, when Fox Company was pinned down by enemy small-arms and mortar fire from a long, narrow ridge running north and south to the west of Hill 330.

Fortunately for the attackers, a parallel valley enabled a platoon of tanks from Baker Company, 1st Tank Battalion, to knock out unusually strong CCF bunkers with direct 90mm fire while Fox Company riflemen followed along the ridgeline with a grenade attack before the enemy had time to recover. Thanks to intelligent planning, not a single Marine was killed or wounded as the battalion dug in for the night on Hill 330.

Adopting the same tactics on the 20th, after artillery preparation and an air strike by VMF–214 and VMF–323 planes, Easy Company of 2/1 advanced along the ridgeline connecting Hills 330 and 381 while tanks moved forward on either side providing direct flat-trajectory 90mm fire. By 1315 the Marines had overrun the enemy's main line of resistance without a casualty.[24]

1st KMC Regiment Returns to Division

As the Eighth Army jumped off on 20 March from Line BUFFALO toward Line CAIRO (Map 9), the 1st KMC Regiment was attached again to the 1st Marine Division. This was the third time that Lieutenant Colonel Charles W. Harrison had been directed to reorganize and reassemble a KMC liaison advisory group. The 3d Battalion of the 11th Marines, commanded by Lieutenant Colonel William McReynolds, was placed in direct artillery support. When the advance

[23] Smith, *Chronicle*, 17 Mar 51.

[24] 1stMarDiv *HD*, Mar 51, 10; VMF–214 and VMF–323 *HD*, Mar 51.

was resumed, the KMCs attacked between the 1st Marines on the right and the 5th Marines on the left.[25]

The high *esprit de corps* of the KMCs shines forth from a comment written in his own English by First Lieutenant Kim Sik Tong: "The KMC ideal is to complete the mission, regardless of receiving strong enemy resistance, with endurance and strong united power, and always bearing in one's mind the distinction between honor and dishonor." [26]

The zone of the KMC Regiment was a roadless wilderness, making it necessary to air-drop ammunition and supplies for the attack on Hill 975. This was the hardest fight of the Division advance to Line CAIRO. Excellent artillery support was provided for the 2d and 3d Battalions as they inched their way forward in three days of bitter combat. Not until the morning of 24 March was the issue decided by maneuver when the 1st Battalion moved around the left KMC flank into a position threatening the enemy's right. Resistance slackened immediately on Hill 975 and the KMCs took their objective without further trouble.

The 1st and 5th Marines were already on Line CAIRO, having met comparatively light opposition from NKPA troops who had relieved the 66th and 39th CCF Armies. Apparently the enemy was using North Koreans as expendable delaying elements while massing in the rear for an offensive that could be expected at any time. A smoke screen, produced by burning green wood, shrouded the front in an almost constant haze.

Although the objectives of Operation RIPPER had been reached, General Ridgway planned to continue the UN offensive for the purpose of keeping the enemy off balance during his offensive preparations. The Eighth Army had been attacking with few and brief pauses for regrouping even since 21 February, and the commanding general wished to maintain its momentum.

An advance of the 1st Marine Division to a new Line CAIRO was ordered by IX Corps on 26 March. This was simply a northeast extension of the old line to the boundary between IX and X Corps (Map 9). There was no need for the 5th Marines to advance, and the 1st Marines and KMC Regiment moved up to the new line on schedule without opposition.

Eighth Army units had made average gains of about 35 miles during the last three weeks while driving nearly to the 38th Parallel.

---

[25] Col C. W. Harrison, *Narrative*, Mar-Apr 51, 2–3.
[26] 1stKMCRegt *POR*, 24 Mar 51.

On 29 March, General Ridgway published a plan for Operation RUGGED. It was to be a continuation of the offensive, with Line KANSAS (Map 8) as the new objective. While other 1st Marine units were being relieved by X Corps elements, the 7th Marines was to be moved up from reserve near Hongchon and attached to the 1st Cavalry Division for the attack beyond Chunchon, evacuated by the retreating enemy.[27]

On 1 April the Marines were informed of sweeping changes in IX Corps plans. Instead of being relieved, the 1st Marine Division was to continue forward with two infantry regiments plus the KMCs. Its new mission called for a relief of the 1st Cavalry Division (with the 7th Marines attached) north of Chunchon. This modification gave General Smith the responsibility for nearly 20 miles of front.[28]

"I visited this front frequently," commented Major General A. L. Bowser, the G–3 of that period, "and it was difficult at times to even locate an infantry battalion. . . . Visitors from the States or FMFPac were shocked at the wide frontages." [29]

38th Parallel Recrossed by Marines

Further IX Corps instructions on 2 April directed that the 1st Marines go into Division reserve near Hongchon while the 5th Marines and 1st KMC Regiment attacked. The deep, swift Soyang-gang, fordable in only a few places, lay squarely in the path of the 5th Marines. Speculations as to the method of crossing became rife just as air mattresses were issued. And though the officers denied any such intent, the troops were convinced that inflated mattresses would be used.

As it happened, the regimental executive officer, Lieutenant Colonel Stewart, worked out a plan that did not include any such novelty. A narrow ford was discovered that would get the 1st and 2d Battalions across while the 3d rode in DUKWs. Light enemy opposition of a rear guard nature was encountered but the regiment completed the operation without casualties. Stewart reported to the regimental CP and learned that a jeep waited to take him on the initial lap of

---

[27] EUSAK *Cmd Rpt*, Mar 51, 18–19.
[28] Smith, *Chronicle*, 1 Apr 51.
[29] BrigGen A. L. Bowser, ltr of 14 Feb 58.

his homeward journey. He was the last man to leave Korea of the 1st Provisional Marine Brigade, which had landed at Pusan on 2 August 1950.[30]

After reaching their prescribed objectives, the 5th Marines and KMC Regiment were relieved on 5 April by elements of the 7th Infantry Division of X Corps. Meanwhile, the 7th Marines, attached to the 1st Cavalary Division, advanced northward with the 7th and 8th Cavalry Regiments. Little opposition developed and on 4 April the Marines were among the first Eighth Army troops to recross the 38th Parallel.

General Ridgway published another operation plan on 6 April 1951 and designated new Eighth Army objectives to the northward. The purpose was to threaten the buildup for the forthcoming CCF offensive that was taking place behind the enemy lines in the so-called "Iron Triangle."

This strategic area, one of the few pieces of comparatively level real estate in central Korea, was bounded by Kumhwa, Chorwon, and Pyongyang (Map 14). A broad valley containing a network of good roads, it had been utilized by the Chinese for the massing of supplies and troops.

Experience had proved that interdictory bombing could not prevent the enemy from nourishing an offensive, even though the FEAF had complete control of the air over roads and rail lines of a mountainous peninsula. The Chinese, though hampered in their efforts, had been able to bring up large quantities of supplies under cover of darkness. General Ridgway determined, therefore, to launch his ground forces at objectives threatening the Iron Triangle, thus forcing the enemy to fight.

On 8 April, in preparation for the new effort, the 1st Marine Division was directed by IX Corps to relieve the 1st Cavalry Division on Line KANSAS and prepare to attack toward Line QUANTICO (Map 8).

Renewal of Division's CAS Problems

By this time, after three months of various sorts of operational difficulties, VMF–311 was riding a wave of efficiency. The distance from the operating base to the combat area emphasized the superior speed

[30] Col J. L. Stewart, ltr of 25 Oct 57.

of the F9Fs. The Panther jets could get into action in half the time required by the Corsairs. The jets were more stable in rocket, bombing, and strafing runs. They were faster on armed reconnaissance and often were pouring it into the enemy before he could disperse. These advantages offset the high fuel consumption of the F9Fs and made them ideal planes for close air support.

On the morning of 8 April an opportunity arose for the Marine jets to help the 7th Marines. It started when 3/7 patrols encountered 120mm mortars, small arms, automatic weapons, and grenades employed by an enemy force dug in on a ridge looming over the road near the west end of the Hwachon reservoir. The battalion forward air controller radioed DEVASTATE BAKER at Hongchon for air support.[31]

At the time Major Roy R. Hewitt, an air officer on General Shepherd's FMFPac staff, was visiting the Air Support Section of Marine Tactical Air Control Squadron–2 (MTACS–2). His blow-by-blow report of events is as follows:

> a. At 0900 a request for an air support strike on an enemy mortar position was received from the 7th Marines. It took the Air Support Section until 0945 to get through to JOC and then it had to be shunted through K–1 in order to get the request in.
>
> b. The G–3 1st Marine Air Wing had arranged with JOC to have four (4) F9F 'scramble alert' for use by the 1st Marine Division. The F9Fs were requested, and JOC authorized their use, but when Marine Aircraft Group–33 was contacted they informed the Air Support Section that JOC had already scrambled the aircraft and sent them to another target.
>
> c. Air Support Section again contacted JOC, and JOC said aircraft would be on station in one (1) hour. At the end of one (1) hour JOC was again contacted concerning aircraft. This time JOC said they would have two (2) flights on station within one (1) hour. At the end of the second one (1) hour period no aircraft were received.
>
> d. Again the Air Support Section contacted JOC and was informed that any air support for the 7th Marines would have to be requested through the 1st Cavalry Division to which the 7th Marines were attached. [In fact, JOC notified DEVASTATE BAKER that any such requests from the 1st Marine Division would not be honored until the Division went back into action.][32]

During all this time ten Marine planes—six from VMF–311 and four from VMF–214—had reported in and out of the area. They had

[31] The following description is from: 7th Mar *HD*, 8 Apr 51, 5; IX Corps *Cmd Rpt*, Apr 51, Bk III, Vol 2, PORs #574 and 575 of 8 Apr; Shepherd Memo, 27 Apr 51, encl (7) "Excerpt from Maj Hewitt rept;" 1stMAW *HD*, Apr 51, App VII, "Staff Journal G–3 Section," 9 Apr.

[32] 1st MAW G–3 Staff Journal, 9 Apr, *op. cit.*

been sent by MELLOW to work under the control of Mosquito STRATEGY, the tactical air controller (airborne) (TACA) of the 1st Cavalry Division. The flights also supported the 6th ROK Division patrols on the Marines' left, hit troops in a small settlement 3 miles to the Marines' front, and aided the 7th and 8th Cavalry regiments which were encountering resistance on the commanding ground to the right. None of the flights supported the Marines.

Meanwhile, the 3/7 Marines employed artillery and tanks on the enemy positions, and late in the day a Mosquito brought in a flight of four Air Force F–80s. Major Hewitt's report continued:

> e. At the end of six (6) hours air support was finally received by the 7th Marines. It was brought in by a Mosquito who would not relinquish control of the aircraft to the Forward Air Controller who could see the target much better than the Mosquito.
>
> f. After having the fighters make a couple of passes the Mosquito took the fighters and went to another target without having completely destroyed the position.

This was the beginning of a deterioration in air support for Marine ground forces that can be charged in large measure to the JOC system of control. Major Hewitt's report was read with great interest by high-ranking Navy and Marine Corps officers. By now they were devoting a lot of thought to the breakup of the Marine air-ground team.

CHAPTER VI

The CCF Spring Offensive

Prisoners Reveal Date of Offensive—Hwachon Occupied by KMC Regiment—CCF Breakthrough Exposes Marine Flank—Marine Air in Support Everywhere—Plugging the Gap on the Marine Left—Repulse of Communist Attacks—Withdrawal to the KANSAS *Line—Enemy Stopped in IX Corps Sector—1st Marine Division Returns to X Corps*

ON 10 APRIL 1951 the 1st Marine Division was poised on Line KANSAS for a drive to Line QUANTICO. Then a new IX Corps directive put on the brakes, and for 10 days Marine activities were limited to patrolling and preparation of defensive works. Boundary adjustments between the Division and the 6th ROK Division on the left extended the Marine zone about 2,000 yards to the west; and General O. P. Smith's CP was advanced to Sapyong-ni, just south of the 38th Parallel (Map 8).

Out of a blue sky came the announcement on the 11th that General MacArthur had been recalled by President Truman for failure to give wholehearted support to the policies of the United States Government and of the United Nations in matters pertaining to his official duties. General Ridgway was appointed to the UN command, and he in turn was relieved on 14 April by Lieutenant General James A. Van Fleet, USA.

The new Eighth Army commander, youthful in appearance for his 59 years, was no novice at fighting Communists. In 1949 and 1950 he had been Director of the Joint Military Aid Group that saved Greece from falling into the clutches of Communism after Moscow fomented a civil war. Van Fleet also brought to his new command a World War II reputation as a vigorous leader with a preference for offensive doctrines.

Prisoners Reveal Date of Offensive

Chinese prisoners taken during the first three weeks of April 1951 told all they knew with no apparent reluctance, just as Japanese captives had given information in World War II. Inconsistent as it may seem that fanatical Asian soldiers should prove so cooperative, such was the penalty the enemy paid for insisting on resistance to the last ditch. Since the possibility of surrender was not considered, CCF prisoners were taught no code of behavior and answered questions freely and frankly.

POW interrogations were supplemented by captured documents revealing that the Chinese prided themselves on a new tactical doctrine known as "the roving defensive," put into effect in the spring of 1951. It meant "not to hold your position to the death, but to defend against the enemy through movement," explained a secret CCF directive dated 17 March 1951. "Therefore, the wisdom of the roving defensive is based on exhausting the enemy without regard for the loss or gain of some fighting area or the immediate fulfillment of our aims." [1]

It was admitted that the CCF soldier must work harder, "because the troops will have to construct entrenchments and field works in every place they move." But the advantages were that "roving warfare can conserve our power, deplete the enemy's strength, and secure for us more favorable conditions for future victory. Meanwhile, the enemy will make the mistake once again, and collapse on the Korean battlefield."

The last sentence evidently refers to the UN advance of late November 1950 that was rolled back by a surprise CCF counteroffensive. Chinese strategists seem to have concluded that their "roving defensive" had made possible another such offensive victory in the spring of 1951. At any rate, prisoners questioned by the 1st Marine Division and other IX Corps units agreed that the CCF 5th Phase Offensive was scheduled to begin on 22 April 1951. The IX Corps zone was said to be the target area for an attempted breakthrough.[2]

Marine G–2 officers recalled that prisoners gave information on the eve of the CCF offensive in November 1950 that proved to be astonishingly accurate in the light of later events. For it was a paradox that

---

[1] This section is based upon the following documents: IX Corps *CmdRpt* 5, Apr 51, Sec 3, Intelligence; 1stMarDiv *HD*, Apr 51; *PIR* 171, 179, 180.

[2] EUSAK, IX Corps and 1stMarDiv G–2 *PIRs*, 1–20 Apr 51.

the Chinese Reds, so secretive in other respects, let the man in the ranks know about high-level strategic plans. In the spring of 1951 it mattered little, since air reconnaissance had kept the Eighth Army well informed as to the enemy buildup.

Prisoners were taken in the IX Corps zone from the following major CCF units during the first three weeks of April:

> 20th Army (58th, 59th, and 60th Divisions), estimated strength, 24,261;
> 26th Army (76th, 77th, and 78th Divisions), estimated total strength, 22,222;
> 39th Army (115th, 116th, and 117th Divisions), estimated total strength, 19,538;
> 40th Army (118th, 119th, and 120th Divisions), estimated total strength, 25,319.

The 20th and 26th, it may be recalled, were two of the CCF armies opposing the 1st Marine Division during the Chosin Reservoir breakout. It was a satisfaction to the Marines that their opponents of December 1950 had evidently needed from three to four months to reorganize and get back into action.

In CCF reserve on 21 April 1951 were the 42d and 66th Armies, both located in the Iron Triangle to the enemy's rear. The former included the 124th, 125th, and 126th Divisions—the 124th being the unit cut to pieces from 3 to 7 November 1950 by the 7th Marines in the war's first American offensive action against Chinese Red adversaries.

Hwachon Occupied by KMC Regiment

At 0700 on the 21st the 1st Marine Division resumed the attack toward Line QUANTICO with the 7th Marines on the left, the 5th Marines in the center, the KMC Regiment on the right, and the 1st Marines in reserve. Negligible resistance awaited the Marines and other IX Corps troops during advances of 5,000 to 9,000 yards. An ominous quiet hung over the front as green wood smoke limited visability to a few hundred yards.

On the Marine left the 6th ROK Division lost touch, opening a gap of 2,500 yards, according to a message from Corps to the 1st Marine Division. The ROK commander was ordered by Corps to restore lateral contact. This incident would be recalled significantly by the Marines when the CCF blow fell.[3]

The KMC Regiment had the mission of finishing the fight for control of the Hwachon Reservoir area. Early in April the 1st Cavalry Division and the 4th Ranger Company, USA, had been repulsed in attempts to fight their way across the artificial lake in rubber boats. The enemy retaliated by opening the penstocks and spillway gates. Considering that the dam was 275 feet high and the spillway 826 feet long, it is not surprising that a wall of water 10 feet high roared down the Pukhan Valley into areas recently occupied by IX Corps units.[4]

Both Army and Marine engineers were on the alert, having been warned by aerial observers. They cut three floating bridges loose from one bank or another, so that they could ride out the crest of the flood. Thanks to this precaution, only temporary damage and interruption of traffic resulted.[5]

The 1st Engineer Battalion, commanded by Lieutenant Colonel John H. Partridge, was given the mission by Corps of jamming the gates of the dam at the open position. Compliance would have to wait, of course, until the KMCs took the dam. Partridge conferred meanwhile with Colonel Bowser, and it was decided to take no action after the anticipated capture until a demolitions reconnaissance could be made.[6]

As early as 18 April a KMC patrol had crossed the Pukhan into the town of Hwachon, which was found abandoned except for 11 Chinese soldiers, who were taken prisoner. Marine engineers installed a floating bridge on the 21st for the advance of one KMC battalion the next morning. The other two battalions were to cross the river several miles downstream by DUKWs.[7]

Corps plans for the attack were made in full realization of air reconnaissance reports for 20 and 21 April indicating that the enemy offensive buildup was in its final stages. This intelligence was gleaned in spite of all enemy efforts to frustrate the airmen. CCF spotters were placed on mountain tops to give the alarm, and relays of men

[3] 1stMarDiv *HD* Apr 51, 47–48; CO 7thMar msg to CG 1stMarDiv, 1830 21 Apr 51.
[4] Capt D. E. Fowler, "Operations at the Hwachon Dam, Korea," *The Military Engineer*, Jan-Feb 1952, 7–8.
[5] 1stMarDiv *HD*, Apr 51.
[6] BrigGen A. L. Bowser, ltr of 14 Feb 58; Col J. H. Partridge, ltr of 17 Mar 58.
[7] CO KMC Regt msg to CG 1stMarDiv, 2335 22 Apr 51, in KMC In&Out #1; Col C. W. Harrison, *Narrative*, Mar-Apr 51.

fired shots to pass on warnings of approaching planes. Antiaircraft defenses were increased at such vital spots as bridges and supply areas. The Communists even went so far as to put out decoys—fake trucks, tanks, and tank cars—to lure UN fighter-bombers within range of antiaircraft guns.

These efforts resulted in 16 Marine planes being shot down from 1 to 21 April 1951. Nine of the pilots were killed, one was captured, three were rescued from enemy territory, one walked back to friendly outposts, and two managed to bail out or crash-land behind the UN lines.[8]

This total was equivalent to two-thirds of the average tactical squadron. Because of the disruption to the 1st MAW pilot replacement program, the Commandant arranged for 20 pilots to be flown to Korea to augment the normal rotation quotas.[9]

Direct opposition from enemy aircraft was also on the increase. CCF flights even reached the EUSAK battle line as unidentified light planes flew over positions or dropped small bombs. Evidently the enemy was using well camouflaged airfields in North Korea.

An air battle took place on 20 April when two VMF–312 pilots from the *Bataan,* Captain Philip C. DeLong and First Lieutenant Harold D. Daigh, encountered four YAK fighters in the heavily defended Pyongyang-Chinnanpo area. They gave chase and shot down three of the enemy planes.[10]

Marine aircraft were on station when Marine ground forces resumed their forward movement at 0830 on the morning of 22 April. A CCF prisoner taken that very afternoon confirmed previous POW statements that the 22d was the opening day of the Fifth Phase Offensive. The front was quiet, however, as the three Marine infantry regiments advanced almost at will.

A motorized patrol of Division Reconnaissance Company, led by the commanding officer, Major Robert L. Autrey, had the initial contact with the enemy while advancing on the Division left flank. The two platoons, supported by Marine tanks, found their first indications when searching a Korean roadside hut. Although the natives denied having seen any Chinese soldiers, Corporal Paul G. Martin discovered about 50 hidden rice bowls waiting to be washed. Upon

[8] Compilation of data from 1stMAW sqdn *HDs*, Apr 51; *PacFlt Interim Rpt* No. 2, II, 1051, 1071.
[9] *Ibid.*
[10] 1st MAW *HD*, 20 Apr 51; VMF–312 *HD*, 20 Apr 51.

634040 O–62–9

being confronted with this evidence, the terrified Koreans admitted that Chinese soldiers had reconnoitred the area just before dawn.

Farther up the road, an ammunition dump of hidden mortar shells was discovered. The enemy had also put up several crude propaganda signs with such sentiments as YOUR FOLKS LIKE SEE YOU HOME and HALT! FORWARD MEANS DEATH.

The patrol dismounted and proceeded with caution, guided by an OY overhead. Although the "choppers" were the favored aircraft of VMO–6, the OYs also earned the gratitude of the troops on many an occasion such as this. The pilot gave the alarm just before hidden Communists opened fire. Thus the Marines of the patrol were enabled to take cover, and the tanks routed the enemy force with well placed 90mm shells.[11]

The KMCs met no resistance worth mentioning when they secured the town of Hwachan and the north bank of the Pukhan just west of the reservoir. Only light and scattered opposition awaited the 5th Marines (Colonel Richard M. Hayward) and the 7th Marines (Colonel Herman Nickerson, Jr.) on their way to the occupation of assigned objectives on Line QUANTICO.

CCF Breakthrough Exposes Marine Flank

For weeks the Communist forces in Korea might have been compared to an antagonist backtracking to get set for taking aim with a shotgun. There could be no doubt, on the strength of daily G–2 reports, about both barrels being loaded. And on the night of 22 April the enemy pulled the trigger.

The KMCs, after taking their objectives, reported a concentration of enemy small-arms fire. At 1800 the command of the 1st Marine Division directed a renewal of the advance at 0700, on the morning of the 23d. This order was cancelled at 2224 by a message calling for all Marine units to consolidate and patrol in zone, pending further instructions.[12]

One of the reasons for the sudden change was the receipt of a message by the 1st Marine Division at 2120, informing that the 6th

[11] Sgt Paul G. Martin, USMC (Ret.), ltr of 2 Jul 56.
[12] CG 1stMarDiv msg to COs 5th, 7th, 11th Mar, 1st KMC, Tk, and EngBns, 2224 22 Apr 51 in Div In&Out #21.

ROK Division was under heavy attack to the west of the Marines. Meanwhile, an on-the-spot questioning of a CCF prisoner just taken by the KMCs convinced the command and staff of the 1st Marine Division that the CCF 5th Phase Offensive was only hours away and gathering momentum. Thanks to this timely interpretation, all forward Marine units were alerted two hours before the main blow fell.

It was on the left of the 1st Marine Division that the situation first became critical. The 6th ROK Division had never quite succeeded in closing up the gap on its right and restoring contact with the Marines. But this failure was trivial as compared to the collapse of the entire ROK division an hour before midnight, leaving a gap wide enough for a major breakthrough.

The 1st Marine Division took prompt measures to cope with the emergency. As early as 2130, the 1st Marines, in reserve just north of Chunchon, were alerted to move one battalion to contain a possible enemy threat to the Division left flank. A second message an hour later called for immediate execution. And at midnight the Division Provost Marshal was directed to stop ROK stragglers and place them under guard. The Division Reconnaissance Company received orders to aid the military police.[13]

Colonel Francis M. McAlister, commanding the 1st Marines, selected Lieutenant Colonel Robley E. West's 1st Battalion to carry out Division orders. "By midnight we were all on trucks and rolling on the roads north," wrote Second Lieutenant Joseph M. Reisler in a letter home. "Mile after mile, all the roads were covered with remnants of the ROKs who had fled. Thousands of them [were] straggling along the roads in confusion." [14]

Despite these preparations for trouble on the left flank, the KMCs on the right and the 5th Marines in the center were first in the Division to come under attack. During the last minutes of 22 April the 2d KMC Battalion had it hot and heavy on Hill 509. To the left the 1st KMC Battalion, partially encircled, notified the 5th Marines of a penetration.

The effects were felt immediately by 1/5, with its CP in Hwachon. Hill 313 was the key to the town, being located at the Hwachon end of a long ridge forming a natural avenue of approach from the

---

[13] CG 1stMarDiv msg to CO 1stMar 2130 and 2232 22 Apr 51 in Div In&Out #12; 1stMarDiv *HD*, Apr 51, 6 and 50.

[14] 2dLt J. M. Reisler, ltr to family of 1 May 51.

northeast. Captain James T. Cronin's Baker Company of 1/5 had the responsibility for protecting the CP and shifting troops to the right flank if necessary. He sent Second Lieutenant Harvey W. Nolan's platoon to run a race with the enemy for the occupation of Hill 313. Attached in excess of T/O for familiarization was Second Lieutenant Patrick T. McGahn.[15]

About 220 yards from the summit the slope was so steep that the Marines clawed their way upward on hands and knees. The company commander posted the attached light machine gun section while Nolan, McGahn, and Sergeant William Piner organized the assault. The three squads of riflemen advanced a few yards, only to be pinned down by well directed CCF machine gun fire. Another rush brought the Marines closer to the enemy but a stalemate ensued in the darkness. Seven of the platoon were killed and 17 wounded.

The situation in the 1/5 area was so serious that Fox Company of 2/5 (Lieutenant Colonel Glen E. Martin) sent reinforcements. At dawn, however, Hill 313 proved to be abandoned by the enemy. A vigorous KMC counterstroke had swept the Communists from Hill 509, so that the front was relatively quiet in this area. The courage and determination of the KMC Regiment were praised by General Smith, who sent this message on the morning of the 23d to Colonel Kim, the commanding officer:

> Congratulate you and your fine officers and men on dash and spirit in maintaining your positions against strong enemy attacks. We are proud of the Korean Marines.[16]

It is taking no credit away from the KMCs and 5th Marines to point out that they appear to have been hit by enemy holding attacks. The main CCF effort was directed at the left of the Division line, held by the 7th Marines.

The heaviest fighting took place in the sector of 1/7 on the extreme left, commanded by Major Webb D. Sawyer. It was obvious that the enemy planned to widen the penetration made at the expense of the 6th ROK Division. The 358th Regiment of the 120th Division, CCF 40th Army, hurled nearly 2,000 men at the Marine battalion. Charlie

---

[15] Sources for this 1/5 action are 5thMar *HD*, Apr 51; LtCol John L. Hopkins, interv of 24 Jan 58; Maj J. T. Cronin, ltr of 30 Jan 58; Capt P. T. McGahn, interv of 27 Jul 56.

[16] CG 1stMarDiv msg to CO KMC Regt, 0910 23 Apr 51.

Company, commanded by Captain Eugene H. Haffey, took the brunt of the assault.[17]

The thin battalion line bent under sheer weight of numbers. But it did not break. It held through three hours of furious fighting, with the support of Marine and Army artillery, until the 1st Battalion of the 1st Marines came up as reinforcements under the operational control of the 7th Marines. The newcomers took a position to the left of 1/7, so that the division flank was no longer completely "in the air."

This was one of the first examples of the Corps and Division maneuvering that played such a large part throughout in the blunting of the CCF offensive. Troops were not left to continue a desperate fight when a shift of units would ease the pressure.

Marine Air in Support Everywhere

At first light on the 23d the FEAF Mosquitoes and fighter-bombers went into action. The Marines had four two-plane flights of Corsairs airborne before sun-up. VMF–323 responded to a call from Baker Company, 1/5, only to find that the enemy had abandoned Hill 313. A low-flying OY of VMO–6, commanded by Major D. W. McFarland, guided the Corsairs to the withdrawing Chinese, who were worked over thoroughly. VMF–214 planes meanwhile supported 1/7 in that battalion's desperate fight at the left of the line.[18]

A pilot's-eye view showed fighting in progress from one coast to another, although the enemy was making his main effort in the IX Corps sector. The U.S. 24th Infantry Division, to the left of the 6th ROK Division, was having to bend its right flank southward to defend against the CCF penetration. Toward the rear the 27th Brigade of the British Commonwealth Division, in IX Corps reserve, was being alerted to meet the Communists head on and bring the breakthrough to a halt.

Elements of the U.S. 24th and 25th Divisions on the edge of the Iron Triangle were giving ground slowly. Seoul was obviously an objective of CCF units that had crossed the Imjin in the moonlight.

[17] This account of the 7th Marines' fight is based on the following sources: 1stMarDiv *HD*, Apr 51; BrigGen A. L. Bowser, ltr of 14 Feb 58; Col R. G. Davis, Comments, n.d.; Col H. Nickerson, comments of 25 Feb 58; Col W. F. Meyerhoff, ltr of 25 Feb 58; Col J. T. Rooney, ltr of 26 Feb 58.

[18] *HDs* of VMF–323, VMF–214, and VMO–6 for Apr 51.

But General Ridgway had decided that the city was not to be abandoned. "Considerable importance was attached to the retention of Seoul," he explained at a later date, "as it then had more value psychologically than its acquisition had conferred when we were still south of the Han." [19]

Near the junction of X Corps and I ROK Corps the 7th ROK Division had been hard hit, although the enemy attack in this area was a secondary effort. Air support helped this unit to hold its own until it could be reinforced.

Of the 205 Marine aircraft sorties on 23 April, 153 went to support the fighting front. The 1st Marine Division received 42 of these CAS strikes; 24 went to the ROK 7th Division; 59 to I Corps to check the advance on Seoul; and 28 to pound the Communists crossings the Imjin.[20]

Only about 66 percent of the landing strip at K–3 (Pohang) could be used; the remainder was being repaired by the Seabees. In order to give the Panther jets more room, VMF–212 shifted its squadrons for two days to K–16 near Seoul. A detachment of VMF–323 planes from K–1 (Pusan) also made the move. Since K–16 was only 30 miles from the combat area along the Imjin, the Corsairs were able to launch their attacks and return for rearming and refueling in an hour or less.[21]

Plugging the Gap on the Marine Left

At first light on 23 April the entire left flank of the 1st Marine Division lay exposed to the Chinese who had poured into the gap left by the disintegration of the 6th ROK Division. IX Corps orders called for the ROKs to reassemble on the KANSAS line, but most of them straggled from 10 to 14 miles behind the positions they held prior to the CCF attack. The 1st Marine Division ordered Reconnaissance Company to stop ROK stragglers at the river crossing, and several groups were turned back.

The reasons for the ROK collapse are variously given. Weak command and low morale have been blamed for the debacle, yet the shattered division did not lack for defenders. No less an authority

[19] Gen M. B. Ridgway, ltr of 29 Jan 58.
[20] Compilation of data from 1st MAW squadrons for 23 Apr 51.
[21] VMF–212, MAG–33, and MAG–12 *HDs* for Apr 51.

than General Van Fleet declared himself ". . . reluctant to criticize the 6th ROK Division too severely. I do not believe they deliberately threw away their equipment—I am inclined to believe such equipment was abandoned due to the terrain, lack of roads and weight. Our check at the time indicated that the Korean soldiers held on to their hand weapons. It is interesting to know that General Chang who commanded 6th ROK Division at the time . . . is today [March 1958] Vice Chief of Staff of the Korean Army."[22]

As a first step toward setting up a defense in two directions, the 1st Marine Division received orders from IX Corps to fall back to Line PENDLETON (Map 10). This was one of the Eighth Army lines assigned to such profusion that they resembled cracks in a pane of glass. PENDLETON ran generally southwest to northeast through the 7th Marines sector, then turned eastward just north of the town of Hwachon.

By occupying this line, the 7th Marines could bend its left to the south in order to refuse that flank. Still farther to the south, the 1st and 3d Battalions of the 1st Marines were to take positions facing west. Thus the line of the 1st Marine Division would face west as much as north. On the center and right the KMCs and 5th Marines would find it necessary to withdraw only about 1,000 yards to take up their new positions.[23]

It was up to 1/1 to make the first move toward plugging the gap. At 0130 on the 23d Captain John Coffey's Baker Company led the way. Moving north in the darkness along the Pukhan and then west along a tributary, the long column of vehicles made its first stop about 1,000 yards from the assigned position. Here the 92d Armored Field Battalion, USA, was stationed in support of the 6th ROK Division and elements of the 1st Marine Division. The commanding officer, Lieutenant Colonel Leon F. Lavoie, was an old acquaintance of 1/1, having supported that battalion during the final days of the Chosin Reservoir breakout. Lavoie was held in high esteem by the Marines, who found it characteristic of him that in this fluid situation his cannoneers were formed into a tight defensive perimeter, ready to fight as infantry if need be.

[22] Gen J. S. Van Fleet, USA (Ret.), ltr of 24 Mar 58.

[23] *HDs* of 1stMarDiv, 1stMar, and 7thMar for Apr 51; MajGen O. P. Smith, *Chronicle*, 23–24 Apr 51; MajGen E. W. Snedeker, ltr of 12 Feb 58; Col J. T. Rooney, ltr of 26 Feb 58; Col H. Nickerson, ltr of 13 Feb 58; Col W. F. Meyerhoff, ltr of 25 Feb 58; Maj R. P. Wray, ltr of 27 Apr 58; LtCol J. F. Coffey and Maj N. B. Mills, interv of 4 Apr 58.

MAP 10
CCF Offensive Starting 22 April
And Subsequent Marine Withdrawals

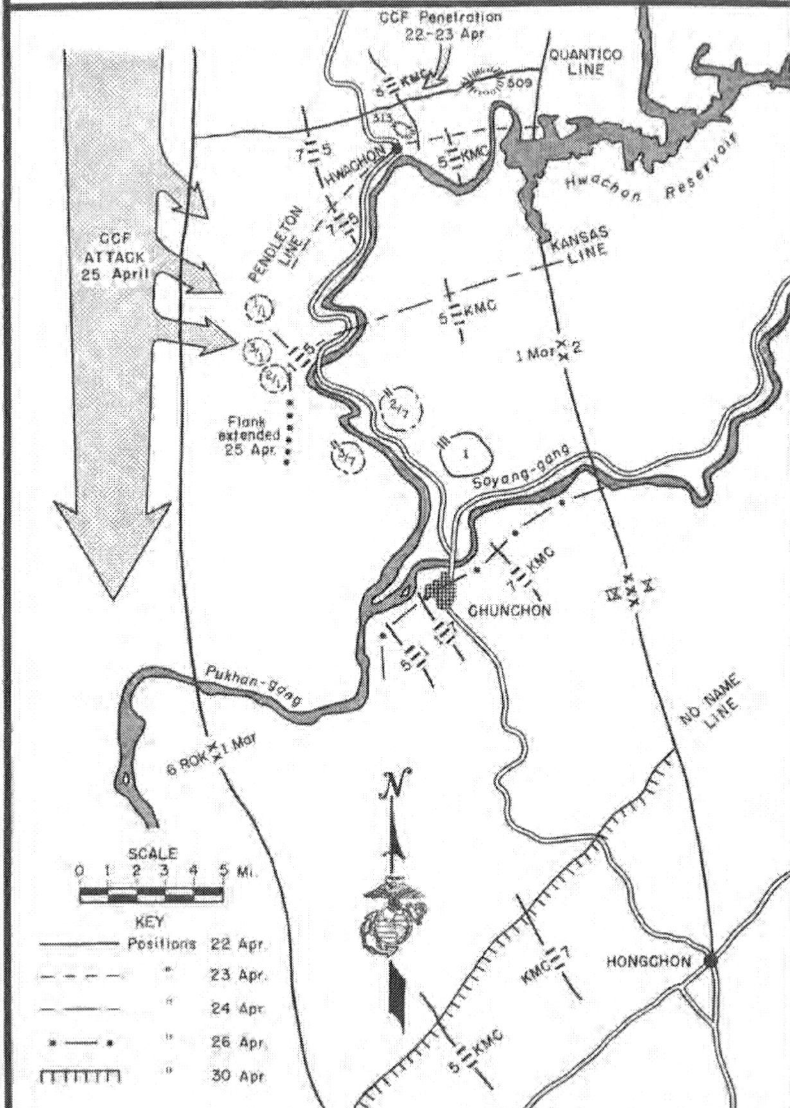

Another Army artillery unit, the 987th Armored Field Artillery Battalion, had been roughly used by the Chinese who routed the ROK division. Losses in guns and equipment had resulted, and Coffey moved with his company about 1,500 yards to the west to assist in extricating from the mud all the 105s that could be saved. Resistance was encountered in the form of machine gun fire from Chinese who had set up a road block.[24]

Upon returning to 1/1, Coffey found it occupying what was in effect an outpost to the southwest of the 7th Marines. Baker Company was assigned to the left of Captain Robert P. Wray's Charlie Company, holding the curve of a horseshoe-shaped ridge, with Captain Thomas J. Bohannon's Able Company on the right. In support, along the comparatively level ground to the immediate rear, was Weapons Company (Major William L. Bates).

With 1/1 facing in three directions to block a CCF attack, 1/7 managed to disengage and withdraw through 3/7, which occupied a position on Line PENDLETON. VMO–6 helicopters and troops of 2/7 helped to evacuate the 1/7 casualties incurred during the night's hard fighting.

During the early morning hours of the 23d the Marines of 3/1 had boarded trucks to the village of Todun-ni (Map 11) on the west bank of the Pukhan. Their assigned position was Hill 902, a 3,000-foot height dominating the surrounding terrain. The Chinese also were interested in this piece of real estate, since it overlooked the river crossing of the 1st Marine Division. Pressure to beat the Communists to the crest mounted as NCOs urged the men to their utmost efforts over steep uphill trails.

The Marines won the race. Once in position, however, it was evident to Lieutenant Colonel Banning that three ridge lines leading up to the hill mass would have to be defended. This necessity imposed a triangular formation, and he placed Captain Horace L. Johnson's George Company at the apex, with First Lieutenant William J. Allert's How Company on the left, and First Lieutenant William Swanson's Item Company on the right. The heavy machine guns of Major Edwin A. Simmons' Weapons Company were distributed among

[24] Sources for operations of the two Army artillery battalions are: Gen W. M. Hoge, USA (Ret.), ltr of 3 Feb 58; LtCol Leon F. Lavoie, USA, ltr of 5 Feb 58; LtCol Roy A. Tucker, USA, ltr of 30 Nov 57; LtCol John F. Coffey, USMC, ltr of 9 Feb 58; Capt Russell A. Gugeler, USA, *Combat Actions in Korea* (Washington, 1954), 162–173.

MAP II
Action of 1/1 at Horseshoe Ridge, 3/1 on 902,
and Subsequent Withdrawals, 23-25 April

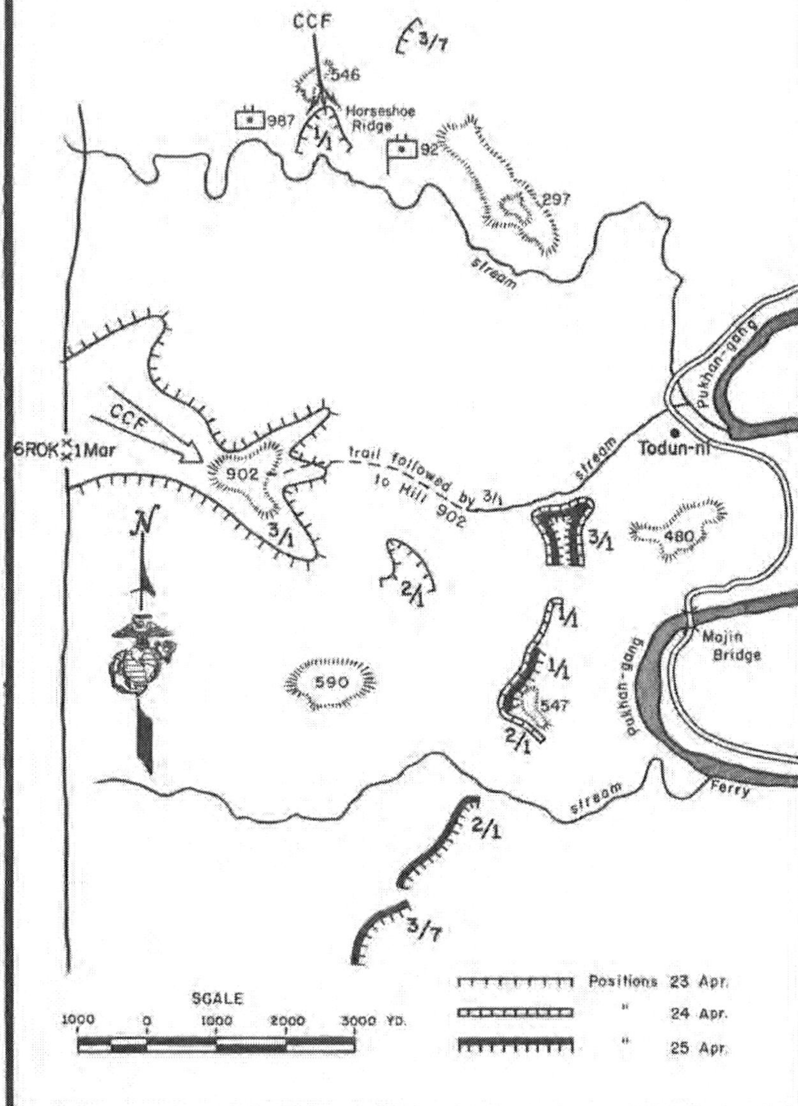

the rifle companies and the 81mm mortars placed only 10 to 20 yards behind the front lines.[25]

The KMCs and 5th Marines completed their withdrawal without interference. Thus the line of the 1st Marine Division on the afternoon of 23 April might have been compared to a fishhook with the shank in the north and the barb curling around to the west and south. The three Marine battalions plugging the gap were not tied in physically. Major Maurice E. Roach's 3/7 was separated by an interval of 1,000 yards from 1/1, and the other two Marine battalions were 5,500 yards apart (Map 11). But at least the 1st Marine Division had formed a new front under fire and awaited the night's attacks with confidence.

Repulse of Communist Attacks

Bugle calls and green flares at about 2000 announced the presence of the Chinese to the west of 1/1 on Horseshoe Ridge.

"They came on in wave after wave, hundreds of them," wrote Lieutenant Reisler, whose platoon held an outpost in advance of Charlie Company. "They were singing, humming and chanting, 'Awake, Marine. . . .' In the first rush they knocked out both our machine guns and wounded about 10 men, putting a big hole in our lines. We held for about 15 minutes, under mortar fire, machine gun fire, and those grenades—hundreds of grenades. There was nothing to do but withdraw to a better position, which I did. We pulled back about 50 yds. and set up a new line. All this was in the pitch-black night with Chinese cymbals crashing, horns blowing, and their god-awful yells." [26]

For four hours the attacks on Horseshoe Ridge were continuous, particularly along the curve held by Wray's company. He was reinforced during the night by squads sent from Coffey's and Bohannon's companies. Wray realized that the integrity of the battalion position depended on holding the curve of the ridge, but his main problem was bringing up enough ammunition. Men evacuating casualties to the rear returned with supplies, but the amount was all too limited until Corporal Leo Marquez appointed himself a one-man committee.

[25] 1stMarDiv, 1stMar, and 7thMar *HDs*, Apr 51; LtCol E. A. Simmons, interv of 12 Jun 57.

[26] 2dLt J. M. Reisler, ltr to family of 1 May 51.

His energy equalled his courage as he carried grenades and small-arms ammunition all night to the men on the firing line. Marquez emerged unhurt in spite of bullet holes through his cartridge belt, helmet, and a heel of his shoe.

About midnight it was the turn of 3/1. These Marines had dug in as best they could, but the position was too rocky to permit much excavation. Ammunition for the mortars had to be hand-carried from a point halfway up the hill.

Several hours of harrassing mortar fire preceded the CCF effort. George Company, at the apex of the ridge, was almost overwhelmed by the first Communist waves of assault. The courage of individual Marines shone forth in the ensuing struggle. Technical Sergeant Harold E. Wilson, second in command of the center platoon, suffered four painful wounds but remained in the fight, encouraging his men and guiding reinforcements from How Company as they arrived.[27]

Steady artillery support was provided by Colonel McAlister, who rounded up a jury-rigged liaison party and three forward observer teams composed mainly of officers from the 987th AFA Battalion. They registered 11th Marines and 987th Battalion defensive fires which had a large part in stopping the CCF attack as it lapped around George Company and hit How and Item on the other two ridges.

Colonel McAlister and Colonel Nickerson paid a visit to the CP of 1/1, which remained under the operational control of the 7th Marines until morning. The two regimental commanders arranged for artillery and tank support to cover the gap between 1/1 and 3/7.[28] The enemy, however, seemed to be wary about infiltrating between the three battalion outposts. This reluctance owed in large part to the deadly flat-trajectory fire of the 90mm rifles of Companies A and B of the 1st Tank Battalion, whose commanding officer, Lieutenant Colonel Holly F. Evans, had relieved Lieutenant Colonel Harry T. Milne that day.

Attacks on 3/1 and 3/7 also continued throughout the night. At daybreak the close air support of Marine aircraft prevented further Communist efforts, though dug-in enemy groups remained within machine gun range. Identification of Chinese bodies at daybreak indi-

[27] 1stMarDiv *HD*, Apr 51; 1stMar *HD*, Apr 51.
[28] Col H. Nickerson Jr., ltr of 13 Feb 58; Col R. E. West, comments, n.d.

cated that the 359th and 360th Regiments, 120th Division, 40th CCF Army, had been employed.

Withdrawal to the KANSAS *Line*

Now came the problem for the three Marine battalions of letting loose of the tiger's tail. Corps orders were received on the morning of 24 April for all units of the Division to pull back to Line KANSAS. This was in accordance with General Ridgway's policy, continued by General Van Fleet, of attaching more importance to destruction of enemy personnel than the holding of military real estate.

Some of the most seriously wounded men of 1/1 required immediate evacuation, in spite of the obvious risks. A VMO–6 helicopter piloted by First Lieutenant Robert E. Matthewson attempted a landing at the base of Horseshoe Ridge. As he hovered over the panel markings, CCF small-arms fire mangled the tail rotor. The machine plunged to earth so badly damaged that it had to be destroyed. Matthewson emerged unhurt and waved off a helicopter flown by Captain H. G. McRay. Then the stranded pilot asked for a rifle and gave a good account of himself as an infantryman.[29]

While First Lieutenant Norman W. Hicks' second platoon fought as the rear guard, First Lieutenant Niel B. Mills' first platoon of Charlie Company led the attack down the hill, carrying the wounded behind. In an attempt to rout the Chinese from a flanking hill, Mills was wounded in the neck by a bullet that severed an artery. Corpsman E. N. Smith gripped the end of the artery between his fingers until a hemostat could be applied, thus saving the lieutenant's life. Just before losing consciousness, Mills looked at his watch. It was 1000 and 1/1 had weathered the storm.[30]

The 3d Battalion of the 7th Marines, which had beaten off probing attacks all night, coordinated its movements with those of the two Marine battalions as they slowly withdrew toward the Pukhan. Despite Marine air attacks, the Communists not only followed but infiltrated in sufficient numbers to threaten the perimeter of Lavoie's cannoneers. The training this Army officer had given his men in infantry tactics now paid off as the perimeter held firm while mowing down the

[29] VMO–6 *HD* for Apr 51.
[30] LtCol R. P. Wray and Maj N. W. Hicks, interv of 16 Dec 59.

attackers with point blank 105mm shells at a range of 1,000 yards. The Marines of Captain Bohannon's company soon got into the fight, and the 92d repaid the courtesy by supporting 1/1 and 3/7 during their withdrawal. Counted CCF dead numbered 179 at a cost to the 92d of 4 KIA and 11 WIA casualties.[31]

As the morning haze lifted, the OYs of VMO–6 spotted for both Army and Marine artillery. DEVASTATE BAKER fed close support to the forward air controllers as fast as it could get planes from K–16 at Seoul, only a 15-minute flight away. Not only 49 Corsairs but also 40 of the Navy ADs and Air Force F–51s and jets aided the Marine ground forces in their withdrawal to Line KANSAS. To speed the fighter-bombers to their targets, some of the Marine pilots were designated tactical air coordinators, airborne (TACA). Their familiarity with the terrain was an asset as they led incoming pilots to ground force units most in need of support.[32]

It was a confusing day in the air. The mutual radio frequencies to which planes and ground controllers were pretuned proved to be inadequate. The consequence was all too often the blocking out of key information at a frustrating moment. Haze and smoke made for limited vision. The planes needed a two-mile circle for their attacks, yet the battalions were at times less than 1,000 yards apart. DEVASTATE BAKER had to deal with this congested and dangerous situation as best it could.

In addition to its strong support of Marine ground forces, the 1st MAW sent 10 sorties to the ROKs in east Korea and 57 to I Corps in its battle along the Imjin. By this time the Gloucestershire Battalion of the 29th British Brigade was isolated seven miles behind enemy lines and receiving all supplies by air-drop. The outlook grew so desperate that officers ordered their men to break up and make their way back to the UN lines if they could. Only 40 ever succeeded.

In the former 6th ROK Division sector units of the 27th Brigade of the British Commonwealth Division had done a magnificent job of stopping the breakthrough. The 2d Battalion of the Princess Patricia's Canadian Light Infantry and the 3d Battalion of the Royal Australian Regiment distinguished themselves in this fight, which won a Distinguished Unit Citation for the division.[33]

---

[31] Capt R. A. Gugeler, USA, *Combat Actions in Korea*, 170–172.

[32] Summary of data from 1st MAW *HDs* for 24 Apr 51.

[33] EUSAK *Cmd Rpt*, Apr51, Sec I, 98, 100, 101; Brig C. N. Barclay, *The First Commonwealth Division* (Aldershot, 1954), 69–70.

Enemy Stopped in IX Corps Sector

Spring had come at last to war-ravaged Korea and the hills were a misty green in the sunshine. Looking down from an aircraft on the warm afternoon of 24 April 1951 the Marine sector resembled a human anthill. Columns of weary men toiled and strained in every direction. Chaotic as the scene may have seemed, however, everything had a purpose. The 1st Marine Division was in full control of all troop movements, despite enemy pressure of the last two nights.

The 5th Marines and KMCs had no opposition as they continued their withdrawal. Marine air reduced to a minimum the harassing efforts of the Chinese following the 1st Marines. As front-line units disengaged and fell back, the length of the main line of resistance was contracted enough for the 7th Marines to be assigned a reserve role. The 1st and 2d Battalions were given the responsibility for the defense of Chunchon as well as the crossing sites over the Pukhan and Soyang Rivers. Major Roach had reached the outskirts of Chunchon when 3/7 was ordered back across the Chunchon, to be attached to the 1st Marines on the left flank.[34]

Throughout the night of 24–25 April the enemy probed the Marine lines, seeking in vain a weak spot where a penetration could be made. It was already evident that the breakthrough in this area had given the Communists only a short-lived advantage. By the third night they were definitely stopped. Only minor patrol actions resulted except for two attacks in company strength on 2/1 at 0050 and 0150. Both were repulsed with total CCF losses of 25 counted dead.

Contrary to the usual rule, the Marines saw more action during the daylight hours. A company-size patrol from 1/1 became heavily engaged at 1350 and three Company A tanks moved up in support. The fight lasted until 1645, when the enemy broke off action and the tanks evacuated 18 wounded Marines.

Early in the afternoon a 3/1 patrol had advanced only 200 yards along a ridgeline when it was compelled to withdraw after running into concentrated mortar and machine gun fire. Sporadic mortar rounds continued until a direct hit was scored on the battalion CP, wounding Colonel McAlister, Lieutenant Colonel Banning, Major Reginald R. Myers, the executive officer, and Major Joseph D. Trompeter, the S–3.

[34] 7thMar *HD*, Apr 51.

Banning and Myers were evacuated and Trompeter assumed command of 3/1.

Losses of 18 KIA and 82 WIA for 24–25 April brought the casualties of the 1st Marines to nearly 300 during the past 48 hours.[85]

A simple ceremony was held at the 1st Marine Division CP on the afternoon of the 24th for the relief of General Smith by Major General Gerald C. Thomas. The new commanding general, a native of Missouri, was educated at Illinois Wesleyan University and enlisted in the Marine Corps in May 1917 at the age of 23. Awarded the Silver Star for bravery at Belleau Wood and Soissons, he was commissioned just before the Meuse-Argonne offensive, in which he was wounded.

During the next two decades, Thomas chased bandits in Haiti, guarded the U.S. mails, protected American interests in China, and served as naval observer in Egypt when Rommel knocked at the gates of Alexandria in 1941. As operations officer and later chief of staff of the 1st Marine Division, he participated in the Guadalcanal campaign in 1942. The next year he became chief of staff of I Marine Amphibious Corps in the Bougainville operation. Returning to Marine Headquarters in 1944 as Director of Plans and Policies, he was named commanding general of the Marines in China three years later.

General Smith had won an enduring place in the hearts of all Marines for his magnificent leadership as well as resourceful generalship during the Inchon-Seoul and Chosin Reservoir campaigns. Speaking of the Marines of April 1951, he paid them this tribute in retrospect:

> The unit commanders and staff of the Division deserve great credit for the manner in which they planned and conducted the operations which resulted in blunting the Chinese counteroffensive in our area. In my opinion, it was the most professional job performed by the Division while it was under my command.[86]

The night of 25–26 April passed in comparative quiet for the Marines. A few CCF probing attacks and occasional mortar rounds were the extent of the enemy's activity. All Marine units had now reached the modified Line KANSAS, but General Van Fleet desired further withdrawals because the enemy had cut a lateral road.

IX Corps also directed that the 1st Marine Division be prepared on

[85] 1stMar *HD*, Apr 51.
[86] Gen O. P. Smith USMC (Ret.), ltr of 11 Feb 58.

the 26th to move back to Chunchon, where it would defend along the south bank of the Soyang until service units could move out their large supply dumps. The Division was to tie in on the right with the lower extension of the Hwachon Reservoir, and contact was made in that quarter with the French battalion of the 2d Infantry Division, X Corps. On the Marine left flank the 5th Cavalry of the 1st Cavalry Division had relieved elements of the British Commonwealth Division.

Marine regimental officers met with Colonel Bowser, G–3, to plan the continued withdrawal. It was decided that four infantry battalions —1/1, 2/1, 3/5, and 3/7—were to take positions on the west bank of the Pukhan to protect the Mojin bridge and ferry sites while the other units crossed. The execution of the plan went smoothly, without enemy interference. After all other Marine troops were on the east side, 3/7 disengaged last of all and forded the chest-deep stream as a prelude to hiking to Chunchon.[37]

The enemy was kept at a discreet distance throughout the night by continuous artillery fires supplemented by ripples from Captain Eugene A. Bushe's Battery C, 1st 4.5″ Rocket Battalion. An acute shortage of trucks made it necessary for most of the troops to hike. Then came the task of organizing the new Division defenses on a line running northeast and southwest through the northern outskirts of Chunchon (Map 10). Planning continued meanwhile for further withdrawals to positions astride the Hongchon-Chunchon MSR.[38]

It was apparent by this time that the enemy had been badly mauled on the IX Corps front. The Communists were now making a supreme effort to smash through in the I Corps area and capture Seoul. It was believed that they had set themselves the goal of taking the city by May Day, the world-wide Communist holiday.

In this aspiration they were destined to be disappointed. They tried to work around the Eighth Army's left flank by crossing the river Han to the Kimpo Peninsula, but air strikes and the threat of naval gunfire frustrated them. Another flanking attempt 35 miles to the southeast met repulse, and before the end of the month it was evident that the Chinese Reds would not celebrate May Day in Seoul.

Generally speaking, the Eighth Army had kept its major units intact

[37] 1stMarDiv *HD*, Apr 51; CO 7thMar msg to CG 1stMarDiv, 2040 27 Apr 51.

[38] CO IX Corps *IXACT 1370;* 1stMar *HD*, Apr 51; 5thMar *HD*, Apr 51; 7thMar *HD*, Apr 51. A "ripple" normally consists of 144 rounds fired simultaneously by six launchers.

634040 O–62—10

and inflicted frightful losses on the enemy while trading shell-pocked ground for Chinese lives. The night of 27–28 April saw little activity on the IX Corps front, adding to the evidence that the enemy had shot his bolt. The next day the 1st Marine Division, along with other Eighth Army forces, continued the withdrawal to the general defensive line designated NO NAME Line (Map 10). Further withdrawals were not contemplated, asserted the IX Corps commander, who sent this message to General Thomas:

> It is the intention of CG Eighth Army to hold firmly on general defense line as outlined in my Operation Plan 17 and my message 9639, and from this line to inflict maximum personnel casualties by an active defense utilizing artillery and sharp armored counterattacks. Withdrawal south of this line will be initiated only on personal direction of Corps commander.[39]

FEAF placed the emphasis on armed reconnaissance or interdiction flights for Marine aircraft during the last few days of April. 1st MAW pilots reported the killing or wounding of 312 enemy troops on the 29th and 30th, and the destruction of 212 trucks, 6 locomotives, and 80 box cars. On the other side of the ledger, the Wing lost a plane a day during the first eight days of the CCF offensive. Of the fliers shot down, five were killed, one was wounded seriously but rescued by helicopter, and two returned safely from enemy-held territory.[40]

The shortage of vehicles slowed the withdrawal of Marine ground forces, but by the 30th the 5th Marines, KMC Regiment, and 7th Marines were deployed from left to right on NO NAME Line. The 1st Marines went into reserve near Hongchon. On the Division left was the reorganized 6th ROK Division, and on the right the 2d Infantry Division of X Corps.[41]

Nobody was in a better position to evaluate Marine maneuvers of the past week than Colonel Bowser, the G–3, and he had the highest praise. "Whereas the Chosin withdrawal was more spectacular than the April 'retrograde,'" he commented seven years later, "the latter was executed so smoothly and efficiently that a complex and difficult operation was made to look easy. The entire Division executed everything asked of it with the calm assurance of veterans." [42]

---

[39] CG IX Corps msg to CG 1stMarDiv with plans for withdrawal, 28 Apr 51.

[40] 1st MAW *HD*, Apr51, Pt #1, Chronology 22–30 Apr and App VI, PORs #46 (23 Apr) and 54 (1 May).

[41] 1stMarDiv *HD*, Apr 51; 5thMar *HD*, Apr 51; 7thMar *HD*, Apr 51.

[42] BrigGen A. L. Bowser, ltr of 14 Feb 58.

1st Marine Division Returns to X Corps

UN estimates of enemy casualties ranged from 70,000 to 100,000. The Fifth Phase Offensive was an unmitigated defeat for the Communists so far, but EUSAK G–2 officers warned that this was only the first round. Seventeen fresh CCF divisions were available for the second.

General Van Fleet called a conference of corps commanders on 30 April to discuss defensive plans. In the reshuffling of units the 1st Marine Division was placed for the third time in eight months under the operational control of X Corps, commanded by Lieutenant General Edward M. Almond. The Marines were to occupy the western sector of X Corps after its boundary with IX Corps had been shifted about 12 miles to the west.[43]

Van Fleet put into effect a reshuffling of units all the way across the peninsula in preparation for the expected renewal of the CCF offensive. Thus on 1 May the UN line was as follows from left to right:

US I Corps—1 ROK Division, 1st Cavalry Division and 25th Infantry Division in line; the 3d Infantry Division and British 29th Brigade in reserve;
US IX Corps—British 27th Brigade, 24th Infantry Division, 5th and 6th ROK Divisions and 7th Infantry Division in line; the 187th Airborne RCT in reserve;
US X Corps—1st Marine Division, 2d Infantry Division, 5th and 7th ROK Divisions;
ROK III Corps—9th and 3d Divisions;
ROK I Corps—Capitol Division and ROK 11th Division.

"I don't want to lose a company—certainly not a battalion," Van Fleet told the corps commanders. "Keep units intact. Small units must be kept within supporting distance. . . . Give every consideration to the use of armor and infantry teams for a limited objective counterthrust. For greater distances, have ready and use when appropriate, regiments of infantry protected by artillery and tanks." [44]

From the foxhole to the command post a confident new offensive spirit animated an Eight Army which only four months previously had been recuperating from two major reverses within two months. The Eighth Army, in short, had been welded by fire into one of the finest military instruments of American military history; and the foreign units attached to it proved on the battlefield that they were picked troops.

---

[43] EUSAK *Cmd Rpt*, Apr 51, 115–118.
[44] *Ibid.*

With the Hwachon dam now in enemy hands, the Communists had the capability of closing the gates, thus lowering the water level in the Pukhan and Han rivers to fording depth. As a countermeasure, EUSAK asked the Navy to blast the dam. It was a difficult assignment, but Douglas AD Skyraiders from the *Princeton* successfully torpedoed the flood gates on 1 May.[45]

An atmosphere of watchful waiting prevailed during the next two weeks as the Marines on NO NAME Line improved their defensive positions and patrolled to maintain contact with the enemy. Eighth Army evolved at this time the "patrol base" concept to deal with an enemy retiring beyond artillery range. These bases were part of a screen, called the outpost line of resistance (OPLR), established in front of the MLR. Their mission was to maintain contact with the enemy by means of patrols, give warning of an impending attack, and delay its progress as much as possible.

When it came to artillery ammunition, the 11th Marines found that it had progressed from a famine to a feast. Where shells had recently been rationed because of transport difficulties, the Eighth Army now directed the cannoneers along NO NAME Line to expend a unit of fire a day. The 11th Marines protested, since the infantry was seldom in contact with the enemy. One artillery battalion submitted a tongue-in-cheek report to the effect that the required amount of ammunition had been fired "in target areas cleared of friendly patrols." [46] The requirement was kept in force, however, until the demands of the renewed CCF offensive resulted in another ammunition shortage for the 11th Marines.

Marine tanks were directed by Division to use their 90mm rifles to supplement 11th Marine howitzers in carrying out Corps fire plans. The tankers protested that their tubes had nearly reached the end of a normal life expectancy, with no replacements in sight. This plaint did not fall upon deaf ears at Corps Headquarters and two Army units, the 96th AFA Battalion and 17th FA Battalion, were assigned to fire the deep missions.[47]

Eighth Army staff officers concluded that the enemy would launch his next effort in the center. Intelligence, according to General Van Fleet, "had noted for some 2 weeks prior to the May attack that

[45] *PacFlt Interim Rpt* No. 2, II, 766.
[46] Col Merritt Adelman, ltr of 10 Feb 58.
[47] 11thMar tel to G–2 1stMarDiv, 4 May 51; CG 1stMarDiv msg to CG X Corps, 9 May 51; X Corps msg X9613, 10 May 51.

the Chinese Communists were shifting their units to the east." Nevertheless, the blow fell "much farther east than [was] expected."[48]

Although the east offered the best prospects of surprise, a rugged terrain of few roads imposed grave logistical handicaps on the enemy. Moreover, UN warships dominated the entire eastern littoral. Despite these disadvantages, an estimated 125,000 Chinese attacked on the morning of 16 May 1951 in the area of the III and I ROK Corps between the U.S. 2d Infantry Division and the coast. Six CCF divisions spearheaded an advance on a 20-mile front that broke through the lines of the 5th and 7th ROK Divisions. Pouring into this gap, the Communists made a maximum penetration of 30 miles that endangered the right flank of the U.S. 2d Infantry Division.

General Van Fleet took immediate steps to stabilize the front. In one of the war's most remarkable maneuvers he sent units of the 3d Infantry Division, then in reserve southeast of Seoul, on a 70-mile all-night ride in trucks to the threatened area.[49]

The 1st Marine Division was not directly in the path of the enemy advance. During the early morning hours of 17 May, however, an enemy column made a thrust that apparently was intended as an end-run attack on the left flank of the 2d Infantry Division. Avoiding initially the Chunchon-Hongchon highway, Chinese in estimated regimental strength slipped behind the patrol base set up by a KMC company just west of the MSR (Map 12).

For several days Colonel Nickerson and his executive officer, Lieutenant Colonel Raymond G. Davis, had been apprehensive over the security of this road on which the 7th Marines depended for logistical support. On the afternoon of the 17th they pulled back Lieutenant Colonel Bernard T. Kelly's 3/7 (less Company G) to establish a blocking position, generally rectangular in shape, at the vital Morae-Kogae pass on the Chunchon road. This move was not completed until sunset and George Company did not rejoin the battalion until midnight, so that the enemy probably had no intelligence of the new position. The main road ran along a shelf on one shoulder of the pass, but the Chinese avoided it and came by a trail from the northwest (Map 12).

The surprise was mutual. A platoon of D/Tanks, a Weapons Com-

[48] Gen J. A. Van Fleet USA (Ret.), ltr of 24 Mar 58.

[49] Unless otherwise specified, accounts of the CCF offensive of 16 May 51 are based on the following sources: EUSAK *Cmd Rpt*, May 51, 12–18; 1stMarDiv *HD*, May 51; CO 7thMar msg to CG 1stMarDiv, 2015 17 May 51.

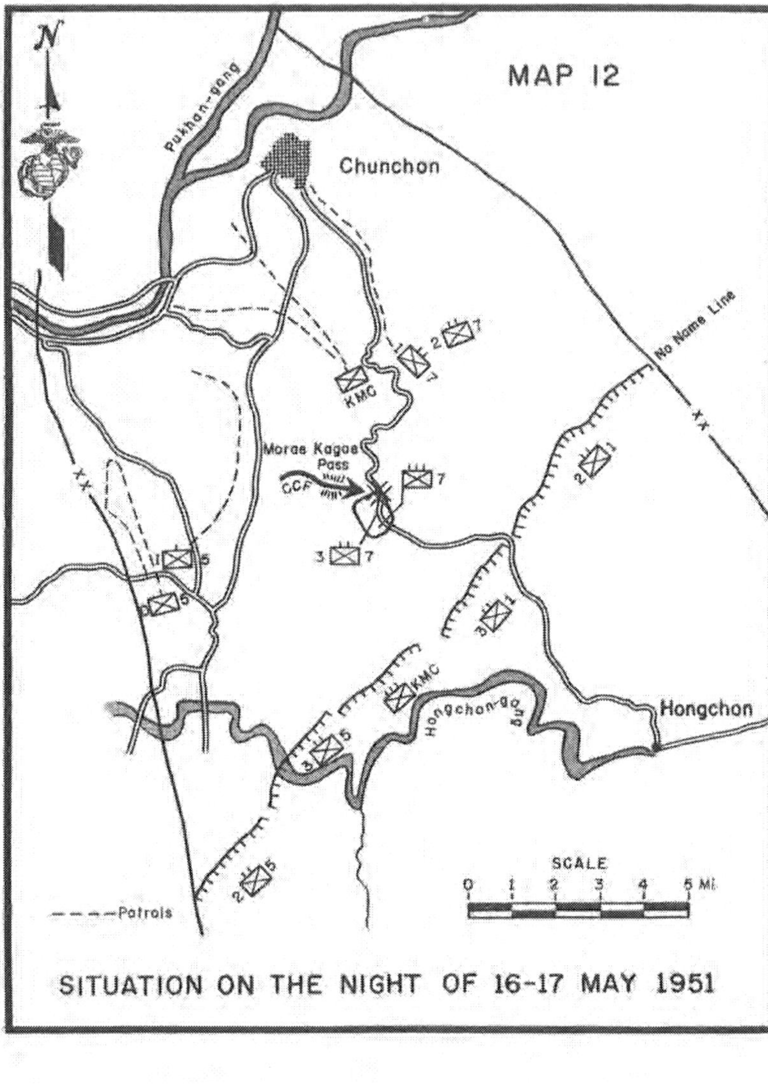

SITUATION ON THE NIGHT OF 16-17 MAY 1951

pany platoon, and an Item Company platoon, defending the northern end of the perimeter, opened up with everything they had. A desperate fire fight ensued as the enemy replied with a variety of weapons—mortars, recoilless rifles, satchel charges, grenades, and machine guns.

Two CCF soldiers were killed after disabling a Marine tank by a grenade explosion in the engine compartment. A satchel charge knocked out another tank, and the enemy made an unsuccessful attempt to kill a third by rolling up a drum of gasoline and igniting it.

Captain Victor Stoyanow's Item Company, at the critical point of the thinly stretched 3/7 perimeter, was hard-pressed. The enemy made a slight penetration into one platoon position but was repulsed by a counterattack that Stoyanow led. Marine infantry and tanks were well supported by artillery that sealed off the Chinese column from the rear. The action ended at daybreak with the routed enemy seeking only escape as Marine artillery and mortars continued to find lucrative targets. Air did not come on station until about 1030, when it added to the slaughter. Scattered enemy groups finally found a refuge in the hills, leaving behind 82 prisoners and 112 counted dead. Captures of enemy equipment included mortars, recoilless rifles, and Russian 76mm guns and machine guns. Friendly losses were 7 KIA and 19 WIA.[60]

1st MAW squadrons were kept busy furnishing close air support to the 2d Infantry Division and the two ROK divisions hit by the enemy's May offensive. Because of the patrolling in the Marine sector, the OYs of VMO–6 took over much of the task of controlling air strikes. They flew cover for the infantry-tank patrols, and in the distant areas controlled almost as many air strikes as they did artillery missions. From the 1st to the 23d of May, VMO–6 observers controlled 54 air strikes involving 189 UN planes—159 Navy and Marine F4Us, F9Fs, and ADs, and 30 Air Force F–80s, F–84s, and F–51s. About 40 percent of the aircraft controlled by the OYs were non-Marine planes.[61]

On the 18th the 1st Marine Division, carrying out X Corps orders, began a maneuver designed to aid the U.S. 2d Infantry Division on the east by narrowing its front. The 7th Marines pulled back to NO NAME Line to relieve the 1st Marines, which side-slipped to the east to take

[60] This account of 3/7's action is derived from 1stMarDiv, 7thMar, 1stTkBn, and 3/7 *HDs*, May 51; Col B. T. Kelly, interv of 28 Dec 57.

[61] VMO–6 *HD*, May 51.

over an area held by the 9th Infantry. The 5th Marines then swung around from the Division left flank to the extreme right and relieved another Army regiment, the 38th Infantry. This permitted the 2d Infantry Division to face east and repulse attacks from that direction.

By noon on 19 May the enemy's renewed Fifth Phase Offensive had lost most of its momentum as CCF supplies dwindled to a trickle along a tenuous line of communications. That same day, when Colonel Wilburt S. Brown took over the command of the 1st Marines from Colonel McAlister, all four Marine regiments were in line—from left to right, the KMCs, the 7th Marines, the 1st Marines, and the 5th Marines. A new NO NAME Line ran more in a east-west direction than the old one with its northeast to southwest slant. Thus in the east of the Marine sector the line was moved back some 4,000 yards while remaining virtually unchanged in the west.

Enough enemy pressure was still being felt by the 2d Infantry Division so that General Van Fleet ordered a limited offensive by IX Corps to divert some of the CCF strength. While the rest of the 1st Marine Division stood fast, the KMC Regiment advanced with IX Corps elements.

At the other end of the line the Marines had the second of their two fights during the CCF offensive. Major Morse L. Holliday's 3/5 became engaged at 0445 on the 20th with elements of the 44th CCF Division. Chinese in regimental strength were apparently on the way to occupy the positions of the Marine battalion, unaware of its presence.

This mistake cost them dearly when 3/5 opened up with every weapon at its disposal while requesting the support of Marine air, rockets, and artillery. The slaughter lasted until 0930, when the last of the routed Chinese escaped into the hills. Fifteen were taken prisoner and 152 dead were counted in front of the Marine positions.[52]

From 20 May onward, it grew more apparent every hour that the second installment of the CCF Fifth Phase Offensive had failed even more conclusively than the first. The enemy had only a narrow penetration on a secondary front to show for ruinous casualties. Worse yet, from the Chinese viewpoint, the UN forces were in a position to retaliate before the attackers recovered their tactical balance. The Eighth Army had come through with relatively light losses, and it was now about to seize the initiative.

---

[52] 5thMar *HD*, May 51.

CHAPTER VII

Advance to the Punchbowl

Plan to Cut Off Communists—Initial Marine Objectives Secured—MAG–12 Moves to K–46 at Hoengsong—Fight of the 5th Marines for Hill 610—1st MAW in Operation STRANGLE*—KMC Regiment Launches Night Attack—1st Marines Moves Up to* BROWN *Line—7th Marines Committed to Attack*

ONLY FROM THE AIR could the effects of the UN counterstroke of May and June 1951 be fully appreciated. It was more than a CCF withdrawal; it was a flight of beaten troops under very little control in some instances. They were scourged with bullets, rockets, and napalm as planes swooped down upon them like hawks scattering chickens. And where it had been rare for a single Chinese soldier to surrender voluntarily, remnants of platoons, companies, and even battalions were now giving up after throwing down their arms.

There had been nothing like it before, and its like would never be seen in Korea again. The enemy was on the run! General Van Fleet, after his retirement, summed up the double-barreled Chinese spring offensive and the UN counterstroke in these words:

> We met the attack and routed the enemy. We had him beaten and could have destroyed his armies. Those days are the ones most vivid in my memory—great days when all the Eighth Army, and we thought America too, were inspired to win. In those days in Korea we reached the heights.[1]

Communist casualties from 15 to 31 May were estimated by the Eighth Army at 105,000. This figure included 17,000 counted dead and the unprecedented total of some 10,000 prisoners, most of them Chinese Reds taken during the last week of the month in frantic efforts to escape. Such results were a vast departure from past occa-

[1] Gen J. A. Van Fleet, USA (Ret), "The Truth About Korea," *Life*, 11 May 53.

sions when Mao Tse-tung's troops had preferred death to surrender.

In all probability, only the mountainous terrain saved them from a complete debacle. If the Eighth Army had been able to use its armor for a mechanized pursuit, it might have struck blows from which the enemy could not recover. As it was, the Communists escaped disaster by virtue of the fact that a platoon could often stand off a company or even a battalion by digging in and defending high ground commanding the only approach. Every hill was a potential Thermopylae in this craggy land of few roads.

It was the misfortune of the 1st Marine Division to have perhaps the least lucrative zone of action in all Korea for the peninsula-wide turkey shoot. A chaos of jagged peaks and dark, narrow valleys, the terrain alone was enough to limit an advance. Even so, the Marines inflicted 1,870 counted KIA casualties on the Communists in May and captured 593, most of them during the last eight days of the month.

General Almond congratulated the Division for its accomplishment of "a most arduous battle task. You have denied [the enemy] the opportunity of regrouping his forces and forced him into a hasty retreat; the destruction of enemy forces and materiel has been tremendous and many times greater than our own losses." [a]

Plan to Cut Off Communists

The 187th Airborne Regimental Combat Team, released from IX Corps reserve, arrived in the Hongchon area on 21 May and took a position between the 1st Marine Division on the left and the 2d Infantry Division on the right. Two days later X Corps gave the Marines the mission of securing the important road center of Yanggu at the eastern end of the Hwachon Reservoir (Map 13). Elements of the 2d Infantry Division, with the 187th Airborne RCT attached, were meanwhile to drive northeast to Inje after establishing a bridgehead across the river Soyang. From Inje the 187th (reinforced) would continue to advance northeast toward its final objective, Kansong on the coast. After linking up with I ROK Corps, the Army regiment might be able to pull the drawstring on a tremendous bag of prisoners —all the CCF forces south of the Inje-Kansong road. There was, however, a big "if" in the equation. The Communists were falling

[a] CG X Corps msg of 1500, 3 Jun 51; 1stMarDiv *HD*, May 51.

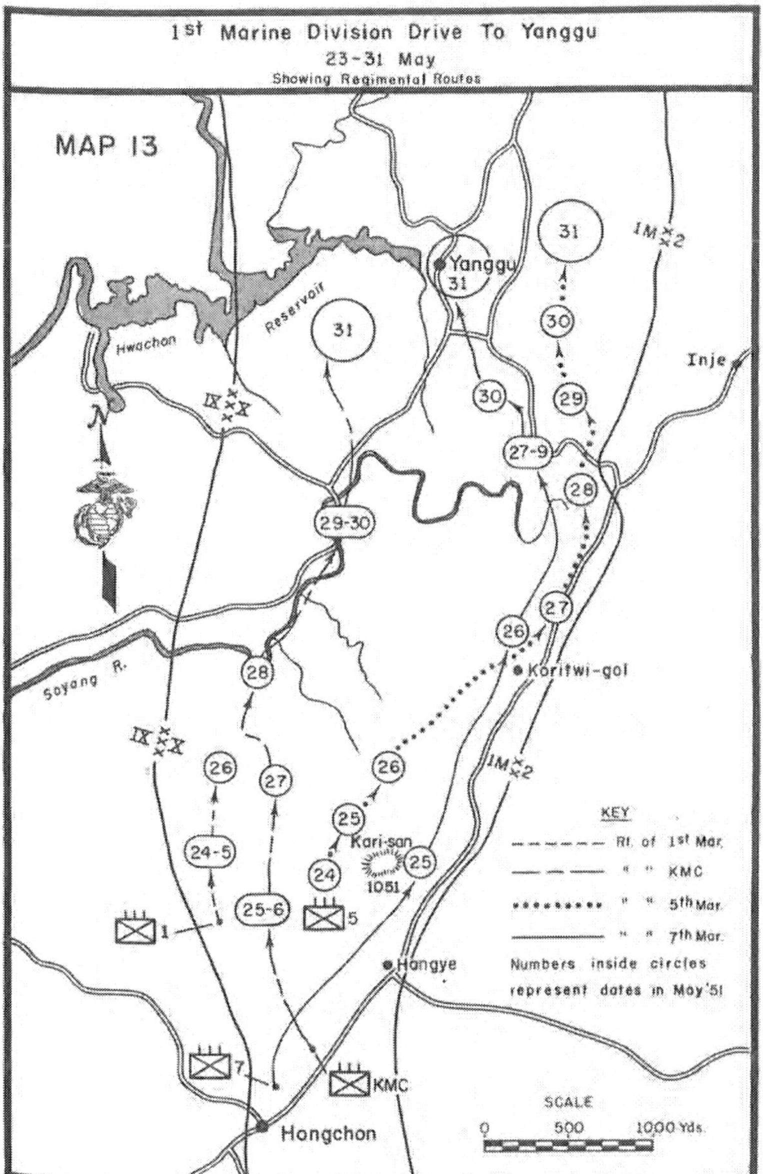

1st Marine Division Drive To Yanggu
23-31 May
Showing Regimental Routes
MAP 13
Hwachon
Reservoir
Yanggu
Inje
Soyang R.
Koritwi-gol
Kari-san
1051
Hangye
Hongchon
KMC
KEY
Rt. of 1st Mar.
" " KMC
" " 5th Mar.
" " 7th Mar.
Numbers inside circles represent dates in May '51
SCALE
0 500 1000 Yds.

back with all haste, and it was a question whether the bag could be closed in time.

The 1st Marine Division jumped off at 0800 on 23 May with the 1st and 5th Marines abreast, the 1st on the left. Both regiments advanced more than 5,000 yards against negligible opposition. During the course of this attack the 1st Marines experimented by calling an air strike in the hope of detonating an entire mine field. The results were disappointing. Live mines were blown to new locations, thus changing the pattern, but few exploded.[3]

The 7th Marines was relieved on the 23d by elements of the 7th Infantry Division (IX Corps) and moved to the east for employment on the Marine right flank. The KMC Regiment, relieved by other IX Corps units, went into Division reserve.[4]

The 1st Marines, advancing on the left, reached its objectives, about two-thirds of the way to the Soyang, by noon on the 26th. The regiment reverted to Division reserve upon relief by the KMCs. In the right half of the Division zone, resistance gradually stiffened. On the 24th, the 2d and 3d Battalions of the 5th Marines ran into trouble as they started their advance toward their initial objective, three hills about 7,000 yards north of Hangye (Map 12). Both battalions were slowed by heavy enemy mortar and machine gun fire. They requested immediate artillery and air support.

Captain John A. Pearson, commanding Item Company, could observe the enemy on Hill 1051, holding up the attack with flanking fire. He directed air and artillery on the crest and on the Communists dug in along the southeastern slopes. Soon the enemy troops were seen retiring northward. This eased the pressure on the center, and Captain Samuel S. Smith's Dog Company managed to work forward and gain the summit of Hill 883 by 1300. Tanks moved up in support and at midnight Colonel Hayward reported his portion of the Division objective secured.[5]

The 7th Marines, moving forward in the right rear of the 5th, veered to the left and drove into the center of the Division zone, reaching the southern bank of the Soyang by nightfall on the 26th. That same day 2/7 overran an enemy ammunition dump and took

[3] 2/1 *HD*, May 51.
[4] 1stMarDiv *HD*, May 51
[5] CO 5thMar msg to CG 1stMarDiv, 2359 24 May 51.

27 CCF prisoners, some of them wounded men who had been left behind. The captured material included the following items:

100,000 rounds of small-arms ammunition;
12,000 rounds of mortar ammunition;
1,000 rounds of artillery ammunition;
6,000 pounds of explosive charges;
9,000 hand grenades.

Five U.S. trucks and jeeps were "released to higher headquarters." Two CCF trucks, two mules, and a horse were "integrated into the battalion transportation system and profitably employed thereafter." [6]

The 187th Airborne RCT reported on the 24th that its advance was being held up by increasing enemy resistance.[7] It was already evident that the CCF flight had frustrated the plan of cutting off decisively large numbers in the X Corps zone. Air observation established, however, that hundreds of Chinese Reds had merely escaped from the frying pan into the fire. By fleeing westward along the south shore of the Hwachon Reservoir, they stumbled into the IX Corps zone. There the remnants of whole units surrendered, in some instances without striking a blow. Along the route they were pitilessly attacked by UN aircraft. 1st MAW units had never before known such good hunting as during the last week in May 1951.[8]

Despite the "murky instrument weather" of 27 May the all-weather fighters of VMF(N)-513 reporting the killing of an estimated 425 CCF soldiers. Two F7F pilots killed or wounded some 200 Chinese Reds in the I Corps zone. On the following day the 1st MAW claimed a total of 454 KIA casualties inflicted on the enemy.[9]

Estimates of enemy dead by pilots are likely to be over-optimistic, but there can be no doubt that UN aircraft slaughtered the fleeing Communists in large numbers. Only poor flying weather saved the enemy from far worse casualties. So intent were the Chinese on escape that they violated their usual rule of making troop movements only by night. When the fog and mist cleared briefly, Marine pilots had

[6] CO 7thMar msg to CG 1stMarDiv, 2050 26 May 51; Col W. F. Meyerhoff, ltr of 8 Aug 58.

[7] CO 5thMar msg to CG 1stMarDiv, 24 May 51, in 5thMar In& Out #13.

[8] James T. Stewart, *Airpower, The Decisive Force in Korea* (Princeton, N.J.: D. Van Nostrand Company, Inc., 1957) 13–15, 84–86; 1st MAW *HD*, May 51, Pts 4 and 5, Fifth Air Force Frag orders (hereafter listed as FAF FragOs), 20–31 May; 1st MAW *HD* May 51, Pt 1, G–3 PORS for 20–31 May; *Ibid.*, Pt 2, Staff Jrn G–3, 25 May, 26 May, 27 May, 31 May; EUSAK *Cmd Rept*, May 51, Sec II, Bk 4, Pts 5 and 6, Encls 20–31, PORs, sections entitled G–3 Air.

[9] *Ibid.*, VMF(N)–513 *HD*, 27 May 51.

glimpses of CCF units crowding the roads without any attempt at concealment. Napalm, bombs, and machine guns left heaps of dead and wounded as the survivors continued their flight, hoping for a return of fog and mist to protect them.

Initial Marine Objectives Secured

As the Marine ground forces advanced, they found fewer and fewer Chinese Reds opposing them. The explanation was given by a prisoner from the 12th Division, V Corps, of the North Korean People's Army (NKPA). His unit had the mission, he said, of relieving troops in the Yanggu-Inje area and conducting delaying actions. The purpose was to allow CCF units to escape a complete disaster and dig in farther north. The North Koreans, in short, were being sacrificed in rear guard delaying actions in order that the Chinese Reds might save their own skins.

U.S. interrogators asked NKPA prisoners why they put up with such treatment. The answer was that they couldn't help themselves. The Chinese had impressed them into service, armed them, and trained them after the NKPA collapse in the fall of 1950. They were under the thumb of political commissars holding life and death authority over them. Any NKPA soldier suspected of trying to shirk his duty or escape was certain to be shot like a dog. At least the man on the firing line had a chance to come out alive; the man who defied the system had none.

This attitude accounts to a large extent for the many occasions when NKPA troops literally resisted to the last man in delaying actions. Marines in general, judging by their comments, considered the Chinese Red the better all-around soldier; but they credited the Korean Red with more tenacity on the defensive.

Because of the stubborn NKPA opposition in East Korea, the Eighth Army staff and command gave some thought to the possibility of an amphibious operation in the enemy's rear by the 1st Marine Division. Plans were discussed on 28 May for a landing at Tongchon (Map 8). The Marines were to drive southward along the Tongchon-Kumhwa road to link up with the IX Corps units attacking toward the northeast along the same route. After meeting, the two forces would systematically destroy the pocketed enemy units. It was decided that 6 June would be D-day. And then, to the great disappointment of Generals

Thomas and Almond, the plan was suddenly cancelled by EUSAK on 29 May after a single day's consideration.[10]

Another scheme for cutting off large enemy forces was abandoned on 28 May when the 187th Airborne got as far as Inje. Most of the CCF units having escaped, this regiment was given a new mission of securing the high ground to the north of Inje.

During the last five days of May the 5th and 7th Marines continued to advance steadily. On the morning of the 31st the 7th faced the task of breaking through a stubbornly contested pass leading into Yanggu. With a battalion on each ridge leading into the pass, Colonel Nickerson found it a slow yet precarious prelude to get the men down. Adding to their trials were some 500 enemy 76mm and mortar shells received by the regiment.

General Van Fleet, an onlooker while visiting the 7th Marines OP, shook his head wonderingly. "How did you ever get the men up those cliffs?" he asked Colonel Nickerson.

The answer was short and simple. "General," said the regimental commander, "they climbed."

As the day wore on, Nickerson called for what his executive officer, Lieutenant Colonel Davis, described as "a through-the-middle play. A company of tanks [Company C, 1st Tank Battalion, commanded by Captain Richard M. Taylor] was launched up the road with infantry on foot hugging the protective cover of the steep road embankments. As the tanks drew fire, the infantry could spot the source and . . . quickly cleaned the enemy out. This rapid thrust caused the enemy defenders to flee as fire was poured into them from our center force as well as the flank attackers." [11]

By nightfall on the 31st the 7th Marines had control of Yanggu, its airfield, and the hills surrounding that burnt-out town. The 5th Marines had reached a point 6,000 yards northeast of Yanggu, astride the north-south ridgeline between that road center and Inje.

Losses for the 1st Marine Division in May added up to 75 KIA, 8 DOW, and 731 WIA. The ratio of wounded to killed, it may be noted, is more than nine-to-one. This proportion, so much more favorable than the usual ratio, rose to an even more astonishing 15-to-1

---

[10] EUSAK *Cmd Rpt*, May 51, 24; Gen G. C. Thomas, USMC (Ret.), interv of 6 Jun 58; LtGen E. M. Almond, USA (Ret.), ltr of 22 May 58.

[11] Col R. G. Davis, comments, n.d.; *HDs* for 1stMarDiv, 5thMar and 7thMar for May 51.

in June. Various explanations have been offered, one of them being the spirit of cool professionalism of Marines who had learned how to take cover and not expose themselves to needless risks. But this doesn't account for the unusual ratio, and it may perhaps be concluded that the Marines were simply lucky in this operation.

The comparatively low death rate has also been credited in part to the alertness with which Marine officers adapted to changing situations. War is a grim business on the whole, but Colonel Wilburt S. Brown took an amusing advantage of enemy propaganda accusing Americans of all manner of crimes against humanity. At the outset he had requested colored smoke shells for signaling. But upon learning from POW interrogations that NKPA soldiers were terrified by what they believed to be frightful new gases, the commanding officer of the 1st Marines had an added reason for using green, red, and yellow smoke. Unfortunately, Lieutenant Colonel Merritt Adelman, commanding officer of the 2d Battalion, 11th Marines, soon had to inform him that the inadequate supply was exhausted.[12] It was never renewed during Brown's command.

Major David W. McFarland, commanding officer of VMO–6, also exploited enemy ignorance. His original purpose in initiating night aerial observation by OY planes was to improve artillery accuracy. Soon he noticed that the mere presence of an OY overhead would silence enemy artillery.

"The aerial observer," McFarland explained, "was often unable to determine the location of enemy artillery even though he could see it firing, because he would be unable to locate map coordinates in the dark—that is, relating them to the ground. Fortunately, this fact was unknown to the enemy. From their observation of the OYs in the daytime, they had found that the safest thing to do whenever an OY was overhead was to take cover. This they continued to do at night." [13]

VMO–6 also put into effect an improvement of 1st Marine Division aerial photographic service at a time when the 1st MAW photo section had missions all over the Korean front. Lieutenant Colonel Donald S. Bush, commanding officer of the section, is credited with the innovation of mounting a K–17 camera on a OY. Only a 6-inch focal length lens could be installed on one of these small planes. This meant that

[12] MajGen W. S. Brown, USMC (Ret.), ltr of 21 Aug 58.
[13] LtCol D. W. McFarland, ltr of 21 Aug 58.

in order to get the same picture as a jet the OY must fly at half the altitude. The pilot would be in more danger but haze problems were reduced.

The experiment was an immediate success. The Division set up a photo laboratory near the VMO–6 CP for rapid processing and printing. A helicopter stood by for rapid delivery to the units concerned.[14]

Not all the variations in tactics were innovations. Lieutenant Colonel Bernard T. Kelly, commanding officer of 3/7, revived an old device on 31 May by using indirect automatic weapons fire with good effect. Four water-cooled heavy machine guns provided long range (2,600 yards) plunging fires on the reverse slopes of hills in support of his leading elements during the final attack on Yanggu.[15]

MAG–12 Moves to K–46 at Hoengsong

Delay and uncertainty were still the two great stumbling blocks to adequate air support for the ground forces under the JOC control system. Marine officers contended that infantry units sometimes took unnecessary casualties as a consequence. Worse yet, there were occasions when the expected planes did not arrive at all.

Statistics kept by the 1st MAW and Navy during the spring of 1951 upheld these conclusions. During the Inchon-Seoul operation, the average delay in receiving air support had been 15 minutes as compared to 80 minutes in May and June of 1951. Approximately 35 minutes of this time was required to process the request through JOC. And only 65 to 70 percent of the sorties requested were ever received by Marine ground forces.[16]

Generals Shepherd and Harris had discussed the problem during the early spring of 1951 with General Partridge of the Fifth Air Force. Several compromises were reached, and for brief periods the 1st Marine Division received more air support than it could use. Unfortunately, these periods were at times of the least need. When the chips were down, the old delays and uncertainties reappeared. General Partridge commented:

> The 1st Marine Air Wing was assigned for operational control by the Fifth Air Force and it was used just as any of the other units of the Fifth

[14] *Ibid.*
[15] Col B. T. Kelly, interv of 9 Jun 58.
[16] *PacFlt Interim Rpt* No. 2, II, 523–537.

634040 O–62–11

> were employed, that is, in support anywhere along the battle front were it appeared to be most urgently needed.
>
> In every action such as took place in Korea when the resources and especially the air resources are far too few, ground commanders inevitably feel that they are being shortchanged. They are trying to accomplish their objectives under the most difficult circumstances and with the minimum number of casualties and they want all the assistance from the air that they can get. I am sure I would feel the same in similar circumstances. However, there was never enough air support to satisfy everyone and I was most unhappy that this was the case.
>
> From time to time I was called upon to denude one section of the front of its close air support in order to bolster some other area where the situation was critical. Sometimes this worked to the advantage of the Marines as in the case of operations near the Chosin Reservoir in December 1950, and at other times it worked to their disadvantage. In retrospect, however, I would estimate that, day in and day out, the Marine ground units had more air support than any other division which was engaged.[17]

With all due respect to General Partridge, Marine officers felt that the discussion should not be limited merely to the amount of air support. It was not so much the amount as the delay and unreliability under JOC control that constituted the problem as the Marines saw it. On 24 May, while on one of his periodic tours of the Far East, General Shepherd brought up the matter of CAS with General Ridgway. He agreed with the UN commander in chief that it would be improper for a Marine division to expect the exclusive support of a Marine air wing in Korea. The main difficulty, he reiterated, lay in the slowness and uncertainty of getting air support when needed.[18]

At this time an extensive reshuffling of Air Force commanders was in progress. On 21 May General Partridge relieved Lieutenant General George E. Stratemeyer, CG FEAF, who had suffered a heart attack. Partridge in turn was relieved by Major General Edward J. Timberlake, who assumed temporary command of Fifth Air Force until Major General Frank E. Everest arrived to take over a few days later.

The 1st MAW was also undergoing changes in command. General Harris was relieved on 29 May by his deputy commander, Major General Thomas J. Cushman. Brigadier General William O. Brice, just arrived from the States, became the Wing's new deputy commander.

---

[17] Gen E. E. Partridge, USAF, ltr of 28 Jun 59.

[18] *FMFPac Visit* 21–31 May 51, 5, 6.

After several "get acquainted" discussions, the new Air Force and 1st MAW generals agreed on a plan to cut down delays in air support. It was a simple solution: the aircraft were merely to be brought nearer to the Marine ground forces. This was to be managed by moving the MAG–12 forward echelon from K–16 at Seoul to K–46 at Hoengsong (Map 16). The new field, if such it could be called, was nothing more than a stony dirt strip. But it was only 40 miles, or a 10- to 15-minute flight, from the firing line. The first missions from the new field were flown on 27 May. VMFs–214 and –323 kept an average of 12 Corsairs at K–46 thereafter, rotating them from K–1.[19]

On the surface this seemed to be a practical solution, especially after a four-plane alert was established at K–46 for use by the 1st Marine Division when needed. DEVASTATE BAKER was permitted to put in an alerting call directly to the field. The rub was that JOC must be called in order to make the original request. Before the planes could take off, the MAG–12 operations officer at the field was likewise required to call JOC and confirm the fact that the mission had been approved.

Communications were poor at first for the 40 miles between the field and the front. DEVASTATE BAKER got better results by calling 1st MAW Headquarters at K–1, 140 miles south, and having the Wing call K–46 and JOC. This meant delays such as General Thomas described in a letter to General Almond. On 29 May, he said, the 5th and 7th Marines were up against severe enemy fire in their attack. The TACPs had enemy targets under observation and were ready to control any aircraft they could get. The Marines requested 92 sorties and received 55. Of these, 20 were flown by Corsairs or Panther Jets, and 35 by Air Force jets and Mustangs. And though 55 sorties were considerably less than optimum air support, practically all arrived from two to four hours late. On the firing line the enemy's resistance, concluded General Thomas, was broken not by air power but by Marine riflemen.[20]

On other days the new plan made a more encouraging showing. There was, for instance, the occasion when the OYs discovered an enemy regiment near the 1st Marine Division right flank. DEVASTATE

[19] MAG–12 *HD*, May 51, 24, 25 and 27 May; 1st MAW *HD*, May 51, Summary and Chronology for 19, 24, 27 and 28 May 51.
[20] CG 1stMarDiv ltr to CG X Corps, 31 May 51.

BAKER called the 1st MAW direct on 31 May for 16 fighters as soon as possible. Wing called JOC for approval to launch the flight and put in a call to K–46 to alert the planes. In just 48 minutes after the initial call from DEVASTATE BAKER, 16 pilots had jumped into their flight gear at K–46, had been briefed, and were airborne on what proved to be a timely strike with excellent results.[21]

A new tactic of night air support was introduced late in May when Marine R4D transports were outfitted to operate as flare planes. Not only did these unarmed aircraft light up targets along the front lines for the VMF(N)–513 night fighters; they were also on call for use by the 1st Marine Division. Later, on 12 June, the Navy provided the 1st MAW with PB4Y–2 Privateers for the nightly illumination missions.[22]

Fight of the 5th Marines for Hill 610

During the heyday of the battleship, every midshipman dreamed of some glorious future day when he would be on the bridge, directing the naval maneuver known as crossing the T. In other words, his ships would be in line of battle, firing converging broadsides on an enemy approaching in column. Obviously, the enemy would be at a disadvantage until he executed a 90° turn under fire to bring his battered ships into line to deliver broadsides of their own.

It was a mountain warfare variation of crossing the T that the Korean Reds were using against the Marines. Whenever possible, the enemy made a stand on a hill flanked by transverse ridgelines. He emplaced hidden machine guns or mortars on these ridgelines to pour a converging fire into attackers limited by the terrain to a single approach. It meant that the Marines had to advance through this crossfire before they could get in position for the final assault on the enemy's main position.

There were two tactical antidotes. One was well directed close air support. The other was the support of tanks advancing parallel to enemy-held ridgelines and scorching them with the direct fire of 90mm rifles and 50 caliber machine guns.

---

[21] 1st MAW *HD*, May 51, Pt 2, Assessment Rpt for 31 May 51.

[22] 1st MAW *HD*, May 51, Pt 1, App II, 2; Chronology, 31 May; MAG–12 *HD* Jun 51, Chronology and 12 Jun.

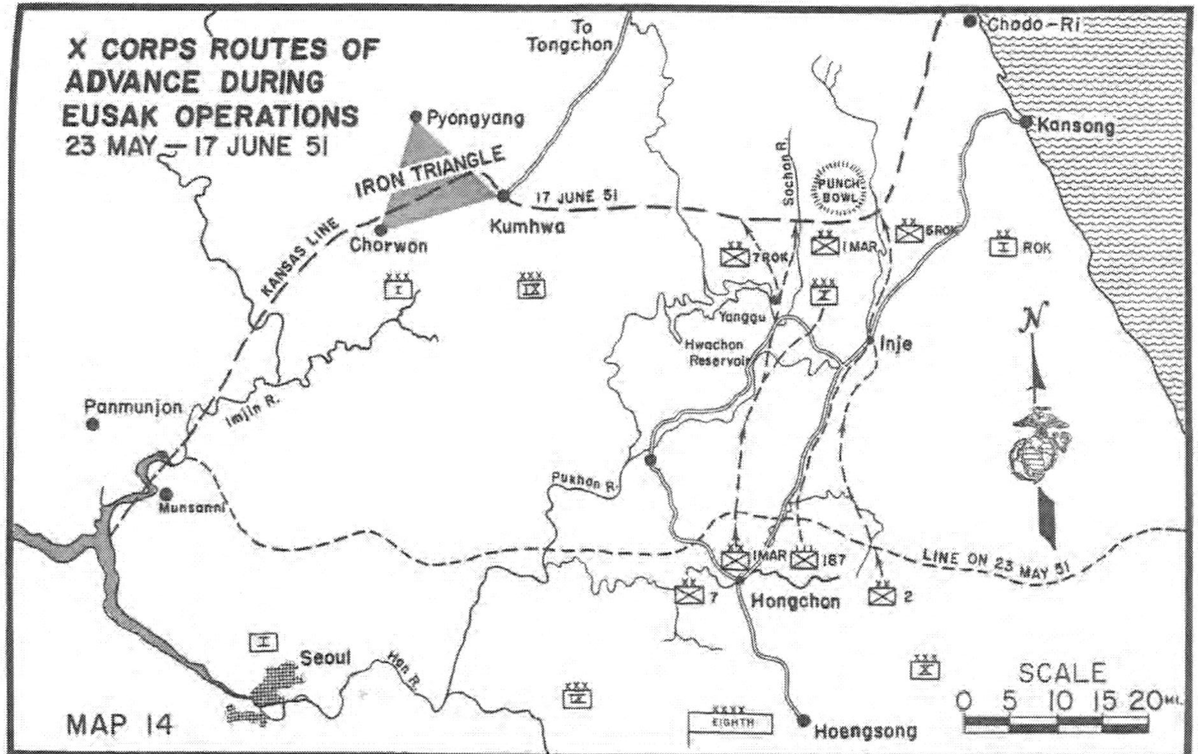
X CORPS ROUTES OF
ADVANCE DURING
EUSAK OPERATIONS
23 MAY — 17 JUNE 51
To
Tongchon
Pyongyang
IRON TRIANGLE
KANSAS LINE
17 JUNE 51
Chorwon
Kumhwa
Sochon R.
PUNCH BOWL
Chodo-Ri
Kansong
7 ROK
1 MAR
5 ROK
ROK
Yanggu
Hwachon
Reservoir
Inje
Panmunjon
Imjin R.
Munsanni
Pukhan R.
1 MAR
187
LINE ON 23 MAY 51
7
Hongchon
2
Seoul
Han R.
SCALE
0 5 10 15 20 MI.
EIGHTH
Hoengsong
MAP 14

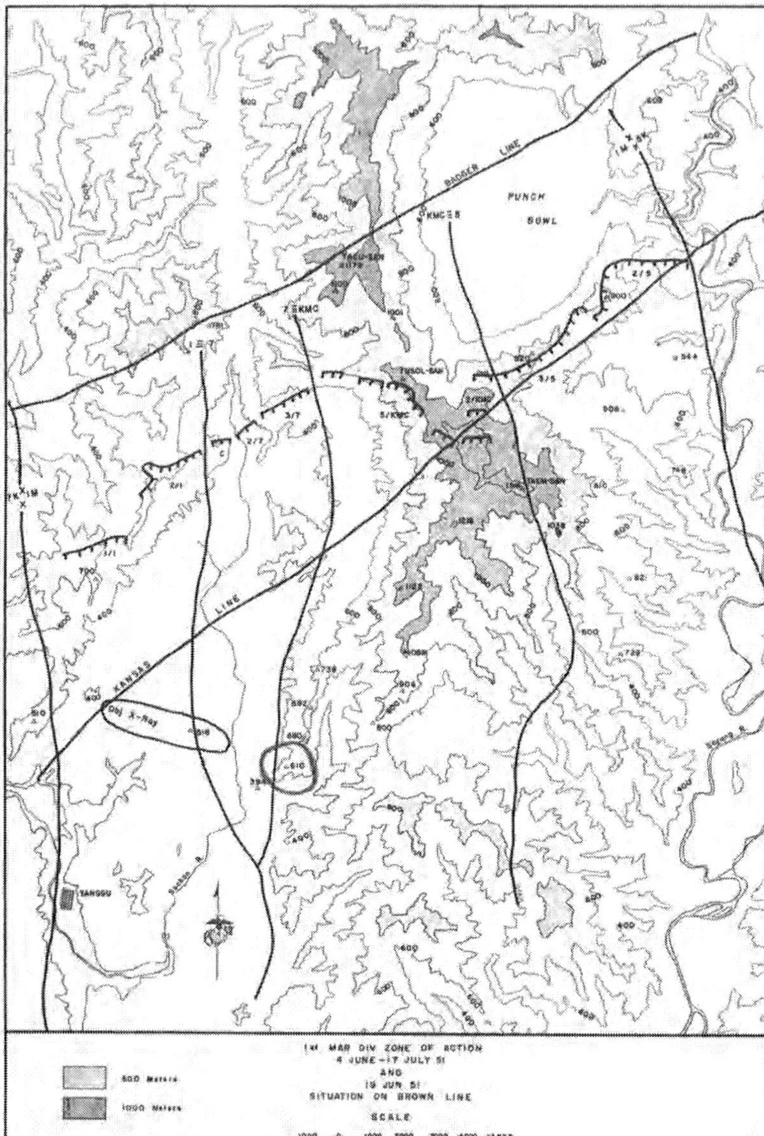

1st MAR DIV ZONE OF ACTION
4 JUNE–17 JULY 51
AND
19 JUN 51
SITUATION ON BROWN LINE

MAP 15

On 1 June the two regiments in assault, the 5th and 7th Marines, found the resistance growing stiffer as they slugged their way forward toward Line KANSAS (Map 15). Within an hour after jumping off, 2/5 was heavily engaged with an estimated 200 enemy defending Hill 651 tenaciously. At noon, after ground assaults had failed, a request was put in for air support. Four VMF–214 planes led by Captain William T. Kopas bombed and strafed the target. This attack broke the back of NKPA opposition, and 2/5 moved in to seize the objective.[23]

Early on the morning of the 2d, Lieutenant Colonel Hopkins' 1/5 moved out to secure the southwest end of the long ridge line that stretched northeast from Yanggu (Map 15) and afforded a natural avenue of approach to Taeam-san and the KANSAS line on the southern rim of the Punchbowl. The Marine advance got under way at 0915. After two four-plane strikes by VMF–214 and a "preparation" by 1/11 and the 1st Rocket Battery, the battalion attacked across a valley with Baker Company (First Lieutenant William E. Kerrigan) on the right and Charlie Company (First Lieutenant Robert E. Warner) on the left to seize the terminal point on the ridge leading to Hill 610 (Map 15). Able Company (Captain John L. Kelly) followed Charlie as Company C (Captain Richard M. Taylor) of the 1st Tank Battalion moved into supporting position.

Converging fire from transverse ridges had the Marine riflemen pinned down until the tankers moved along the valley road running parallel. Direct 90mm fire into NKPA log bunkers enabled C/1/5 to advance to the forward slope of Hill 610. The enemy fought back with machine guns and grenades while directing long-range rifle fire against 2/5, attacking along a parallel ridge across the valley.

By 1945 the last bunker on Hill 610 had been overrun. Meanwhile, 2/5 had pushed ahead some 5,000 yards to the northeast.

The capture of Hill 610 will never have its glorious page in history. It was all in the day's work for Marines who could expect a succession of such nameless battles as they clawed their way forward. That night the weary men of 1/5 were not astonished to receive a counterattack in the darkness. It was all part of the job, too. After driving off the unseen enemy, the new tenants of Hill 610 snatched a few

---

[23] This section, unless otherwise specified, is based on the following sources: X Corps *Cmd Rpt*, Jun 51; *HDs* of 1st MarDiv, 1stMar, 5thMar, 7thMar, and VMF–214 for Jun 51.

hours of sleep. They were on their feet again at dawn, ready to go up against the next key terrain feature in a rocky area that seemed to be composed entirely of Hill 610s.

The next knob along the ridge happened to be Hill 680, about 1,000 yards to the northeast. VMF–214 planes from K–46 napalmed and strafed the enemy, and Able Company led the 1/5 attack. During the air strike the Koreans had taken to cover in their holes on the reverse.

They were back in previously selected forward slope firing positions by the time the Marines came in sight. Close-in artillery support enabled the attackers to get within grenade range and seize the last NKPA bunker by 1400. Able Company pushed on.

Midway from Hill 680 to the next knob, Hill 692, the advance was stopped by enemy small-arms and mortar fire. An air strike was requested on the bunkers holding up the assault, but fog closed in and the planes were delayed more than two hours.

At 1600, after Able Company had renewed the assault without air support, four VMF–214 Corsairs started a target run controlled by a liaison plane from VMO–6. The foremost Marines, almost at the summit by this time, had to beat a hasty retreat to escape the napalm and 500-pound bombs being dumped on Hill 692. Fortunately, there were no friendly casualties. Some were caused indirectly, however, when hostile mortar fire caught Marines withdrawing along a connecting saddle to the comparatively safe reverse slope of Hill 680. When the danger passed, Able Company returned to the attack on 692 and routed the remaining defenders.[24]

The 1st Marine Division made it a policy thereafter that only the forward air controllers on the ground were to direct close air support along the front. Control of air strikes farther behind the enemy lines was reserved for the OYs.

1st MAW in Operation STRANGLE

Sightings of enemy vehicles during the month of May totaled 54,561—seven times those of January. This increase prompted General Van Fleet to ask the Fifth Air Force and Seventh Fleet to initiate a

[24] 5thMar *UnitReport* (*URpt*), Jun 51, 35.

program of cutting off all possible enemy road traffic between the latitudes 38° 15′ N and 39° 15′ N.

Earlier in 1951 the interdiction program had been aimed chiefly at the enemy's rail lines and bridges. The Communists had countered by using more trucks. The new program, known as Operation STRANGLE, was to be concentrated against vital road networks. Flight leaders were briefed to search out critical spots where truck and ox cart traffic could be stopped. Roads skirting hills were to be blocked by landslides caused by well placed bombs. Where cliffside roads followed the coast, as they so often did in East Korea, naval gunfire started avalanches of dirt and rocks which sometimes reached a depth of 20 feet. Roads running through a narrow ravine or rice paddy could often be cut by a deep bomb crater.[25]

The 1st MAW was given the assignment of stopping traffic on three roads in East Korea—from Wonsan to Pyonggang, from Kojo to Kumhwa, and along a lateral route linking the two (Map 16). Since Kumhwa and Pyonggang were two of the three Iron Triangle towns, these roads were of more than ordinary importance.

The Communists reacted to the new UN pressure by increasing their flak traps. UN pilots were lured with such bait as mysterious lights, tempting displays of supposed fuel drums, or damaged UN aircraft that called for investigation. The cost of the UN in planes and pilots showed an increase during the first two months of Operation STRANGLE. From 20 May to the middle of July, 20 Marine planes were shot down. Six of the pilots returned safely; two were killed and 12 listed as missing.[26]

The demands of Operation STRANGLE added to the emphasis on interdiction and armed reconnaissance by the Fifth Air Force. Statistics compiled by the 1st Marine Division for 1–17 June 1951 show that 984 close air support sorties had been requested and 642 received—about 65 percent. The ratio of Marine planes to other UN aircraft reporting to the Division was about four to one.[27]

The statistics of the 1st MAW indicate that out of a total of 1,875 combat sorties flown from 1 to 15 June 1951, about a third were close

---

[25] Descriptions of Operation STRANGLE are based on *Pac Flt Interim Rpt* No. 3, Chapter 10, 10–45 to 10–47; and on 1st MAW *HDs*, May to Jul 51, G–3 PORs, G–3 Journal entries, Assessment Rpts.

[26] 1st MAW *HDs* May-Jul 51, Summaries; MAG–12 and MAG–33 *HDs* May-Jul 51, Summaries.

[27] Summarization from DivAirO memo of 26 Jun 51 to CG 1stMarDiv.

air support—651 day CAS and 19 night CAS. Of this number, 377 sorties went to the 1st Marine Division, which received more than half. Next in line were the 7th Infantry Division (41 sorties), the 3d Infantry Division (31 sorties), and the 25th Infantry Division (28 sorties).[28]

The effect of Operation STRANGLE on the enemy must be left largely to conjecture. There can be no doubt that it added enormously to the Communists' logistical problem. It is equally certain that they solved these problems to such an extent that their combat units were never at a decisive handicap for lack of ammunition and other supplies. Operation STRANGLE, in short, merely added to the evidence that interdictory air alone was not enough to knock a determined adversary out of the war, as enthusiasts had predicted at the outbreak of hostilities in Korea.

KMC Regiment Launches Night Attack

On the night of 1–2 June, Colonel Nickerson was notified that the 7th Marines would be relieved next day by the 1st Marines, which would pass through and continue the attack. The 1st Marines moved into assembly areas at 0630. Lieutenant Colonel Homer E. Hire, commanding officer of 3/1, went forward at 0800 with his command group to make a reconnaissance of the area. As his staff paused for a conference in a supposedly enfiladed location, a Communist mortar barrage hit the group by complete surprise. The artillery liaison officer was killed instantly. His assistant, two forward observers, four company commanders, the S–3 and 32 enlisted men were wounded. So hard hit was the battalion that its attack had to be postponed until the following day.[29]

The first Division objective was designated X-RAY. 2/1 had the mission of taking the high point, Hill 516 (Map 15). Across the valley 3/1 advanced up a parallel ridge. Planes from VMF–214 and VMF–323 cleared the way for the securing of this battalion's objective at 1900. Aircraft from these same squadrons also aided 2/1 in over-

---

[28] 1st MAW *HD*, Jun 51, Pt 1, Chronology, 15 Jun.
[29] CO 1stMar msg to CG 1stMarDiv, 1915 2 Jun 51.

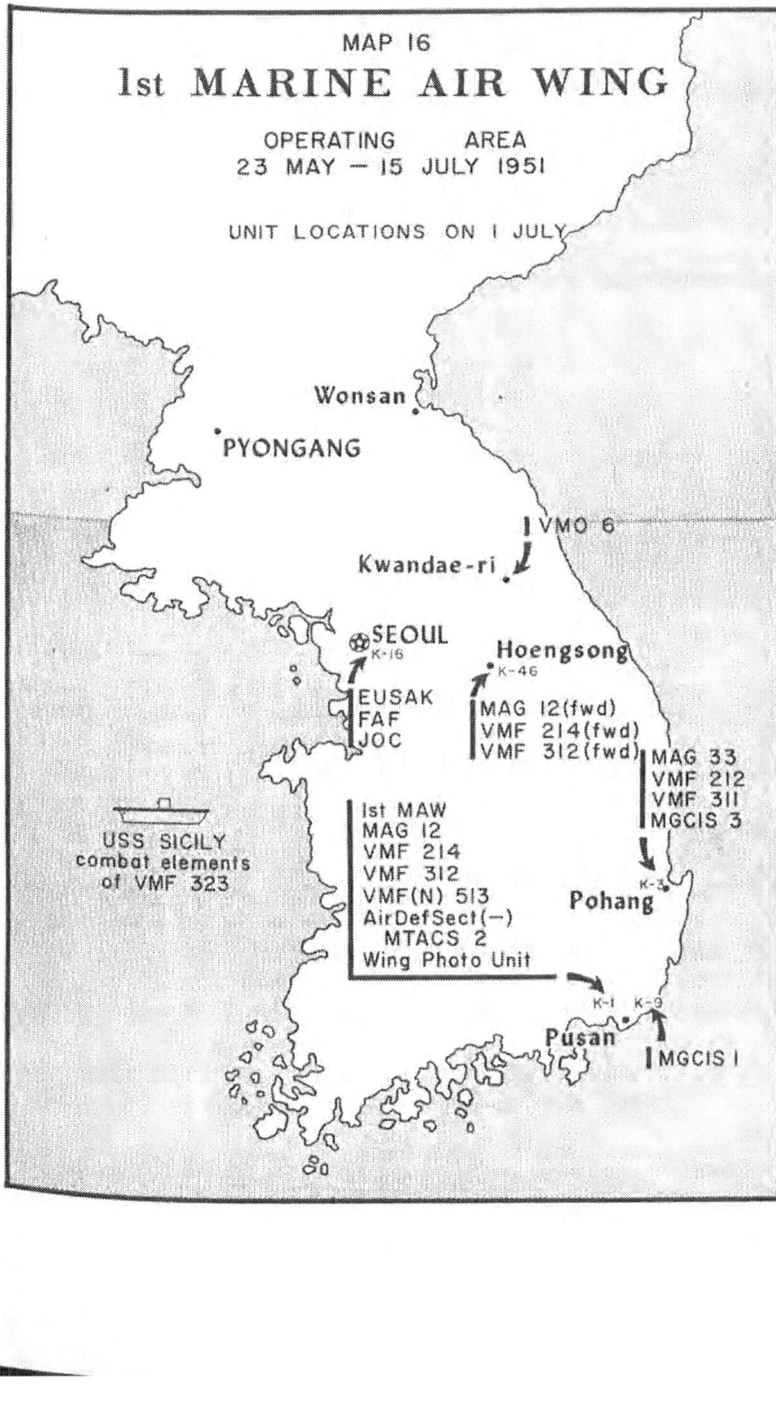
MAP 16
1st MARINE AIR WING
OPERATING AREA
23 MAY — 15 JULY 1951
UNIT LOCATIONS ON 1 JULY
Wonsan
PYONGANG
VMO 6
Kwandae-ri
SEOUL
K-16
Hoengsong
K-46
EUSAK
FAF
JOC
MAG 12(fwd)
VMF 214(fwd)
VMF 312(fwd)
MAG 33
VMF 212
VMF 311
MGCIS 3
USS SICILY
combat elements
of VMF 323
1st MAW
MAG 12
VMF 214
VMF 312
VMF(N) 513
AirDefSect(–)
MTACS 2
Wing Photo Unit
K-3
Pohang
K-1
K-9
Pusan
MGCIS 1

running the last opposition on Hill 516, where 80 NKPA dead were counted.[30]

The KMC regiment, in reserve only two days, was ordered to relieve the 5th Marines on 4 June. This would permit Colonel Hayward to shift over to the right flank, thus extending the 1st Marine Division zone 5,000 yards to the east with a north-south boundary of the Soyang river valley (Map 15). The purpose of this maneuver was to free 2d Infantry Division troops for a mission of mopping up in the X Corps rear area.

Three Marine regiments were now in line, the 1st on the left, the KMCs in the center, the 5th on the right, and the 7th in reserve. A reshuffling of units also took place in the 1st MAW when VMF–312 ended its tour of duty on the CVL *Bataan*. The replacement involved a change of carriers when VMF–323 was alerted for west coast duty on the CVE *Sicily* a week later.[31]

Ahead of the KMCs stretched the most difficult of the regimental zones of action—the main mountain range extending northeast from Yanggu to Hill 1316, known to the Koreans as Taeam-san. Along these ridges the Chinese had placed North Korean troops with orders to "hold until death." [32]

From the air, the ground in front of the KMCs resembled a monstrous prehistoric lizard, rearing up on its hind legs. The 1st Battalion was to ascend the tail and the 2d the hind legs. The two would meet at the rump, Hill 1122 (Map 15). From this position the backbone ran northeast to the shoulders, Hill 1218. Still farther northeast, along the neck, was the key terrain feature—Taeam-san, the head of the imagined reptile.

The 1st and 2d Battalions ran immediately into the opposition of an estimated NKPA regiment. In an effort to outflank the enemy, the 3d Battalion swung over to the east and attacked up the ridge forming the forelegs. Seizure of the shoulders (Hill 1218) would render enemy positions along the back, rump, hind legs, and tail untenable. Major General Choe Am Lin, commanding the 12th NKPA Division,

---

[30] CO 1stMar msg to CG 1stMarDiv, 1830 3 Jun 51; *HDs* of VMF–214 and VMF–323, Jun 51.

[31] *PacFlt Interim Rpt* No. 3, VI, 6–6, 6–7; 1stMarDiv *Special Action Report* (*SAR*), Jun 51.

[32] The account of the KMC attack is based upon these sources: 1stMarDiv *HD*, Jun 51; "KMC Operations in Korea, Jun 51," n.d., by Col C. W. Harrison, then KMC senior adviser.

was quick to recognize the tactical worth of this height and exact a stiff price for it.

That the KMCs could expect little mercy from their fellow countrymen was demonstrated when the bodies of ten men reported missing were found. All had been shot in the back of the head.

For five days the fight raged with unabated fury. The terrain limited the advance to a narrow front, so that the attack resembled the thrust of a spear rather than a blow from a battering ram. When the KMCs did gain a brief foothold, the enemy launched a counterattack.

At 2000 on 10 June, after six days of relatively unsuccessful fighting, the KMCs decided to gamble on a night attack. This had heretofore been the enemy's prerogative, and the Korean Reds were caught unaware in a devastating surprise. Most of the NKPA troops were attending to housekeeping duties at 0200 when all three KMC battalions fell upon them like an avalanche. Hill 1122, the rump of the lizard, was seized; and under pressure the enemy withdrew from the shoulders. This made the fall of Taeam-san inevitable, and only mopping-up operations remained for KMCs who had suffered more than 500 casualties. General Thomas sent the regiment this message on 12 June:

> Congratulations to the KMC on a difficult job well done. Your seizure of objectives on the KANSAS Line from a determined enemy was a magnificent dash of courage and endurance. Your courageous and aggressive actions justify our pride in the Korean Marines.

Logistical support of the three regiments in the attack presented a problem to the Division supply echelons. The KMCs in the center and the 1st Marines on the left could be supplied over a narrow, winding mountain road that scaled a high pass before dropping down into an east-west valley giving relatively easy access to the center and left. The 5th Marines had to receive its supplies over another mountain road leading north of Inje, then west into the regimental zone.[aa]

Both of the Division supply routes needed a good deal of engineering work before trucks could move over them freely. Landslides were frequent and many trucks skidded off the slippery trail while rounding the hairpin turns.

The 1st Marines moved northward on north-south ridges, and the

[aa] The KMC's drew fuel and ammunition from the 1st Marine Division and rations from the ROK Army. Other classes of supplies were obtained generally on a catch-as-catch-can basis with some aid from KMC Headquarters in Pusan.

KMCs in the center had spurs leading to their objectives. It was the misfortune of the 5th Marines to have a topographical washboard effect ahead. The axis of advance was south to north, but the ground on the way to the final objectives on the KANSAS Line consisted of five sharply defined ridgelines running northwest to southeast. Instead of attacking along the ridgelines Colonel Hayward's men had to climb some 1,200 feet, then descend 1,200 feet, five separate times while covering an advance of 8,000 yards (Map 15).

Artillery fired for more than two hours on the morning of 6 June to soften defenses on the next regimental objective, Hill 729. An air strike was attempted but fog with low-hanging clouds forced the flight leader to abort the mission. At 1300 the assault battalions moved across the LD against small-arms and machine gun fire. The fog lifted sufficiently at 1400 to allow four F9Fs from VMF-311 to deliver an effective attack. And by 2100 both 2/5 and 3/5 were consolidating their positions on the first of the five ridges.

This assault is typical of the fighting as the 5th Marines took the remaining four ridges, one by one, in a slugging assault on an enemy defending every commanding height. The advance resolved itself into a pattern as the Korean Reds probed the Marine lines at night and continued their tough resistance by day. For 10 days the regiment plugged ahead, step by step, with the support of artillery, air, mortars, and 75mm recoilless rifles.[34]

1st Marines Moves Up to BROWN *Line*

On the left flank, the 1st Marines devoted several days to consolidating its position and sending out reconnaissance patrols in preparation for an attack on the ridge just north of the Hwachon Reservoir. From this height the Communists could look down the throats of Colonel Brown's troops.

From 6 to 8 June, Lieutenant Colonel Hire's 3d Battalion led the attack against moderate but gathering resistance. A gain of 1,500 yards was made on the right flank by 2/1, commanded by Major Clarence J. Mabry after the evacuation of Lieutenant Colonel McClellan, wounded on the 5th. On the left, Lieutenant Colonel Robley E. West's 1/1 held fast as the 5th ROK Regiment, 7th ROK Division,

[34] *HDs* of 1stMarDiv and 1stMar, Jun 51.

X Corps, passed through on its way to a new zone of action to the west.

Early on the 9th, as 2/1 was preparing to launch its attack, an intense artillery and mortar barrage fell upon the lines, followed by the assault of an estimated NKPA company. The Korean Reds were beaten off with heavy losses. And though the enemy fire continued, 2/1 jumped off on schedule, fighting for every inch of ground. Colonel Brown committed 1/1 on the left. It was an all-day fight for both battalions. After taking one ridge in the morning, it was used as the springboard for an assault on the second objective. The weapons of the regimental Anti-Tank Company built up a base of fire that enabled this ridge to be secured by 1600.

The 5th ROK Regiment took its objectives by the morning of the 10th. The 1st Marines provided additional fire support by diverting all its antitank guns and tank rifles to the aid of the ROKs.

The pressure, which had been building up for several days, reached a new high on 10 June. Late that morning Colonel Brown met General Almond and the Division G–3, Colonel Richard G. Weede, at a conference. By 1100 the entire 2d Battalion of the 1st Marines was committed. On the left, Lieutenant Colonel West had to hold up the 1st Battalion until 1330, when the ROKs completed the occupation of the high ground dominating the route of advance.

For several hours it appeared that the Marines had met their match this time. A tenacious enemy defended log bunkers expertly, refusing to give ground until evicted by grenade and bayonet attacks. At every opportunity the Communists counterattacked. So effective was their resistance that at dusk the two Marine battalions were still short of their objectives in spite of casualties draining the strength of both units.

Colonel Joseph L. Winecoff, commanding officer of the 11th Marines, remained on the telephone for hours with Colonel Brown. He gave all possible artillery support, not only of his own regiment but also nearby Corps units. By nightfall, with the attacking battalions still held up, the atmosphere was tense in the regimental forward CP. Lieutenant Colonel Adelman, commanding the supporting artillery battalion, 2/11, helped to coordinate air strikes and artillery with Lieutenant Colonel Donald M. Schmuck, executive officer of the 1st Marines, and the air liaison officers.

"Everything I had ever hoped to see in years of teaching such co-ordination of fires seemed to come true that night," commented Colonel Brown at a later date. "I stayed in my regular CP until I was sure all I could do through Winecoff was done, and then went forward to see the finale. It was a glorious spectacle, that last bayonet assault. In the last analysis 2/1 had to take its objective with the bayonet and hand grenades, crawling up the side of a mountain to get at the enemy. It was bloody work, the hardest fighting I have ever seen." [35]

This was no small tribute, coming from a veteran officer whose combat service included three major wars, not to mention Nicaragua and China. It was nearly midnight before Mabry's battalion took its final objective. Casualties for the day's attack were 14 KIA and 114 WIA exclusive of slightly wounded, who were neither counted nor evacuated. West's battalion, which seized Hill 802, overlooking the Soyang River, had won its all-day fight at a cost of 9 KIA and 97 WIA.

Unfailing support had been given throughout the daylight hours by aircraft of VMF–214. VMF(N)–513 took over on the night shift, and planes came screeching in as late as 2200 to attack moonlit targets a hundred yards ahead of the leading infantry elements.

The 1st Marines had outfought and outgamed a tough enemy. Never again, after the 10th, was the NKPA resistance quite as determined. The 3d Battalion led the other two during the next few days. There was plenty of fighting for all three, but the result was never again in doubt.

By the late afternoon of 14 June the regiment was in position on the BROWN Line. This was the unofficial name for an extension of the KANSAS Line some 3,000 yards north. It had been requested by Colonel Brown when he realized that positions along the KANSAS Line were completely dominated by the next ridge to the north.

The change made necessary a continued advance by the KMCs on the right to tie in with the 1st Marines. The so-called BROWN Line was then officially designated the modified KANSAS Line.

7th Marines Committed to the Attack

For several days General Thomas had been concerned over the heavy casualties suffered by his command. In order to give greater impetus

---

[35] MajGen W. S. Brown, USMC (Ret.), ltr of 8 Jun 58. Other sources for this section are the *HDs* of 1stMarDiv, 1stMar, 1/1, 2/1, 3/1, and VMF–214.

to the Division effort, he decided to commit the reserve infantry regiment, the 7th Marines (minus one battalion held back as Division reserve) to complete the occupation of the modified KANSAS Line.

On 8 June, Colonel Nickerson's regiment (minus 3/7) moved into an assembly area between the 1st Marines and the KMCs, ready to attack in the morning. Ahead stretched a narrow but difficult zone of advance up the valley of the So-chon River (Map 15). Tank-infantry patrols went forward to select favorable positions for the jumpoff, and engineers worked throughout the daylight hours to clear the valley roads of mines. Despite their best efforts, 10 Marine tanks were lost to mines during the first week.[30]

As the two battalions advanced on the morning of the 9th they came under heavy enemy artillery and mortar fire. Nevertheless, they secured Hill 420 and dug in before nightfall.

On the 10th Rooney's 1/7 advanced along the ridgeline to support the attack of Meyerhoff's 2/7 up the valley floor. The maneuver was carried out successfully in spite of NKPA automatic weapons and mortar opposition. Contact was established with KMC forward units at dusk. Sixteen POWs were taken by the 7th Marines and 85 North Korean dead were counted on the objectives.

The two battalions continued the attack throughout the next week. The 3d Battalion of the 7th Marines remained General Thomas' sole Division reserve until he committed it on the afternoon of 18 June.

The newcomers got into the fight just in time for the enemy's all-out effort to defend the steep east-west ridge marking the BROWN Line. The nature of the terrain made maneuver impossible—a frontal assault was the only answer. Defending the ridge was the 1st Battalion, 41st Regiment, 12th NKPA Division. Waiting on the reverse slope, the enemy launched a counterattack when the Marines neared the crest. George Company, commanded by First Lieutenant William C. Airheart, met five successive repulses at the hands of superior numbers. Item Company (First Lieutenant Frank A. Winfrey) also took part in the fifth assault, and both companies held their ground near the summit when the fighting ended at dusk. They expected to resume the attack at dawn, but the enemy had silently withdrawn during the night. All three 7th Marines battalions occupied their designated positions on the BROWN Line without further interference.

---

[30] Unless otherwise noted, this section is based on the *HD*s of the 1stMarDiv, 7thMar, 1/7, 2/7, and 3/7 for Jun 51.

By early afternoon on the 20th, the Division was in complete control of the modified KANSAS Line and construction of defenses began in earnest. The next day the 1st Marines and KMCs extended their right and left flanks respectively and pinched out the 7th Marines, which dropped back into reserve.

Thus ended two months of continual hard fighting for the 1st Marine Division, beginning on 22 April with the great CCF offensive. Few and far between were the interludes of rest for troops which saw both defensive and offensive action. After stopping the enemy's two drives, they launched a month-long counterstroke that had the enemy hardpressed at times for survival. Only the ruthless sacrifice of NKPA troops in defensive operations enabled the Chinese Reds to recover from the blows dealt them in late May and early June.

The cost in Marine casualties had been high. Throughout the entire month the 1st Marines alone suffered 67 KIA and 1,044 WIA, most of them being reported during the first 2 weeks. This was a higher total than the regiment incurred during the Chosin Reservoir operation. Reflecting on the caliber of these men, their regimental commander had this to say:

> They were war-wise when I got command; I contributed nothing to their training because they were in battle when I joined them and I left them when they came out of the lines for a rest. They used cover, maneuvered beautifully, used their own and supporting arms intelligently, were patient and not foolhardy; but when it came to the point where they had to rely on themselves with bayonet, hand grenade and sheer guts, they could and did do that too. I have long ago given up telling people what I saw them do on many occasions. Nobody believes me, nor would I believe anyone else telling the same story of other troops.[37]

Colonel Brown, of course, paid this tribute to the troops of his regiment. But it is safe to say that any commanding officer of the 1st Marine Division would have felt that these sentiments applied equally to his own men. All the combat Marines of the 60-day battle had shown themselves to be worthy heirs of the traditions of Belleau Wood, Guadalcanal, Iwo Jima, and the Chosin Reservoir.

---

[37] MajGen W. S. Brown, USMC (Ret.), ltr to Maj W. T. Hickman, 22 Apr 57.

CHAPTER VIII

The Truce Talks at Kaesong

Communists Ask for Truce Talks—Patrol Bases on BADGER *Line—Red Herrings at Kaesong—1st Marine Division in Reserve—Marine Helicopters Take the Lead—Marine Body Armor Tested in Korea—MAG–12 Moves to K–18 —The Division Back in Action Again*

IT IS NOT LIKELY that the date 25 June 1951 meant much to the Marines on the KANSAS Line. In all probability few of them recalled that it was the first anniversary of the Communist aggression which started the war in Korea.

Since that surprise attack on a June Sunday morning in 1950, some 1,250,000 men had been killed, wounded or captured in battle—a million of them from the Communist forces of Red China and the North Korean People's Republic. This was the estimate of J. Donald Kingsley, Korean reconstruction agent general for the United States. He reckoned the civilian victims of privation, violence, and disease at two million dead. Another three million had been made homeless refugees.[1]

On 25 June 1951 the Communists held less territory by 2,100 square miles than they occupied when they began their onslaught with an overwhelming local superiority in arms and trained troops. Losses of Communist equipment during the first year included 391 aircraft, 1,000 pieces of artillery, and many thousands of machine guns, automatic rifles, and mortars. North Korea, formerly the industrial region of the peninsula, lay in ruins. Cities, factories, and power plants had been pounded into rubble.

In short, the thrifty conquest planned by the Koreans and their

[1] This section is based on by Peter Kihss, "One Year in Korea," *United Nations World*, Vol. 5, No. 7, July 1951, 21–23.

Soviet masters had backfired. Not only had the Communist offensives of April and May been stopped; the United Nations forces had rebounded to win their greatest victory of the war's first year. While X Corps was advancing to the Punchbowl, other major Eighth Army units had also gained ground. Perhaps the most crushing blow was dealt by I Corps in its attack on the Iron Triangle. Units of two U.S. infantry divisions fought their way through extensive mine fields into Chorwon and Kumhwa on 8 June. By the end of the month, I Corps held defensive positions about midway between the base and apex of the strategic triangle that had been the enemy's main assembly area for the troops and supplies of his spring offensives.[2]

On the east-central front, units of IX Corps pushed within 10 miles of Kumsong while I ROK Corps advanced along the east coast to Chodo-ri. Thus the UN forces occupied the most favorable line they had held since the great CCF offensive early in January. From the mouth of the Imjin this line ran northeast to the middle of the Iron Triangle, eastward across the mountains to the southern rim of the Punchbowl, then northeast to the coast of Chodo-ri (Map 14).

Communists Ask for Truce Talks

The first anniversary of the Korean conflict was overshadowed two days earlier by the news that the Communists had taken the initiative in proposing truce talks. The suggestion was made in a New York radio address of 23 June by a Soviet delegate to the United Nations—Jacob Malik, Foreign Minister of the USSR. On the 25th the idea was unofficially endorsed in a radio broadcast by the Chinese Communist government. UN officials immediately indicated their willingness to discuss preliminary terms. The outcome was an agreement that representatives of both sides would meet on 7 July at Kaesong, then located between the opposing lines in west Korea.

Why had the Communists been first to ask for a truce conference? Both Generals Van Fleet and Almond believed that the answer might have been traced to military necessity rather than any genuine desire for peace. "I felt at that time that the Chinese Communists and the North Korean armies were on the most wobbly legs that they had been on to that date," said General Almond when interviewed shortly

[2] EUSAK *Cmd Rpt*, Jun 51.

after his retirement in 1953. "They were punch drunk and ineffective, and I, personally, thought at that time that it was the time to finish off the effort." [3]

Raymond Cartier, representing a Paris newspaper, probably spoke for most of the correspondents at the front when he suspected that the proposal for truce talks "was possibly just a crafty trick devised by the Communists to gain time and build up again the badly mauled Chinese armies." [4]

It might have been recalled at this time that the Communists had used truce negotiations for military purposes during the Chinese Civil War. In 1945 and 1946, when prospects for a Nationalist victory were bright, the enemy took advantage of American peace efforts by agreeing on several occasions to meet for truce conferences. And while prolonging the talks by all manner of subterfuges, the Communists profited from the breathing spells by regrouping their forces and planning new offensives. Their final triumph, in fact, owed in no small measure to interludes when the conference table served a military purpose.[5]

History repeated itself in June and July 1951 when events of the next two years were shaped by the political decisions of a few summer weeks. Indeed, Admiral C. Turner Joy believed that the war was actually prolonged rather than shortened as a result of the negotiations.

"Military victory was not impossible nor even unusually difficult of achievement," wrote the Senior Delegate and Chief of the UN Command delegation at the truce talks. "Elimination of the artificial restraints imposed on United States forces, coupled with an effective blockade on Red China, probably would have resulted in military victory in less time than was expended on truce talks." [6]

Mao Tse-tung's forces had lost face by the failure of their long heralded 5th Phase Offensive. They had been badly beaten during the UN counteroffensive. Pretensions of high CCF morale could no longer be maintained when troops were laying down their arms without a

[3] *U.S. News and World Report*, 13 Feb 53, 40–41.

[4] *UN World*, Vol. 5, No. 10, Oct 51, 10.

[5] U.S. State Department Publications 3573, Far East Series 30, pp. 352–363.

[6] Admiral C. Turner Joy, USN (Ret.), *How Communists Negotiate* (New York: Macmillan, 1955), 176, hereafter Joy, *How Communists Negotiate*. One of Admiral Joy's last services to his country before his death in 1956 was the writing of this book. Other sources for this section are William H. Vatcher, Jr., "Inside Story of Our Mistakes in Korea," *U.S. News and World Report*, 23 Jan 1953, 35–36; E. Weintal, "What Happened at Kaesong and What is in Prospect," *Newsweek*, 23 Jul 1951, 38; Comments n.d., Col J. C. Murray.

fight. Nor could charges of low UN morale be supported when the fighting spirit of the Eighth Army was being shown every day at the front.

In view of these circumstances, it would appear that the Communists had poor cards to play against United Nations trumps at a truce conference. But they played them so craftily, with such a sly sense of propaganda values, that the victors of the May and June battles were soon made to appear losers begging for a breathing spell.

To begin with, the Chinese knew that the mere public announcement of the possibility of truce talks would have a tremendous appeal in the United States, where the war was unpopular. Pressure would be brought upon Washington to meet the enemy immediately for negotiations. And while a cease fire remained even a remote prospect, American public opinion would demand a slackening of offensive military operations with their attendant casualties.

From the outset it was apparent that the United Nations Command was no match for the Communists in low cunning. The UN suggested, for instance, that the truce teams meet on the Danish hospital ship *Jutlandia*. Here, surely, was neutral ground, since the Danes had no combat forces in Korea. Moreover, the ship was to be anchored in Wonsan harbor within range of CCF shore batteries.

The Reds won the first of many such concessions with their refusal. They insisted that the talks be held at Kaesong, and the UN Command let them have it their way. The reason for the Communist decision was soon made evident. Kaesong was in the path of the advancing Eighth Army, which meant that an important road center would be immune from attack. And though the ancient Korean town was originally in no man's land, the Communists soon managed to include it within their lines.

All delegates were requested to display white flags on their vehicles for identification. Communist photographers were on hand to snap countless pictures of UN delegates which convinced Asia's illiterate millions at a glance that the beaten United Nations had sent representatives to plead for terms. If any doubt remained, other photographs showed the unarmed UN delegates being herded about Kaesong by scowling Communist guards with burp guns.

No detail of the stage setting was too trivial to be overlooked. Oriental custom prescribes that at the peace table the victors face south and the losers face north. Needless to add, the UN delegates were seated at Kaesong with a view to enhancing Communist prestige.[7]

Some of the propaganda schemes bordered on the ridiculous. "At the first meeting of the delegates," Admiral Joy related, "I seated myself at the conference table and almost sank out of sight. The Communists had provided a chair for me which was considerably shorter than a standard chair. Across the table, the senior Communist delegate, General Nam Il, protruded a good foot above my cagily diminished stature. This had been accomplished by providing stumpy Nam Il with a chair about four inches higher than usual. Chain-smoking Nam Il puffed his cigarette in obvious satisfaction as he glowered down on me, an obviously torpedoed admiral. This condition of affairs was promtply rectified when I changed my foreshortened chair for a normal one, but not before Communist photographers had exposed reels of film." [8]

Patrol Bases on BADGER *Line*

The war went on, of course, during the negotiations. But the tempo was much reduced as the UN forces consolidated their gains, and the enemy appeared to be breaking off contact at every opportunity. Generally speaking, the Eighth Army had shifted from the offensive to the defensive. In keeping with this trend, the 1st Marine Division occupied the same positions for nearly three weeks after fighting its way to the BROWN Line.

On 22 June all three infantry regiments were directed to establish battalion-size patrol bases on the BADGER Line—1½ to 2½ miles forward of their present positions. In the 1st Marines sector 3/7 was attached to Colonel Brown and ordered to relieve 3/1 on the left flank of the regiment. The purpose was to free 3/1 to move forward and establish a patrol base on Hill 761, about 1,000 yards forward of the MLR.

While these arrangements were being carried out, General Almond called at the 1st Marines CP. He expressed surprise that the establishment of patrol bases was being contemplated by EUSAK when some of the front-line units were still in contact with the enemy.[9]

Execution of these orders was accordingly suspended. The follow-

[7] Joy, *How Communists Negotiate*, 4–5.
[8] *Ibid.*
[9] 1stMarDiv *HD*, Jun 51, 55.

ing day, however, Division again alerted the infantry regiments to be prepared to occupy patrol bases on order. This was by direction of Corps, which in turn had been directed by EUSAK.

The Marine regimental and battalion commanders were not happy about this turn of affairs. The patrol base concept had been tried out early in May, during the lull between the enemy's two offensives, and found wanting. In theory it was a good means of keeping contact with an enemy who had pulled back out of mortar and light artillery range. In practice the enemy had shown that he could bypass patrol bases at night for probing attacks on the MLR. The bases themselves ran the constant risk of being surrounded and overwhelmed. As a final objection, a regiment was often deprived of its reserve battalion, which was the logical choice for such duty.

In compliance with orders, 3/1 moved out on 26 June and established a patrol base on Hill 761. This position received such a bombardment of large caliber mortar fire that Colonel Brown pulled the battalion back to the MLR the following day.[10]

General Thomas gave his opinion of the patrol base concept after his retirement when he summed it up as "an invitation to disaster." [11] He could only carry out orders, however, when Corps directed early in July that a patrol base be established on Taeu-san.

This 4,000-foot peak, located some 2 miles north of the MLR, afforded excellent observation eastward into the Punchbowl and westward into the So-chon River Valley. The enemy, of course, was aware of these advantages and had made Taeu-san a strongpoint of his MLR. This was clearly indicated by the stiff resistance encountered by KMC reconnaissance patrols.[12]

Nevertheless, Division G–3 was suddenly alerted on the morning of 7 July by the Marine Liaison Officer with X Corps to expect an order directing the setting up of a patrol base on Taeu-san the following day. The KMC Regiment, warned by telephone, had little time for planning and organizing an attack. Since the KMCs could not be relieved for responsibility for their sector, it was necessary to form a composite battalion of the three companies that could most conveni-

---

[10] CO 1stMar msg to CG 1stMarDiv, 0815 27 Jun 51.

[11] Gen G. C. Thomas interv, 6 Feb 58. It is interesting to note that there was no mention of the patrol base concept in the then current *Field Service Regulations, Operations, FM 100–5,* published by the Department of the Army in August 1949.

[12] Unless otherwise specified, the remainder of this section is based on 1stMarDiv *HD,* Jul 51, 7–11; Col C. W. Harrison's account, "KMC Attack on Taeu-san, 8–11 July 1951;" Col G. P. Groves, ltr of 9 Apr 58.

ently be relieved. Unfortunately, they contained a large proportion of recruits, and the battalion commander was a new arrival.

There were two avenues of approach. One was along an open, fairly level, ridgeline that extended from the KMC positions. The other called for a descent into the stream-bed generally paralleling the MLR and a steep climb up a ridge leading directly north to Taeu-san.

Both routes of approach were used. One company advanced on the right by way of the stream bed and two companies took to the ridgeline on the left. The assault was to have been preceded by air strikes and an artillery bombardment, but bad weather kept the aircraft grounded.

The attack jumped off at 1030 on 8 July. All three companies were greeted by enemy mortar and machine gun fire that pinned down the company on the right. The two companies on the left won a foothold on Hill 1100, about a mile in front of Taeu-san. Here the advance ground to a halt.

These KMCs dug in for the night and repulsed a series of counterattacks. On the morning of the 9th the KMC regimental commander, Colonel Kim Tai Shik, committed the entire 1st Battalion to the attack on the right. It had no better success than the company of the day before. Meanwhile, the two companies were driven off Hill 1100.

Colonel Gould P. Groves, senior liaison officer with the KMCs, recommended that the remnants of the two companies be withdrawn. The 1st Battalion had managed to capture Hill 1001, but it was plain that the KMC regiment could not come close to Taeu-san. On 12 July the 1st Marine Division informed X Corps that the position held by the KMCs just forward of Hill 1001 fulfilled the requirements of an advance patrol base. As far as the Marines were concerned, the sad affair was permitted to rest there.

As evidence of the valiant effort made by the KMCs, they suffered 222 casualties. A sequel to this story was written late in July after the 2d Infantry Division relieved the Marines. X Corps again ordered the capture of Taeu-san as a patrol base, and it required the commitment of the major part of the division to accomplish the task.[38]

Although the fighting had not been severe for other units of the 1st Marine Division during the first two weeks of July, the casualties (including KMC losses) were 55 KIA, 360 WIA, and 22 MIA—a total of 437. Relief of the Marines was completed by the 2d Infantry

[38] X Corps *Cmd Rpt*, Jul 51, 13; 2dInfDiv *HD*, Jul 51, 13–19.

Division on 15 July, and by the 17th all units were on their way back to assembly areas in X Corps rear.

It was the second time since the landing of the 1st Provisional Marine Brigade on 2 August 1950 that the Marines had been away from the firing line for more than a few days.

Red Herrings at Kaesong

It is not changing the subject to switch to the truce talks. Kaesong was actually a second UN front.

After the preliminaries had been settled—most of them to Communist satisfaction—the UN delegation, headed by Admiral Joy, held a first meeting on 10 July 1951 with his opposite number, NKPA Major General Nam Il, and the Communist truce team. This was the first of the talks that were to drag on for two dreary years.

Nam Il, a Korean native of Manchuria, born in 1911, had been educated in Russia and had served with the Soviet army in World War II. His career in Korea began when he arrived as a captain with Soviet occupation troops in 1945. Rising to power rapidly, he took a prominent part in the creation of a Soviet puppet state in North Korea.

An atmosphere of sullen hatred surrounded the UN delegates at Kaesong. The CCF sentinel posted at the entrance to the conference room wore a gaudy medal which he boasted had been awarded to him "for killing forty Americans." When Admiral Joy tried to send a report to General Ridgway, the messenger was turned back by armed Communist guards. These are samples of the indignities heaped upon the UN truce team. After several UN delegates were threatened by guards with burp guns, Joy protested to Nam Il, "demanding prompt elimination of such crudities."

In order to give their battered armies more time for recuperation, the Communist delegates met every issue with delaying tactics. They proved themselves to be masters of the ancient art of dragging a red herring across the trail. Going back on their word did not embarrass them in the least if they found it to their advantage to renege.[14]

The truce negotiations were bound to have an immediate effect on military operations. In the United States it seemed a pity to newspaper

[14] This section, except when otherwise noted, is derived from the following sources: Joy, *How Communists Negotiate*, 6–10, 129, 140; Carl Berger, *The Korean Knot* (University of Pennsylvania Press, 1957), 141–151; Comments n.d., Col J. C. Murray.

readers that American young men should have to die in battle at a time when headlines were hinting at the possibility of peace. Mothers wrote to their congressmen, requesting a halt in Korean operations.

General Van Fleet minced no words after his retirement when he commented on the effect of the truce talks on strategy:

> Instead of getting directives for offensive action, we found our activities more and more proscribed as time went on. Even in the matter of straightening out our lines for greater protection, or capturing hills when the Reds were looking down our throats, we were limited by orders from the Far East Command in Japan, presumably acting on directives from Washington.[15]

It was the opinion of Admiral Joy that more UN casualties were suffered as a consequence of the truce talks than would have resulted from an offensive taking full advantage of Red China's military weaknesses in June 1951.

"As soon as armistice discussions began," he wrote, "United Nations Command ground forces slackened their offensive preparations. Instead, offensive pressure by all arms should have been increased to the maximum during the armistice talks. . . . I feel certain that the casualties the United Nations Command endured during the two long years of negotiations far exceed any that might have been expected from an offensive in the summer of 1951." [16]

1st Marine Division in Reserve

Most of the 1st Marine Division units were in X Corps reserve during the last two weeks of July 1951. The 5th Marines, however, remained in "ready reserve" near Inje under the operational control of X Corps. Toward the end of the month, the 3d Battalion of the 11th Marines passed to the operational control of the 2d Infantry Division. Meanwhile, the 7th Marines and Division Reconnaissance Company displaced to the Yanggu area to aid in the construction of defensive positions and undergo special training.

1st Marine Division Training Order 2–51, covering the period from 23 July to 20 August 1951, provided for a stiff daily schedule of general and specialist military subjects. The objectives were "to maintain each individual and unit of the command at a very high state

---

[15] Gen J. A. Van Fleet, USA (Ret.), "The Truth About Korea," *Life*, 11 May 53, 133.
[16] Joy, *How Communists Negotiate*, 166.

of proficiency, while emphasizing rest and rehabilitation of personnel and repair and maintenance of equipment. . . . A minimum of 33% of all technical training was to be conducted at night, stressing individual and unit night discipline. Formal unit schools and on-the-job training were utilized extensively."[17]

Most thoroughly covered among general military subjects were mechanical training, capabilities, tactical employment, and firing of individual and infantry crew-served weapons. Lectures and demonstrations were combined to good effect with instruction in basic infantry tactics.

"The prescribed periods of physical conditioning," the Division report continued, "were supplemented by extensive organized athletic programs outside of training hours, resulting in the maintenance of a high degree of battle conditioning of all hands. Special military subjects encompassed the whole range of activities necessary to the accomplishment of any mission assigned the Division. Building from the duties of the individual Marine, infantry, artillery, engineer, and tank personnel progressed through small unit employment and tactics as it applied to their respective specialities. Meanwhile such diverse training as tank repair and watch repair was conducted in various units."[18]

Fortification came in for study after a tour of the KANSAS Line by Major General Clovis E. Byers, who had relieved General Almond as X Corps commander. He listed the weaknesses he found and directed that "special attention [be] given to the thickness, strength and support of bunker overheads, and to the proper revetting and draining of excavations."[19]

The KMC Regiment received the most thorough training it had ever known, considering that it had been in combat continually since its organization. Each of the Division's three other regiments sent four training teams consisting of a lieutenant, an NCO, and an interpreter to the KMCs on 22 July. The 12 teams had orders to remain until 20 August. Attached to various KMC companies, they acted as advisers for the entire training period.

Another organization of Koreans that had won its way to favorable recognition was the newly formed Civil Transport Corps (CTC). The

---

[17] 1stMarDiv *HD,* Jul 51, 18.
[18] *Ibid.*
[19] CG X Corps, CITE X 21568.

use of indigenous labor for logistical purposes dated back to March 1951, when the Eighth Army's advance was slowed up by supply problems caused by muddy roads. Plans were made to equip and train a special corps to assist in the logistical support of combat troops in areas inaccessible to normal motor transportation.[20]

The project began on 29 March with 720 South Koreans—all from the Korean National Guard—being assigned to I Corps. Plans were developed for a Civil Transport Corps of 82 companies, each containing 240 men. The CTC was to be supervised by a staff of eight U.S. Army officers and four enlisted men under the operational control of the Transportation Section, EUSAK.

The ROK Army had the added responsibility for logistical support, of hospitalization and medical services other than emergency treatment in forward areas. Support for the CTC from UN units was to be provided in a manner similar to that in effect for the ROK forces.[21] No difficulty was found in filling the CTC ranks, for the pay meant food and clothing to a Korean and his family.

The Marines were always astonished at the heavy loads the Korean cargadores could carry uphill on their "A-frames," which looked like sturdy easels with a pair of arm-and-shoulder carrying straps. Humble and patient, these burden bearers were the only means of supply in remote combat areas.

Marine Helicopters Take the Lead

The truce talks continued to be front-page news in August. Some of the more impulsive newspaper and radio commentators hinted at the possibility of a cease fire before the end of summer. As for the Marine command and staff, they were not so optimistic, judging from this sentence in a report:

"All Division units were notified on 14 August that requisitions had been sent to EUSAK for cold weather clothing and equipment."

The training period afforded an opportunity to glance back over the first year of fighting in Korea and evaluate the results. There could be no doubt that the war's foremost tactical innovation so far was the combat helicopter. The Marine Corps had taken the lead in its de-

[20] EUSAK *Cmd Rpt*, Apr 51, 1080110.
[21] *Ibid.*

velopment when VMO–6, made up of OYs and Sikorsky HO3S–1 helicopters in roughly equal numbers, got into action with the 1st Provisional Marine Brigade in the Pusan Perimeter. Brigadier General Edward A. Craig had the historical distinction, insofar as is known, of being the first commanding general to see the advantages of a "chopper" as a command vehicle.

Evacuation of casualties was the principal job of the rotary-wing aircraft, and 1,926 wounded Marines were flown out during the first year. No less than 701 of these mercy flights took place during the three months from 1 April to 30 June 1951, covering the period of the two CCF 5th Phase offensives and the UN counterstroke. By that time the Bell HTL–4, with its built-in litters on both sides sheltered by plexiglas hoods, had taken over most of the evacuation missions from the HO3S–1.

The zeal of the pilots contributed substantially to the successful results. Captain Dwain L. Redalen gave a demonstration of the VMO–6 spirit at the height of the first CCF offensive in the spring of 1951. During the 13½ hours from 0600 to 1930 on 23 April, he was in the air constantly except for intervals of loading or unloading casualties. Logging a total of 9.6 flight hours, he evacuated 18 wounded men under enemy fire that left bullet holes in the plexiglas of his HTL–4.[22]

Practically all the helicopter techniques put into effect by VMO–6 had originally been developed by the Marine experimental squadron, HMX–1, organized late in 1947 at Quantico. Despite the enthusiasm for rotary-wing aircraft then prevailing, HMX–1 decided that an observation squadron should combine OYs with helicopters. The wisdom of this conclusion was proved in Korea, where the test of combat showed that both types were needed. The OYs were the superiors at reconnaissance and artillery spot missions, while the helicopters excelled at transportation and liaison and evacuation flights.

VMO–6 as a whole was the only Marine organization linking the ground and air commands. An administrative unit of the 1st MAW, the squadron was under the operational control of the 1st Marine Division.[23]

---

[22] VMO-6 Daily Flight Log, 23Apr51.

[23] This section, except when otherwise noted, is derived from the following sources: Elizabeth L. Tierney, Historical Branch, G–3, HQMC, statistics compiled from VMO–6 reports of Aug 50 to Jul 51; HMR–161 *HD*, Sep 51; 1stMarDiv type "C" rpt on assault helicopters, 4 Oct 51; Lynn Montross, *Cavalry of the Sky* (Harper, 1954), based on Marine records, 151–158.

Thanks to the ability of the helicopter to land "on a dime," staff liaison missions and command visits were greatly facilitated. The helicopter had become the modern general's steed, and the gap between staff and line was narrowed by rotary wings.

The importance of wound evacuation missions can hardly be overestimated. Surgeons stressed the value of time in treating the shock resulting from severe wounds. The sooner a patient could be made ready for surgery, the better were his chances of survival. Definitive care had waited in the past until a casualty was borne on a jolting stretcher from the firing line to the nearest road to begin a long ambulance ride. Such a journey might take most of a day, but there were instances of a helicopter evacuee reaching the operation table only an hour after being wounded at the front, 15 or 20 miles away.

Captain J. W. McElroy, USNR, commanding the famous hospital ship *Consolation*, asserted that his experience had "proved conclusively the superiority of the helicopter method of embarking and evacuating casualties to and from the ship."[24] A helicopter loading platform was installed on the *Consolation* in July 1951, during an overhaul at the Long Beach Naval Shipyard in California. Marine helicopter pilots advised as to landing requirements, and eventually all the hospital ships had similar platforms.

At a conservative estimate, the 1,926 wounded men flown out by VMO–6 helicopters during the squadron's first year in Korea included several hundred who might not have survived former methods of evacuation.

Marine Body Armor Tested in Korea

Another far-reaching tactical innovation was being launched at this time as Lieutenant Commander Frederick J. Lewis (MSC) USN, supervised a joint Army-Navy three-month field test of Marine armored vests made of lightweight plastics.

A glance at the past reveals that body armor had never quite vanished from modern warfare. European cavalry lancers wore steel cuirasses throughout the 19th century. During the American Civil War two commercial firms in Connecticut manufactured steel breastplates purchased by thousands of Union soldiers. So irksome were

[24] CO USS *Consolation* rpt to ComNavFe, 26 Jan 52.

the weight and rigidity of this protection, however, that infantrymen soon discarded it.

World War I dated the first widespread adoption of armor in the 20th century. The idea was suggested when a French general noted that one of his men had survived a lethal shell fragment by virtue of wearing an iron mess bowl under his beret. France led the way, and before the end of 1915 steel helmets were being issued to all armies on the Western Front.

When the United States entered the war, General John J. Pershing put in a request for body armor. Some 30 prototypes using steel or aluminum plates were submitted but rejected. In every instance the weight and rigidity were such that too high a price in mobility would be paid for protection.[26]

During the 1930's new possibilities were opened up by developments in lightweight plastics. The Japanese attack at Pearl Harbor interrupted experiments that were not resumed until 1943. Then a new start was made with the formation of a joint Army-Navy committee headed by Rear Admiral Alexander H. Van Kueren and Colonel George F. Doriot.

Wound statistics indicated that the great majority of fatal wounds were received in a comparatively small area of the body. The following table shows the regional frequency:

NON-FATAL	Percent	FATAL	Percent
Head	10	Head	20
Chest	10	Chest	50
Abdomen	10	Abdomen	20
Upper Extremity	30	Upper Extremity	5
Lower Extremity	40	Lower Extremity	5

Shell, mortar, or grenade fragments caused 60 percent of the fatal wounds, the statistics revealed, with the remainder being charged to rifle or machine gun fire. It was futile to hope for lightweight protection against high-velocity bullets. But researchers hoped that plastic body armor could stop enough shell or mortar fragments to reduce serious wounds to light wounds while preventing light wounds altogether.

[26] This section, except when otherwise noted, is derived from the following sources: Rpt of Joint Army-Navy Mission at HQMC, 9 Nov 51, in G–4 Files; *Instructional Information, Vest, Armored,* M–1951, G–4 Files, HQMC; LCdr F. J. Lewis (MSC) USN, ltr of 21 Jun 54; Capt Louis Kirkpatrick (MC) USN, ltr of 22 Jun 54; Capt D. G. McGrew, ltr of 2 Jul 54; LtCol G. A. Hardwick, USMC, ltr of 30 Jun 54.

Doron and nylon were the materials approved by the joint Army-Navy committee. The first, named in honor of Colonel Doriot, consisted of laminated layers of glass cloth filaments, bonded under heavy pressure to form a thin, rigid slab. That a ⅛-inch thickness could stop and partially flatten a submachine gun bullet with a muzzle velocity of 1,150 feet per second was demonstrated by ballistic tests at a range of eight yards.

The committee recommended 12-ply, laminated, basket-weave nylon for use where flexibility was required. Both the doron and nylon protected the wearer by offering enough resistance to absorb the energy of the missile, which spent itself at the impact. Thus the shock was spread out over too large a surface for a penetration, although the wearer could receive a bad bruise. If a penetration did result from a missile of higher velocity, its effects would be much reduced in severity.

Aircraft pilots and crewmen, who could tolerate more weight than foot-sloggers, were first to benefit. Flak suits and curtains were being manufactured in quantity for airmen by 1944, and the Eighth Air Force claimed a 50 percent reduction in casualties as a result.

The infantry stood most in need of protection. Statistics from 57 U.S. divisions in the European theater of operations during World War II indicated that foot soldiers, comprising 68.5 percent of the total strength, suffered 94.5 percent of the casualties. It was further established that shell or mortar fragments caused from 61.3 to 80.4 percent of the wounds.

Unfortunately, progress lagged for the ground forces, owing to conflicting requirements. Several prototype armored vests were submitted and rejected. The Marine Corps planned to conduct combat tests in the spring of 1945 by providing the ordinary utility jacket with sheaths to hold slabs of doron. A battalion of the 2d Marine Division had been selected to wear the garment on Okinawa, but the experiment was interrupted by the end of the campaign.

The Navy and Marine Corps renewed their research in 1947 at Camp Lejeune. There a new ballistics center, established for the development and evaluation of body armor, was set up by the Naval Medical Field Research Laboratory (NMFRL). Lieutenant Commander Lewis was placed in charge of experiments.

Scientific precision seemed more important than haste in time of peace, and the NMFRL was not ready with an armored vest when

634040 O-62-13

Communism challenged the free world to a showdown in Korea. Five hundred of the armored utility jackets of the proposed Okinawa test were available, however, and were air-shipped to the 1st Marine Division during the Inchon-Seoul operation.

Many of them went astray during the sea lift to Wonsan and subsequent Chosin Reservoir operation. Only the 50 garments issued to the Division Reconnaissance Company were worn in combat. And though this unit kept no records, the doron slabs were credited by Major Walter Gall, the commanding officer, with saving several lives.

By the summer of 1951, Lieutenant Commander Lewis and his researchers had designed a new Marine armored vest, weighing about 8½ pounds, combining curved, overlapping doron plates with flexible pads of basket-weave nylon. This garment, according to the official description, was capable of "stopping a .45 caliber USA pistol or Thompson submachine gun bullet; all the fragments of the U.S. hand grenade at three feet; 75 percent of the U.S. 81mm mortar at 10 feet; and full thrust of the American bayonet."

Only 40 vests were available for field tests in the summer of 1951. Lewis rotated them among as many wearers as possible in the three regiments selected for the test, the 5th Marines and the 23d and 38th regiments of the U.S. 2d Infantry Division. There was, as he saw it, a psychological question to be answered—would body armor win the acceptance of troops in combat? The hackneyed phrase "bullet-proof vest," for instance, put the wearer in a class with the buyer of a gold brick. Nylon was associated in the minds of the men with alluring feminine attire rather than protection from shell fragments. Finally, there could be no denying that undesired weight had been added, that doron plates hampered movement to some extent, and that nylon pads were uncomfortably warm for summer wear.

Despite these drawbacks, Lewis found that troop acceptance was all that could be asked. The locale of the tests was the Inje area and the approaches to the KANSAS Line in June and early July. "By keeping these few vests almost constantly in use," the Medical Service Corps officer commented, "the maximum amount of troop wear was obtained. Included in the wide sampling were company aid men, riflemen, BAR men, mortar (60mm) men, radio (backpack type) men—each carrying his basic weapon, ammunition load and a one-meal ration."

When Lewis returned to Camp Lejeune, he reported "that body armor, protection of some type for the vital anatomic areas, is almost unanimously *desired by all combat troops,* particularly the combat veteran of several actual fire fights with the enemy." [26]

Infantry body armor had at last made the transition from a dream to a reality. The M–1951 was put into production by a Philadelphia sportswear firm. And it was estimated that by the spring of 1952 nearly all Marines would be protected by the vest in combat.

Saving of American lives, of course, was a primary consideration. But there was a tactical as well as humanitarian advantage to be gained. For if body armor could reduce fatal and serious wounds by as much as 50 percent, as NMFRL researchers hoped, it would mean that a large percentage of the enemy's best antipersonnel weapons had in effect been silenced.

MAG–12 Moves to K–18

There was no respite for 1st MAW while the 1st Marine Division remained in reserve. Operation STRANGLE was at its height, and interdiction flights called for nearly all the resources of Marine aviation during the summer of 1951.

Close air support missions were made secondary. This principle was upheld by Air Force Major General Otto P. Weyland:

> I might suggest that all of us should keep in mind the limitations of air forces as well as their capabilities. Continuous CAS along a static front requires dispersed and sustained fire power against pinpoint targets. With conventional weapons there is no opportunity to exploit the characteristic mobility and fire power of air forces against worthwhile concentrations. In a static situation close support is an expensive substitute for artillery fire. It pays its greatest dividends when the enemy's sustaining capability has been crippled and his logistics cut to a minimum while his forces are immobilized by interdiction and armed reconnaissance. Then decisive results can be obtained as the close-support effort is massed in coordination with determined ground action.[27]

Marine aviation officers, of course, would have challenged some of these opinions. But General Weyland insisted that in the summer and fall of 1951 "it would have been sheer folly not to have concentrated the bulk of our air effort against interdiction targets in the

[26] Quotations are from *Instructional Information, Vest Armored, M–1951.* The italicized words were in the original.

[27] Quoted in James T. Stewart, *Air Power, The Decisive Force in Korea* (Princeton, N. J.: D. Van Nostrand Company, 1957), 22–23.

enemy rear areas. Otherwise, the available firepower would have been expended inefficiently against relatively invulnerable targets along the front, while the enemy was left to build up his resources to launch and sustain a general offensive."[28]

The UN interdiction program was costly to the Communists. Yet it remained a stubborn fact that the enemy had not only maintained but actually increased his flow of supplies in spite of bombings that might have knocked a Western army out of the war. That was because CCF and NKPA troops could operate with a minimum of 50 short tons per day per division—an average of about 10 pounds per man. It was about one-fifth of the supply requirements for an equal number of U.S. troops.

Try as they might, the UN air forces could not prevent the arrival of the 2,900 tons of rations, fuel, ammunition, and other supplies needed every day by the 58 Communist divisions at the front.

The enemy during this period was increasing his own air potential. On 17 June the Fifth Air Force warned that the Communists had stepped up their number of planes from an estimated 900 in mid-May to 1,050 in mid-June. Their Korean airfields were being kept under repair in spite of persistent UN air attacks.

In June enemy light planes made night raids along the UN front lines and even into the Seoul area. VMF(N)–513 pilots, flying the nightly combat patrol over Seoul, had several fleeting contacts with these black-painted raiders. The Marines were unable to close in for the kill, since the opposing planes were nonmetal and difficult to track by radar. Soon, however, the VMF(N)–513 pilots had better hunting. On 30 June Captain Edwin B. Long and his radar operator, CWO Robert C. Buckingham, shot down a black, two-place PO–2 biplane. And on 13 July Captain Donald L. Fenton destroyed another.[29]

Despite the Air Force emphasis on interdiction, better close air support remained a major objective of the 1st MAW. One of the requirements was a shorter flying distance from air base to combat area. K–46, the MAG–12 field near Hoengsong, had qualified with respect to reduced flying time. Maintenance problems caused by the dusty, rocky runway of this primitive strip led to its abandonment.

---

[28] *Ibid.*

[29] MAG–12 *HD*, Jun 51, Summary and Chronology, 30 Jun; MAG–12 *HD*, Jul 51, Chronology, 13 Jul.

On 14 July the squadrons pulled back temporarily to K–1, and on the 26th MAG–12 withdrew its maintenance crews.

The Group's new field was K–18, a 4,400-foot strip on the east coast near Kangnung and just south of the 38th Parallel. Situated only 40 miles behind the 1st Marine Division and on the seacoast, the new field seemed to be ideally located. The runway, reinforced with pierced steel planking, extended inland from a beach where water-borne supplies could be delivered, as at K–3.[30]

The Division Back in Action Again

Political causes had a good deal to do with the renewal of activity for the 1st Marine Division late in August 1951. Apparently the Communist armed forces had been given enough time to recuperate from their hard knocks in May and June. At any rate, the Red delegates walked out on the truce talks after falsely charging on 22 August that UN planes had violated the neutrality of the Kaesong area by dropping napalm bombs. Although the Reds were unable to show any credible evidence, the negotiations came to an abrupt end for the time being.[31]

On the 26th all Marine units received a Division warning that offensive operations were to be initiated in the immediate future. The effective strength, of the Division (including the KMCs) had been reported as 1,386 officers and 24,044 enlisted men on 1 August 1951. Attached to the Division at that time were 165 interpreters and 4,184 Korean CTC cargadores.

On the 26th the regiments were disposed as follows: the 1st Marines near Chogutan; the 5th Marines near Inje; the 7th Marines near Yanggu; and the 1st KMC Regiment at Hangye. Service units and the Division CP were located along the Hongchon-Hangye road in the vicinity of Tundong-ni.

The 11th Marines (–), with the 196th FA Battalion, USA, attached, constituted the 11th Marine Regiment Group, an element of X Corps artillery. Throughout the training period 2/11 remained under the control of the 1st Marine Division and 3/11 was attached to the 2d Infantry Division.

---

[30] "Rpt of Visit to Far East by CG, FMFPac, and his staff during the period 27 August to 12 September 1951," 17 *ff.*

[31] Berger, *The Korean Knot, op. cit.* 144–145.

The 5th Marines, 7th Marines, and KMCs were alerted to be prepared to move up to the combat areas south and west of the Punchbowl on 27 August. The 1st Marines was to remain in Division reserve, and the 11th Marines reverted to parent control.[32]

It was only about a five hour motor march from Tundong-ni to the forward assembly area under normal road and weather conditions. But recent rains had turned roads into bogs and fordable streams into torrents. Bridges were weakened by the raging current in the Soyang, and landslides blocked the road in many places.

The 1st Marine Division was back in action again. But it would have to fight its first battles against the rain and the mud.

[32] 1stMarDiv *HD*, Aug 51, 3–5.

CHAPTER IX

Renewal of the Attack

Crossing the Soyang in Flood—Light Resistance at First—Supply Problems Cause Delay—Resumption of Division Attack—The Mounting Problem of CAS—First Helicopter Supply Operation of History—The Fight for Hill 749—5th Marines Attack Hill 812—The Struggle for the "Rock"

IT WAS to a large extent a new 1st Marine Division on 27 August 1951. Very few veterans of the Reservoir campaign were left, and even the Marines of the hard fighting in April and May had been thinned by casualties and rotation. Whatever the new arrivals lacked in experience, however, they had made up as far as possible by intensive and realistic training while the Division was in reserve.

The new Marine zone of action, in the Punchbowl area, was as bleak and forbidding as any expanse of terrain in Korea. Dominating the Punchbowl from the north and blocking any movement out of it was YOKE Ridge, looking somewhat like an alligator on the map (See Map 17). Hill 930 represented the snout. Hill 1000 was the head, and the body extended eastward through Hills 1026 and 924.

Two smaller hills, 702 and 602, spread off southeast and northeast respectively to the Soyang River and its unnamed tributary from the west. On either side of YOKE Ridge were numerous sharp and narrow ridges. Some of the hills were wooded with enough scrub pine to afford concealment for outposts and bunkers. Altogether, it was an area eminently suited to defense.

The defenders were identified by Division G–2 as troops of the 6th Regiment, 2d Division, II NKPA Corps. Apparently they did not lack supporting weapons, for 3/7 positions on Hill 680 were hit by an estimated 200 mortar and artillery rounds during daylight hours of the 30th.

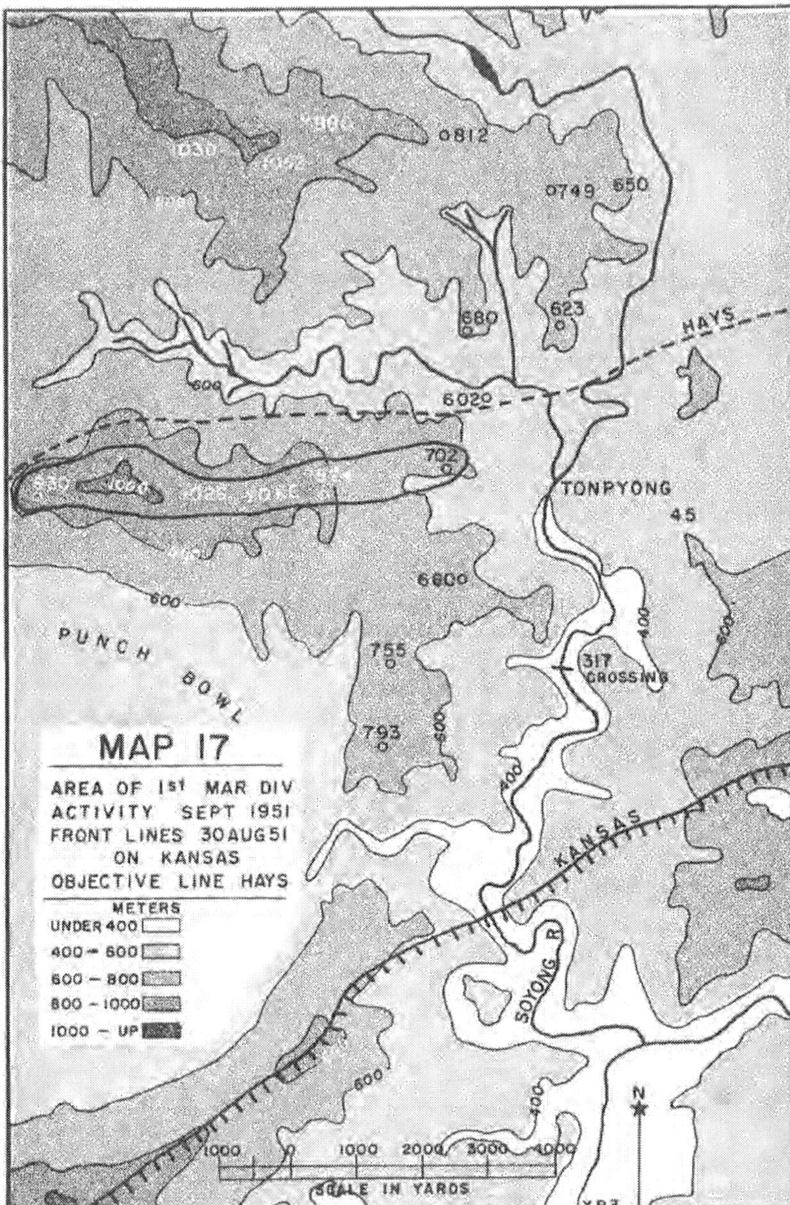
MAP 17
AREA OF 1st MAR DIV
ACTIVITY SEPT 1951
FRONT LINES 30AUG51
ON KANSAS
OBJECTIVE LINE HAYS
METERS
UNDER 400
400 – 600
600 – 800
800 – 1000
1000 – UP
HAYS
KANSAS
TONPYONG
PUNCH BOWL
317 CROSSING
SOYONG R
SCALE IN YARDS
1000
0
1000
2000
3000
4000
N
X83
812
749
650
680
623
602
702
680
755
793
600
400

Crossing the Soyang in Flood

The 7th Marines and KMC Regiment, ordered to relieve U.S. and ROK Army units on the KANSAS Line, started their march in a downpour on 27 August. The 5th Marines (less 1st Battalion) at Inje had orders to follow the 7th up the narrow Soyang valley.

Typical of the wet weather difficulties were those experienced by 3/7. Scheduled to depart early for the forward positions, the companies struck tents. Trucks failed to arrive and they remained to eat the noon meal, a gustatorial bonus of all food the galley crew could not carry with them. Unfortunately, the trucks were delayed further and the men shivered in the rain as they ate an evening meal of "C" rations.

When the vehicles finally arrived at 2100 the rain had reached torrential proportions. Progress was so slow over muddy roads that it took until 0330 on the 28th to reach the CP of the 7th Marines at Sohwari (Map 18), just southeast of the junction of the Soyang and a tributary from the east.

The bivouac area assigned to 3/7 for the night proved to be a foot deep in water, and Lieutenant Colonel Kelly directed his men to catch what sleep they could in the trucks while he and his staff attempted to straighten out the snarled traffic situation.[1]

It took the rest of the night for the 3/7 officers to walk the length of the convoy, cutting out trucks with less essential cargo. With only a small space available for a turn-around, the 3/7 vehicles were ordered to back into it, unload their troops and equipment, and return along a narrow road, which had been churned into a quagmire.

The battalion assembly area was on the other side of the rain-swollen Soyang. How Company and the command group managed to cross over a waist-deep ford, but the crossing was so perilous that DUKWs were requested for the other two rifle companies. Lieutenant Colonel Louis C. Griffin's 2/7 also found the river crossing an operation requiring DUKWs. By the afternoon of the 29th all elements of the two 7th Marine battalions were on the west bank, occupying their assigned assembly areas.

The relief proceeded slowly. Two KMC battalions on the left of the 7th Marines took over the zone formerly held by elements of the

[1] Sources are 1stMarDiv *HD*, Aug 51, 3–5; Col B. T. Kelly's contemporary "Notes on my Service in Korea, 14 Apr–13 Sep 1951" (hereafter Kelly, *Notes*).

2d Infantry Division and the 8th ROK Division. The cosmopolitan character of the Eighth Army was revealed when 2/KMC relieved the French Battalion of the 2d Infantry Division. Linguistic chaos was averted only by the best efforts of the exhausted interpreters.

By the 30th, the 1st and 3d KMC Battalions were behind the line of departure on Hill 755, ready to attack in the morning. The 2d Battalion assumed responsibility for the regimental zone on the KANSAS Line.

The 2d and 3d Battalions of the 7th Marines had meanwhile completed the relief of elements of the 8th ROK Division. On the other side of the river Lieutenant Colonel James G. Kelly's 1/7 had relieved units of the ROK division on the hill mass a mile and a half north of Tonpyong (Map 17). These Marines were first to come under fire as the enemy sent over a few mortar rounds after dark on the 29th.

Division OpnO 22–51 directed the two assault regiments, the 7th Marines and KMCs, to attack at 0600 the following morning and seize their assigned positions on Corps Objective YOKE, the ridgeline running from Hill 930 on the west through Hills 1026 and 924 on the east (Map 17). Objective 1, the hill mass 1½ miles northeast of Tonpyong, was already occupied by 1/7.

The 3d Battalion, 7th Marines, was ordered to seize Objective 2, generally that part of YOKE Ridge east of Hill 924. The KMC Regiment was assigned Objective 3, consisting of Hills 924 and 1026.

Other 1st Marine Division units had the following missions on 31 August:

5th Marines—to patrol the Division zone along the KANSAS Line and protect defensive installations;
1st Marines—to remain in the rear in the Hongchon area in X Corps reserve;
1st Tank Battalion—to move up in readiness to support the assault regiments;
Division Reconnaissance Company—to continue to patrol the Punchbowl and mop up bypassed enemy.

Land mines were a constant menace to troop movements as the assault regiments adjusted positions in preparation for the attack. As usual, neglected "friendly" mines were encountered as well as those planted by the enemy.[2]

POW information and air reports indicated a southward movement of two to three enemy regiments with artillery and supplies. Prisoners

[2] This section, except when otherwise specified, is based on 1stMarDiv *HD*, Aug and Sep 51; X Corps *Cmd Rpt*, Sep 51; 2/7 and 3/7 *HD*, Aug and Sep 51; Kelly, *Notes*; Col G. P. Groves, ltr of 8 Apr 58.

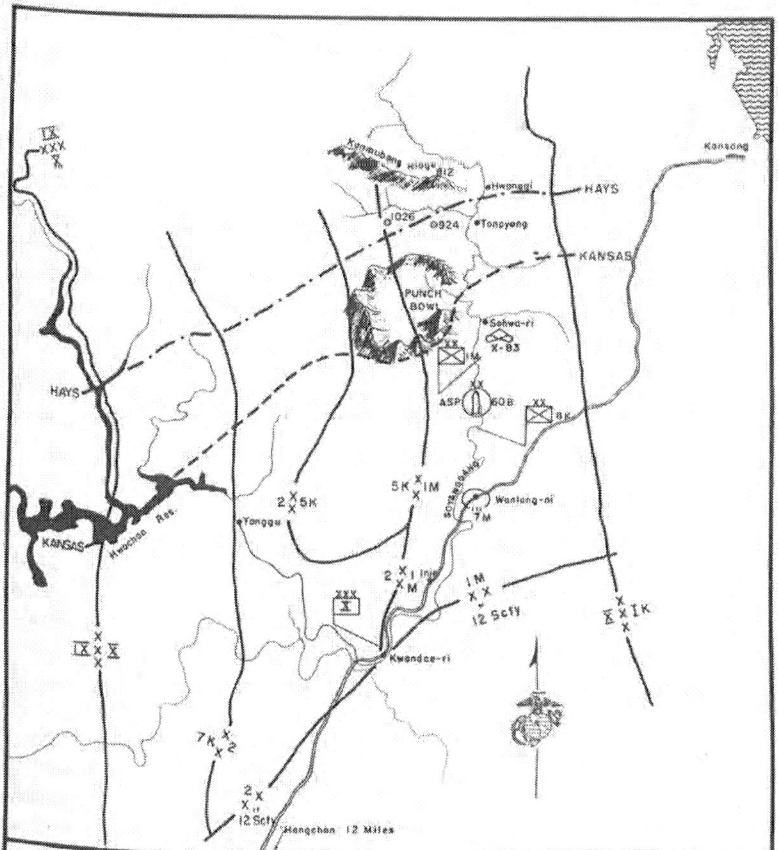

X CORPS ZONE
OF ACTION
SEPTEMBER 51

FRONT LINES ON 20 SEPT 51 WERE GENERALLY ALONG HAYS LINE

SCALE

5 0 5 Miles
(Approx)

MAP 18

stated that an attack was due on 1 September, leading to the G–2 conjecture that the enemy's Sixth Phase Offensive might be about to start.

Light Resistance at First

Priority of air support on 31 August was assigned to the two KMC battalions. They jumped off in column against light to moderate resistance, with Hill 924 as their first objective. Mine fields gave the KMCs more trouble at first than scattered NKPA mortar and machine gun fire. Forward movement and maneuver were restricted as 1/KMC passed through 3/KMC at 1445 to continue the attack against stiffening resistance.

On the right 3/7 also encountered light resistance in the morning which increased as the assault troops neared the objective. The slopes of Hill 702 proved to be heavily mined, and forward elements of 3/7 were hit by a concentration of mortar and artillery fire.

East of the river, on the regimental right flank, where Objective 1 had been occupied without a fight, 1/7 supported the attack of 3/7 with mortar fire. Both 3/7 and the KMCs were within 1,000 yards of their objectives late in the afternoon when a halt was called for the day. Casualties had been light, thanks in large measure to excellent air and artillery support.

When the attack was resumed on 1 September, 3/KMC moved through positions of 3/7 to reach a ridgeline on the flank of the regimental objective. While 3/KMC advanced from the northeast, 1/KMC closed in from the southeast. Both battalions took heavy losses from enemy mines and mortars as well as machine guns and automatic weapons fired from hidden bunkers. The converging attack made slow but steady progress, however, until one company of 3/KMC drove within 200 meters of the top of Hill 924 at 1700. Even so, it took four more hours of hard fighting to secure the objective. That evening 2/KMC was relieved of its defensive responsibility along the KANSAS Line by 3d Battalion, 5th Marines, enabling the KMC battalion to join in the attack.

Throughout the day 3/7 slugged it out in the vicinity of 702 with an NKPA battalion. Four counterattacks were launched from Hill 602, the northeastern fork of YOKE Ridge. More than 500 men were employed in this effort, some of them penetrating briefly into 3/7 positions. Two air strikes, called by patrols of 1/7 from across the river, helped to break up the main NKPA attack, and the 11th Marines (Colonel Custis Burton, Jr.), poured in a deadly concentration of artillery fire. Lieutenant Colonel B. T. Kelly's battalion continued to be engaged until dusk.

The tenacity of the NKPA defense was demonstrated at the expense of the KMCs when they were driven from the top of Hill 924 by a surprise enemy counterattack at midnight. The Korean Marines came back strongly at daybreak and a terrific fight ensued before the North Koreans were in turn evicted shortly before noon. As a measure of the artillery assistance rendered, Major Gordon R. Worthington's 1st Battalion, 11th Marines, fired 1,682 rounds of 105 ammunition in support of the KMC's during the 24 hours ending at 1800 on 2 September. During the same period Lieutenant Colonel William McReynold's 3/11 fired 1,400 rounds in support of 3/7. The other battalions of the Marine artillery regiment, reinforced by the 196th, 937th, and 780th Field Artillery Battalions, USA, brought the number of rounds to a grand total of 8,400 for this 24-hour period.

After the securing of Hill 924, the 2d Battalion of the KMC Regiment passed through the 1st and 3d Battalions to spearpoint the attack west toward Hill 1026. In the zone of 3/7, an NKPA counterattack was repulsed at 0700 on 2 September. Two hours later George Company, supported by How Company with mortar and machine gun fire, moved out to resume the attack on Hill 602. Lieutenant Colonel B. T. Kelly ordered his battalion heavy machine guns set up in battery to deliver overhead supporting fires.

In slightly less than two hours the Marines of 3/7 swept the crest of Hill 602, securing Division Objective 2. Three company-size enemy counterattacks were repulsed before the North Koreans withdrew to the north at 1500.

The 2d KMC Battalion fought its way to a point within 800 yards of Hill 1026 before dusk. So aggressive and persistent was the NKPA defense that several light enemy probing attacks were launched during the night of 2–3 September, not only against forward Marine elements but also against the 5th Marines units on the KANSAS Line, 5 miles to the rear. The front was where you found it.

While 3/7 constructed emplacements and obstacles on Hill 602, the KMCs continued their attack on the morning of 3 September toward Hill 1026. With the extending of the 7th Marines zone to the left to decrease the width of the KMC front, 2/7 was brought up from regimental reserve to help cover a new sector that included Hill 924.

The attack led by 2/KMC collided with a large-scale enemy counterattack. It was nip and tuck for 3½ hours before the North Koreans broke, but, by midmorning, the KMCs were in possession of Division Objective 3 and consolidating for defense. They were not a moment too soon in these preparations, for the enemy counterattacked at 1230 and put up a hot fight for two hours before retiring.

This action completed the battle for Corps Objective YOKE. At 1800 on 3 September, the 1st Marine Division was in full possession of the HAYS Line, dominating the entire northern rim of the Punchbowl (Map 18). Reports from the U.S. 2d Infantry Division and 5th ROK Division, attacking in sectors to the west, indicated that the pressure exerted by the Marines was assisting these units. Large gains had been made on the west side of the Punchbowl against comparatively light resistance.

On 4 September, with all objectives consolidated, 1st Marine Division units patrolled northward from defensive positions. Plans were being formed for the second phase of the Division attack—the advance to seize the next series of commanding ridgelines, 4,000 to 7,000 yards forward of the present MLR.

The victory in the four-day battle had not been bought cheaply. A total of 109 Marine KIA and 494 WIA (including KMCs) was reported. NKPA casualties for the period were 656 counted KIA and 40 prisoners.

As evidence that the enemy had profited by the breathing spell during the Kaesong truce talks, it was estimated that NKPA artillery fire in the Punchbowl sector almost equalled the firepower provided by the organic Marine artillery and the guns of attached U.S. Army units. NKPA strength in mortars and machine guns also compared favorably with that of Marines.

Supply Problems Cause Delay

Logistical shortages made it necessary for the 1st Marine Division to call a six-day halt and build up a new reserve of artillery and mortar ammunition.

During the first phase of the Division attack, the main burden of transport and supply had fallen upon three Marine units—the 1st Ordnance Battalion (Major Harold C. Borth), the 1st Motor Transport Battalion (Lieutenant Colonel Howard E. Wertman), and the 7th Motor Transport Battalion (Lieutenant Colonel Carl J. Cagle). The extraordinary expenditure of artillery shells for these four days posed a resupply problem that was aggravated by an almost impassable supply route. The three Marine battalions had to strain every resource to meet minimal requirements.

Ammunition Supply Point (ASP) 60–B, a U.S. Army installation manned by elements of the Marine 1st Ordnance Battalion, was located about five miles behind the gun positions. From this dump it was 48 miles to Hongchon, the source of supplies for ASP 60–B. A well maintained, two-lane dirt road led from that base to Inje, but northward it deteriorated into a narrow, twisting trail following the Soyang valley. Recent rains, resulting in earth slides and mudholes, had reduced the road to such a condition that the round trip between ASP 60–B and Hongchon took 25 hours.[3]

As an added complication, it was necessary to build up a 10-day reserve of ammunition at ASP 60–B so that Division transport would be available for lifting 2,000 rotated troops to Chunchon some time between 3 and 15 September. This meant that 50 to 60 Marine trucks must be employed daily to haul ammunition, with the result of a drastic shortage of motor transport for other purposes.

Only human transport was available for supplying Marines on the firing line. X Corps started the month of September with 20,070 Korean Service Corps, the successor to CTC, and civilian contract laborers—the equivalent in numbers of a U.S. Army infantry division. Even so, 14 air drops were necessary during the month, only one of which went to a Marine unit. This took place on 1 September, when 20 Air Force cargo planes from Japan dropped ammunition and rations to the KMCs. A 90 percent recovery was reported.[4]

It generally took a full day in the 1st Marine Division zone during the first week of September for a cargador to complete the trip from a battalion supply point to the front lines and return. This made it necessary to assign from 150 to 250 Korean laborers to each infantry

[3] 1stMarDiv *HD*, Sep 51, 4, 7.
[4] X Corps *Cmd Rpt*, Sep 51, 41–42; 1stMarDiv *HD*, Sep 51, 5–6.

battalion. And as the Marines advanced farther into the rugged Korean highlands, the logistic problem was increased.

Resumption of Division Attack

Enemy groups moving southward into the zone of the 1st Marine Division during the six-day lull were sighted by air observation. POW interrogations and other G–2 sources established that the 2d NKPA Division, II Corps, had been relieved by the 1st NKPA Division, III Corps. Accurate 76mm fire from well-hidden guns was received by the Marines throughout the interlude, and patrols ran into brisk mortar fire when they approached too near to enemy bunkers on Hill 673.

For the second time, during the night of 4–5 September, 5th Marines units were assailed on the KANSAS Line, 5 miles to the rear of the 7th Marines troops similarly deployed along the HAYS Line. Yet a large 7th Marines patrol ranged forward some 2,000 yards the next day without enemy contacts. A like result was reported by a patrol representing almost the entire strength of the Division Reconnaissance Company (Major Robert L. Autry) after it scoured the area north of the Punchbowl.[5]

1st Marine Division OpnO 23–51, issued on the morning of 9 September, called for the 7th Marines to jump off at 0300 on the 11th and attack Objectives ABLE and BAKER—Hills 673 and 749 respectively—while maintaining contact with the 8th ROK Division on the right. Other Division units were given these missions:

1st Marines—to be released from X Corps reserve near Hongchon to Division control; to be prepared to pass through the 7th Marines, when that regiment secured its objectives, and continue the attack to seize Objective CHARLIE, the ridgeline leading northwest from Hill 1052.

5th Marines—to maintain one company on KANSAS Line while occupying positions in Division reserve along HAYS Line in rear of 7th Marines.

KMC Regiment—to patrol aggressively on Division left to exert pressure on enemy defenses south and southeast of Objective CHARLIE.

11th Marines—to displace forward to support attack of the 7th Marines.

Division Reconnaissance Company—to patrol northward in the Soyang valley as far as Hwanggi to deny the enemy this area.

[5] This section, except when otherwise specified, is based on the following sources: EUSAK *Cmd Rpt*, Sep 51, 38–53; X Corps *Cmd Rpt*, Sep 51, 9–12; 1stMarDiv *HD*, Sep 51, 8–14; 7th Mar *HD*, Sep 51; 1st, 2d, and 3d Bns of 7th Mar, *HD*s for Sep 51.

The area ahead of the 7th Marines was ideal for defense. From YOKE Ridge the assault troops had to descend into a narrow valley formed by a small tributary of the Soyang-gang, cross the stream, and climb Kanmubong Ridge on the other side. This formidable piece of terrain was dominated by three enemy positions, Hills 812, 980, and 1052 (Map 17). Thus the attack of the 7th Marines had as its primary purpose the securing of initial objectives on Kanmubong Ridge that would give access to the main NKPA defense line, some 4,000 yards to the north.

The 7th Marines was to seize the eastern tip (Objective ABLE) of this commanding terrain feature and "run the ridge" to Hill 749, Objective BAKER. While Lieutenant Colonel Louis G. Griffin's 2/7 maintained its patrolling activities on the left, tied in with the KMCs, Lieutenant Colonel B. T. Kelly's 3/7 in the center and Lieutenant Colonel J. G. Kelly's 1/7 on the right were to attack.

As an intermediate regimental objective on the way to Kanmubong Ridge, the 680-meter hill directly north of B. T. Kelly's position on Hill 602 was assigned to his battalion. He ordered How Company to move forward under cover of darkness and be prepared to attack at dawn. Rain and poor visibility delayed the attempt until surprise was lost, and after a fierce fire fight How Company was stopped halfway up the southeast spur.

In order to relieve the pressure, the battalion commander directed Item Company to attack on the left up the southwest spur. This maneuver enabled How Company to inch forward under heavy mortar and machine gun fire to a point with 50 yards of the topographical crest. Item Company became confused in the "fog of war" and finally wound up on How's spur at 1245.

Twice the two companies made a combined assault after artillery and mortar preparation and air strikes with napalm, rocket, and strafing fire. Both times the North Koreans swarmed out of their bunkers to drive the Marines halfway back to the original jump off line. It was anybody's fight when the two battered companies dug in at dusk.

Across the valley to the east, J. G. Kelly's 1/7 had no better fortune in its attack on Hill 673. Heavy enemy mortar and machine gun fire kept the assault troops pinned down until they consolidated for the night.

With both attacking battalions in trouble, Colonel Nickerson

634040 O-62-14

ordered 2/7 to advance up the narrow valley separating them. His plan called for the reserve battalion to move under cover of darkness around the left flank of 1/7 and into a position behind the enemy before wheeling to the northeast to trap the North Koreans defending Hill 673.

The maneuver succeeded brilliantly. Griffin's troops were undetected as they filed northward during the night, making every effort to maintain silence. By daybreak on 12 September 2/7 had two platoons in position behind the enemy to lead the attack.[8]

The assault exploded with complete surprise as 2/7 swept to the crest of Hill 673 against confused and ineffectual opposition. Griffin's battalion and 1/7 had the enemy between them, but the jaws of the trap could not close in time because of NKPA mine fields. Thus 1/7 continued to be held up on the forward approaches to Hill 673 by NKPA mortar and small-arms fire. Grenades were the most effective weapons as J. G. Kelly's men slugged their way to the summit at 1415 while 2/7 was attacking Objective BAKER, Hill 749.

On the other side of the valley, 3/7 had seized its initial objective. While How and Item Companies attacked up the southeast spur, where they had been stopped the day before, George Company launched a surprise assault up the southwest spur. This was the blow that broke the enemy's will to resist. George Company knocked out seven active enemy bunkers, one by one, thus taking the pressure off the troops on the other spur. At 1028 all three companies met on the summit.

The 2d Battalion, 7th Marines, radioed that Objective BAKER had been secured at 1710 after a hard fight, but this report proved to be premature. Enough NKPA troops to give the Marines a good deal of trouble were still holding the wooded slopes of Hill 749, and it would take the attack of a fresh battalion to dislodge them. Along the ridgeline from Hill 673 to Hill 749, an undetermined number of enemy soldiers had been caught between 2/7 and 1/7, and events were to prove that they would resist as long as a man remained alive.

Casualties of the 1st Marine Division on 11 and 12 September were 22 KIA and 245 WIA, nearly all of them being suffered by the assault regiment. Enemy losses included 30 counted KIA and 22 prisoners.

---

[8] LtCol E. G. Kurdziel interv, 13 Jun 58.

The Mounting Problem of CAS

With the Division in reserve from 15 July until the latter part of August, close air support (CAS) was not a vital problem; however, upon return to the Punchbowl area the situation became serious. The difficulties arose from the time lag between the request for air support to the time the planes arrived over target. The 1st Marine Aircraft Wing operating under the control of the Fifth Air Force was busily employed on interdiction missions. On 30 August, a tactical air observer, spotting what appeared to be a division of NKPA troops moving toward the Marines, hurriedly flashed back a request for a multiplane strike. The enemy troops were beyond artillery range, but they were bunched up—a good target for a concentrated air strike. It was more than three hours later that four fighter bombers arrived on the scene; by that time, the enemy formation had dispersed and the desired number of casualties could not be inflicted.[7]

The reason for this lack of timely air support was apparent. Most of the UN air power was being funneled into Operation STRANGLE, the interdiction operation designed to cut off the enemy's vehicular and rail traffic in the narrow waist of North Korea. With the emphasis on air interdiction, close air support sorties were limited to only 96 per day for the entire Eighth Army.[8] The 1st Marine Division received only a proportionate share.

Marine close air support was needed because of the enemy's determined resistance to the Division's attack. The Reds hurled frequent night counterattacks and pounded the Marine positions with artillery and mortars hidden in the precipitous Punchbowl area. At one time it was estimated that the enemy was using 92 pieces of artillery. The Marines had only 72 field pieces, but in one 24-hour period they expended more than 11,000 rounds of artillery ammunition on a 6,000-yard frontage. The enemy emplacements, hewn out of solid rock, were hard to knock out.

To support the hard-working infantrymen, Marine Aircraft Group 12 (MAG–12) had moved VMF–214 and VMF–312 from the Pusan area to K–18, an airfield on the east coast at Kangnung. By moving closer to the Division area, planes were able to extend their

[7] CG 1stMarDiv ltr to CinCPacFlt, 4 Oct 51, enclosure (1) "Observations on Close Air Support for the 1st Marine Division during 5–23 September 1951."

[8] *PacFlt Interim Rpt* No. 3, VI, 6–6, 6–7; 1stMarDiv *SAR*, Jun 51.

time over the target area and render more effective support to the infantry. Also, Marine Air Support Radar Team One (MASRT–1) was sent to Korea and established positions to support the Division. Using its support radar the team began to evaluate its capability of guiding unseen fighter-bombers at night or under conditions of poor visibility.[9]

Even though the Corsairs at K–18 were less than 50 miles from the 1st Marine Division, very few were available to the Marines. Operation STRANGLE, in full swing, was not achieving the desired results. Since sightings of enemy vehicles were increasing, more and more Marine and Navy air sorties were channeled into interdiction. During 18 days of rugged fighting from 3 to 21 September, forward air controllers made 182 tactical air requests. Fighter-bombers were provided on 127 of these requests; however, in only 24 instances did the planes arrive when needed. The average delay time in getting CAS in response to requests during September was slightly less than two hours, but in 49 cases the planes were more than two hours late.[10] As a consequence, General Thomas reported, many of the 1,621 casualties suffered by the 1st Marine Division during the hard fighting in September were due to inadequate close air support. Furthermore, he said, the tactical capabilities of his battalions were strongly restricted.

During the planning of attacks, infantry commanders almost always desired and requested close air support. It was also desirable to have planes on station overhead should an immediate CAS need arise, for the lack of an air strike when needed could jeopardize success. However, with restricted availability of CAS planes due to participation in STRANGLE, many times desired air cover was not to be had. Attacks under those circumstances were often costly.

First Helicopter Supply Operation of History

The relief of the three battalions of the 7th Marines by their corresponding numbers of Colonel Thomas A. Wornham's 1st Marines took place during the night of 12–13 September. By daybreak 3/1

[9] *PacFlt Interim Rpt* No. 3, Chap. 9, 9–18; Chap. 10, 10–12, Chap. 15, 15–20, 60–61; Gen G. C. Thomas interv, 21 Jan 59.
[10] *PacFlt Interim Rpt* No. 3, Chap. 9, 9–14.

and 1/1 had assumed responsibility for the zones of 3/7 and 1/7, which were on their way to Division reserve at Wontong-ni at the junction of the Inje and Kansong roads. In the center, however, 2/1 could not complete the relief of 2/7. Not only was that battalion engaged most of the day with the enemy, but the units were separated—one company south of Hill 749 being unable to join the other two companies on separate spurs northwest of that height. All three were under persistent NKPA mortar and 76mm fire.[11]

The attack of the 1st Marines, originally scheduled for 0500 on 13 September, had been changed to 0900 by Division orders. One reason for the postponement was the serious shortage of ammunition and other supplies after the urgent demands of the last two days. Another reason was the inability of VMO–6 helicopters, lifting two wounded men at most, to cope with the mounting casualty lists. Enemy interdiction of roads added in several instances to the complications of a major logistical problem, particularly in the zone of Lieutenant Colonel Franklin B. Nihart's 2d Battalion, 1st Marines.

The hour had struck for HMR–161, and the world's first large-scale helicopter supply operation in a combat zone would soon be under way. It was not the development of a day. On the contrary, its roots went all the way back to 1945, when the atomic bomb of Hiroshima rendered obsolescent in 10 seconds a system of amphibious assault tactics that had been 10 years in the making. Obviously, the concentrations of transports, warships, and aircraft carriers that had made possible the Saipan and Iwo Jima landings would be sitting ducks for an enemy armed with atomic weapons.

The problem was left on the doorstep of the Marine Corps Schools, which had reared the Fleet Marine Force from infancy to maturity during the 1930's. A Special Board and Secretariat were appointed for studies. They assigned two general missions to Marine Helicopter Experimental Squadron 1 (HMX–1), organized late in 1947 before the first rotary-wing aircraft had been delivered. These missions were:

(1) Develop techniques and tactics in connection with the movement of assault troops by helicopter in amphibious operations;

(2) Evaluate a small helicopter as a replacement for the present OY type

---

[11] Sources for this section are as follows: EUSAK *Cmd Rpt*, Sep 51, 35–53; X Corps *Cmd Rpt*, Sep 51, 9–12; 1stMarDiv *HD*, Sep 51, 10–16; 1st Marines *HD*, Sep 51; 1/1, 2/1, and 3/1 *HD*, Sep 51; Class "C" Rpt, *Employment of Assault Helicopters*, 1–6; Lynn Montross, *Cavalry of the Sky* (New York, 1954), 159–162, (hereafter *Cavalry of the Sky*).

aircraft to be used for gunfire spotting, observation, and liaison missions in connection with amphibious operation.[12]

The second mission resulted in the small Sikorsky and Bell helicopters of VMO–6 which landed in Korea with the 1st Provisional Marine Brigade in August 1950. Although it was originally believed that rotary wing aircraft might replace the OYs, combat experience soon demonstrated that the best results were obtained by retaining both types in fairly equal numbers.

Landing exercises under simulated combat conditions were conducted by HMX–1 in fulfillment of the first mission. At first the squadron had only three-place helicopters. Later, when the usefulness of the helicopter was fully realized, even the new 10-place "choppers" were never available in sufficient numbers. The capacity designations of these machines, however, were more ideal than real, for the helicopters could lift only four to six men in addition to the pilot, copilot, and crewman. Despite such drawbacks, HMX–1 developed tactical and logistical techniques for helicopter landings to be made from widely dispersed carriers against an enemy using atomic weapons.

Belated deliveries of aircraft delayed the commissioning of the world's first transport helicopter squadron, HMR–161, until 15 January 1951 at El Toro. Lieutenant Colonel George W. Herring was designated the commanding officer and Lieutenant Colonel William P. Mitchell the executive officer.

Nearly three months passed before the first three transport helicopters arrived. The squadron was gradually built up to a strength of 43 officers and 244 enlisted men with a full complement of 15 HRS–1 helicopters. These Sikorsky aircraft, designed to Marine specifications, were simply an enlarged three-place HO3S in configuration, with a similar main rotor and vertical tail rotor. About 62 feet long with maximum extention of rotor blades, the HRS–1 was 11½ feet wide with the blades folded. Following are some of the other statistics:

Gross weight at sea level, 7,000 pounds; cruising speed, 60 knots; payload at sea level, 1,420 pounds; troop-lifting capacity, four to six men with full combat equipment or three to five casualties in litters.[13] Capabilities varied, of course, according to such factors as altitude, temperature, and pilot experience.

Marine Transport Helicopter Squadron 161 arrived in Korea on

[12] CMC ltr to CO MCAS, Quantico, 3 Dec 47.
[13] *Cavalry of the Sky*, 157.

the last day of August, and by the 10th of September it had moved up to the front, sharing Airfield X-83 (see Map 18) with VMO-6.[14] The 11th was devoted to reconnaissance flights in search of landing sites, and on the 12th the transport squadron was ready for its first combat mission. A new means of logistical and tactical support that was to revolutionize operations and create front page headlines had arrived in Korea.

Prior to the squadron's arrival, the Division chief of staff, Colonel Victor H. Krulak, had held numerous planning conferences with Division staff officers, and preparations for the employment of HMR-161 had made noteworthy progress. Then General Thomas ordered executed the first operation of the squadron under combat conditions, and the major logistical problem of moving supplies and evacuating casualties was well on the way to being solved. At 1600 on 13 September 1951—a date that would have historical significance—Operation WINDMILL I was set in motion.

Lieutenant Colonel Herring had attended the final planning conference at Division headquarters at 0830 on the 13th, and he was told that the operation would involve a lift of one day's supplies to 2/1 over a distance of seven miles. The commanding officer of 2/1 was to select suitable landing points and the commanding officer of 1st Shore Party Battalion had the responsibility of providing support teams to operate at the embarkation and landing points.[15]

Only two days had been available for training and rehearsals, but not a minute was wasted. All morning on the 13th the embarkation point section separated the supplies into balanced loads of about 800 pounds per helicopter. Loading commenced at 1520. Half an hour later, seven aircraft were ready to depart while four others went ahead to carry the landing point section to the previously reconnoitered site.

The route followed the valleys as much as possible, so that the helicopters were in defilade most of the way. Smoke was laid down by the 11th Marines for concealment.

The landing point section managed in 20 minutes to clear an area of 20 x 40 feet (later enlarged to 100 x 100 feet) and mark it with fluorescent panels. At 1610 the first HRS-1 hovered with cargo nets suspended from a hook released by manual control. A few minutes

[14] Auxiliary airstrips in Korea had an "X" designation and fields in the "K" category were major installations. Those in proximity to U.S. Army centers were designated "A."

[15] LtCol H. W. Edwards, interv of 20 Feb 61.

later it took off with five walking wounded and two litter cases.

Each helicopter carried out as many casualties as possible, depending on the amount of gasoline in the fuel tanks. Only 30 minutes passed from the time one Marine was wounded and the time of his arrival at a hospital clearing station 17 miles behind the firing line.

Radio provided communications between helicopters in flight, HMR–161 headquarters, 2/1 CP, and the Shore Party team at the landing site.

Fifteen aircraft were employed for one hour, three for two hours, and one for two hours and 45 minutes—a total of 28 flights in over-all time of 2½ hours. The helicopters landed at intervals of two minutes and took off as soon as the landing point section could put the casualties aboard. And though an altitude of 2,100 feet restricted loads, 18,848 pounds of cargo had been lifted into the area and 74 casualties evacuated when the last "chopper" returned to X–83 at 1840.

To even the most pessimistic observer Operation WINDMILL I was a complete success, so successful that a similar operation, WINDMILL II was conducted on the 19th. Two days later the first helicopter lift of combat troops was completed. A new era of military transport had dawned.

The Fight for Hill 749

Although 2/1 alone had 240 Korean cargadores attached, the 7½ tons of helicopter-borne supplies, largely ammunition, were vitally needed by the two assault battalions of the 1st Marines. After relieving Fox Company of 2/7 south of Hill 749 at 1100 on the 13th, Lieutenant Colonel Nihart's 2/1 jumped off to the attack an hour later. Stiff opposition was encountered from the beginning. The relief of the remaining two companies of 2/7 was complicated by the fact that they were some 400 yards from the position reported, on the reverse slope of Hill 749. Throughout the day these Marines were heavily engaged with the enemy.[16]

On the left of 2/1, the 3d Battalion (Lieutenant Colonel Foster C. La Hue) could not make much progress toward its regimental objective, Hill 751, while the enemy was active on Hill 749. A second

[16] Sources for this section are the same as for the previous section except when otherwise noted.

attack of 2/1 at 1500 drove to the summit of that height after fierce fighting with small arms, automatic weapons, and hand grenades. There was still much fighting to be done before the entire objective would be secured since many enemy bunkers hidden among the trees remained to be neutralized.

At 1600 a gap of about 300 yards separated 2/1 from the two 2/7 companies. So fierce was enemy resistance in this area that it took until 2025 for Nihart's men to complete the relief after fighting for every foot of ground.

Air and artillery support had been excellent on the 13th despite the fact that neither could be called by 2/1 in some instances because of the danger of hitting elements of 2/7. Even so, 2/11 (Lieutenant Colonel Dale H. Heely) and other artillery units fired 2,133 rounds and Company C of the 1st Tank Battalion (Lieutenant Colonel Holly H. Evans) contributed 720 rounds of 90mm fire which knocked out six enemy bunkers. The 4.2" mortars had a busy day firing 261 HE and 28 WP rounds, and Company C of the 1st Engineer Battalion (Lieutenant Colonel John V. Kelsey) supported the attack by clearing mine fields.

Mortar fire was received by the 1st Marines throughout the night, and 3/1 repulsed a series of counterattacks by an estimated 300 enemy. Colonel Wornham's regiment continued the attack at 0800 on 14 September. Both the 2d and 3d Battalions inched their way forward against a heavy volume of well-aimed enemy mortar, artillery, and automatic weapons fire.

NKPA resistance persisted on the wooded northern slope of Hill 749, where hidden bunkers had to be knocked out, one by one. It took constant slugging for 2/1 to advance 300 meters before dusk, enabling 3/1 to fight its way to the summit of Hill 751. Again the flat trajectory fire of Company C tanks had been helpful as 400 rounds were directed again NKPA bunkers, while the 11th Marines fired 3,029 rounds.

The 15th was a relatively quiet day as compared to the previous 48 hours. In preparation for an expected passage of lines, the action took a slower tempo as units consolidated their positions. The principal fight of the day was a continuation of the attack by 2/1 north of Hill 749. Although the battalion commander had arranged for a heavy artillery preparation, the attack, which jumped off at 1710, was stopped at 1800 by a terrific pounding from NKPA mortars and artillery coupled with a crossfire of machine guns from concealed bunkers. The assault troops withdrew under effective covering fire by the 11th Marines to positions occupied the previous night. Objective BAKER yet remained to be secured.

The Marines could not help paying reluctant tribute to the skill as well as obstinacy of the NKPA defense. Enemy bunkers were so stoutly constructed that the North Koreans did not hesitate to direct well aimed mortar fire on their own positions when the Marines closed in for the final attack.

NKPA fields of fire were laid out for the utmost effect. Marines with recent memories of college football referred to the enemy's effective use of terrain as the "North Korean T Formation." On Hill 749, for example, the main ridgeline leading to the summit was crossed by another wooded ridgeline at right angles. Attackers fighting their way up the leg of the "T" came under deadly crossfire from the head of the imaginary letter—a transverse ridgeline bristling with mortars and machine guns positioned in bunkers.

In accordance with Division OpnO 25–51, the 5th Marines (Colonel Richard C. Weede) moved up to assembly areas on 15 September in preparation for passing through 3/1 on the 16th to continue the attack. The 3d Battalion, 1st Marines in turn would relieve 1/1 (Major Edgar F. Carney, Jr.), so that it could pass through 2/1 and carry on the assualt to complete the securing of Hill 749.

The KMCs and Division Recon Company were to relieve the 5th Marines of responsibility for the HAYS Line, while the 7th Marines remained in reserve at Wontong-ni.

The comparative quiet of the 15th was shattered a minute after midnight when the enemy launched a savage four-hour attack to drive 2/1 off Hill 749. The NKPA hurricane barrage that preceded the attempt, according to the Division report, "reached an intensity that was estimated to surpass that of any barrage yet encountered by the 1st Marine Division in Korea." [17]

The thinned companies of 2/1 took a frightful pounding from 76mm, 105mm, and 122mm artillery supplemented by 82mm and 120mm mortars. Bugles and whistles were the signal for the onslaught. It was stopped by weary Marines who demonstrated at NKPA expense that they, too, could put up a resolute defensive fight.

Wave after wave of attackers dashed itself at the thinned Marine

[17] 1stMarDiv *HD*, Sep 51, 19–20.

platoons, only to shatter against a resistance that could be bent but not broken. The fight was noteworthy for examples of individual valor. When one of the forward Marine platoons was compelled to give ground slowly, Corporal Joseph Vittori of Fox Company rushed through the withdrawing troops to lead a successful local counterattack. As the all-night fight continued, "he leaped from one foxhole to another, covering each foxhole in turn as casualties continued to mount, manning a machine gun when the gunner was struck down and making repeated trips through the heaviest shell fire to replenish ammunition." [18]

Vittori was mortally wounded during the last few minutes of the fight, thus becoming the second Marine of 2/1 within a 48-hour period to win the Medal of Honor. His predecessor was Pfc Edward Gomez of Easy Company. When an enemy grenade landed in the midst of his squad on 14 September, he "unhesitatingly chose to sacrifice himself and, diving into the ditch with the deadly missile, absorbed the shattering violence of the explosion in his own body." [19]

Not until 0400 on the 16th did the enemy waves of attack subside on Hill 749. NKPA strength was estimated at a regiment. A combined assault by an estimated 150 enemy on 3/1 positions to the west in the vicinity of Hill 751 was repulsed shortly after midnight, as were three lesser efforts during the early morning hours of the 16th.

When the 1st Battalion, 1st Marines moved out at 0830 to pass through 2/1 and continue the fight, it was the first day of command for Lieutenant Colonel John E. Gorman.[20] The passage of lines was slowed by enemy mortar fire, and NKPA resistance stiffened as 1/1 attacked along the ridgeline leading toward Hill 749. At 1800, after a hard day's fighting, Objective BAKER was occupied and defensive positions were organized for the night.

Thus was the attack of the 1st Marines terminated. Around Hill 751, 3/1 remained in control. The regiment's other two battalions, 1/1 and 2/1, held a defensive line about 1,500 yards long on both sides of Hill 749.

Hill 749 had finally been secured. A number of mutually supporting

---

[18] Jane Blakeney, ed., *Heroes, U.S. Marine Corps, 1861–1955* (Washington, 1957), Joseph Vittori Medal of Honor Citation, 45.

[19] *Ibid.*, Pfc Edward Gomez citation, 38.

[20] On 14 September, LtCol Horace E. Knapp, Jr., the previous commanding officer of 1/1, was severely wounded while reconnoitering forward positions. He was evacuated, and the executive officer, Major Edgar F. Carney, Jr., commanded until LtCol John E. Gorman assumed command at noon on the 16th.

hidden enemy bunkers had been knocked out in a ruthless battle of extermination, and veterans of the World War II Pacific conflict were reminded of occasions when Japanese resistance flared up in similar fashion after ground was thought to be secure.

Casualties of the 1st Marine Division during the four-day fight for Hill 749, most of them suffered by the attacking regiment, were 90 KIA, 714 WIA, and 1 MIA. Enemy losses for the same period were 771 counted KIA (although more than twice that number were estimated KIA) and 81 prisoners.

5th Marines Attack Hill 812

Division OpnO 25–51 assigned the 5th Marines the mission of passing through 3/1 in the vicinity of Hill 751 and attacking to secure Objective DOG, the bare, brown hill mass which loomed approximately 1,000 yards ahead. The last few hundred yards were certain to be long ones, for the main east-west ridgeline leading to Hill 812 was crossed by a north-south ridgeline—the leg and head of another "T" formation. Again, as on Hill 749, the attackers had to fight their way through a vicious crossfire.

Lieutenant Colonel Houston Stiff's 2/5 on the right had the main effort. The 3d Battalion, 5th Marines (Lieutenant Colonel Donald R. Kennedy) was to advance on Stiff's left with the mission of supporting his attack on Objective DOG, prepared to seize Hill 980 on order. Lieutenant Colonel William P. Alston's 1/5 remained in regimental reserve.[21]

Fox Company spearheaded the 2/5 attack by moving initially up the low ground between Hill 673 on the right hand and 680 on the left. Owing to delays in completing the relief of 1st Marines elements, it was early afternoon on 16 September before the assault got underway. Progress was slow against heavy mortar and machine gun fire, and a halt came at 1700 for regrouping and evacuation of casualties.

Dog Company, in support on the ridge to the left, sighted troops approaching the objective and requested that the positions of the assault company be identified. In order to pinpoint the locations,

[21] Sources for this section, unless otherwise specified are as follows: 1stMarDiv *HD*, Sep 51, 19–23; 5thMar *HD*, Sep 51, 14–19; 1st, 2d, and 3dBn, 5th Mar, *HD*, Sep 51; LtCol Houston Stiff, interv of 25 Jun 58; Maj G. P. Averill, "Final Objective," *Marine Corps Gazette*, vol. 40, no. 8 (Aug 56), 10–16.

a white phosphorous grenade was used as a mark. It attracted the attention of aircraft summoned by 3/5 against Hill 980 (Map 17), from which fire had been received. The planes, assuming that another target had been designated, attacked the forward platoons of Fox Company with napalm and machine guns. By a miracle, recognition panels were put out before a single casualty resulted, but the men found it a harrowing experience.

Darkness fell before the attack could be resumed, and Fox Company pulled back along the ridgeline to set up a perimeter defense and evacuate the wounded. The night passed without enemy action. Bright moonlight made for unusual visibility which discouraged enemy attacks and permitted the Marine assault platoons more sleep than might otherwise have been expected.

Regimental orders called for 2/5 to resume the attack at 0400 on the 17th, supported by the fires of 3/5, while 1/5 continued in reserve. Fox Company of 2/5 had some difficulty in orienting itself after the confusion of the night before and was delayed until 0700 in jumping off. This proved to be a stroke of luck, for dawn gave the Marines a good view of unsuspecting enemy troops eating breakfast and making ready for the day's fighting. Fox Company called artillery on them with good effect.

Surprise gave the attack an opening advantage and rapid progress was made at first along the main ridgeline leading west to Hill 812. Then Fox Company was stopped by the cross-fire from the head of the "T." Easy Company passed through at 0830 to continue the assault, reinforced by a platoon of Fox Company that had become separated from its parent unit, although it kept in touch by radio.

An air strike was called but did not materialize. After waiting for it in vain, Easy Company drove toward the summit with the support of artillery and mortars.

Two hours after passing through Fox Company, the attackers had advanced only about a hundred yards against the NKPA cross-fire. At 1100, Lieutenant Colonel Stiff ordered an all-out drive for the objective, following a preliminary barrage of everything that could be thrown at the enemy—artillery, 75mm recoilless, rockets, and 81mm and 4.2" mortars. As soon as the bombardment lifted, Easy Company was to drive straight ahead along the ridgeline while the 2d Platoon of Fox Company made a flank attack.

This maneuver turned the trick. The blow on the flank took the enemy by surprise, and in just 36 minutes the assault troops were on the summit after a hard fight at close quarters with automatic weapons and grenades. Since regimental orders had specified "before nightfall," Objective DOG had been seized ahead of schedule.

With scarcely a pause, Easy Company continued along the ridgeline leading west from Hill 812 toward Hill 980. Remarkably fast progress was made against an enemy who appeared to be thrown off balance. Permission was asked to seize Hill 980. The regimental commander refused because of instructions from Division to the effect that this position could not be defended while the enemy remained in possession of Hill 1052, the key terrain feature. Easy Company was directed to withdraw 600 yards toward Hill 812.

Late in the evening of 17 September, Colonel Weede directed his two assault battalions to consolidate on the best ground in their present locations and prepare to hold a defensive line.

When the brakes were put on the attack, 3/5 was strung out over a wide area to the north of Hill 751. This battalion was not tied in with 2/5, which occupied positions coordinated for the defense of Hill 812—Easy Company to the west, on the ridgeline leading to 980; Dog and Weapons Company to the south, protecting the left flank; and Fox Company to the east.

Both Easy and Fox Companies were under fire from Hills 980 and 1052, and daytime movement on 812 was restricted to the northern slope. Even so, sniping shots from well aimed North Korean 76mm mountain guns inflicted a number of casualties.

The Struggle for the "Rock"

An abrupt change in the enemy's strategy became evident throughout these September operations. Where he had previously contented himself with an elastic defense, every position was now bitterly fought for and held to the last man. When it was lost, counterattacks were launched in efforts to regain it.

One of these attempts hit the western outpost of 2/5's Easy Company at 0430 on 18 September, compelling the Marines to give ground. A second counterattack at 0840 was repulsed. Enemy fire from Hills 980 and 1042 continued all day long, and Colonel Stiff's battalion suffered most of the 16 KIA and 98 WIA casualties reported by the Division for 18 September.

The night of 18–19 September passed in comparative quiet, but at daylight the enemy on Hills 980 and 1052 was still looking down the throats of the 2/5 Marines. None of the participants will ever forget a landmark known simply as "the Rock"—a huge granite knob athwart the ridgeline approximately 700 yards west of Hill 812. Only 12 feet high, its location made it visible from afar. The Marines outposted the top and eastern side, while the enemy held tenaciously to the western side. Along the northern slope of the ridge leading west to the Rock were the only positions affording protection to the dug-in forward elements of the battalion.

The need for fortification materials such as sand bags, barbed wire, and mines aggravated the already serious supply problems of 2/5. A request for helicopter support was sent at 1100 on the 19th and approved immediately by General Thomas. Loading commenced early the same afternoon, and Operation WINDMILL II was launched. A total of 12,180 pounds were lifted by 10 HRS–1 aircraft in 16 flights during the overall time of one hour.[22]

Again, on 19 September, 2/5 incurred most of the casualties reported by the Division. During the day 1/5, after relieving the 1st and 2d Battalions of the 1st Marines, moved up on the right of 2/5 to occupy a defensive line stretching two miles east along the ridge almost to the Soyang-gang.

NKPA action was confined to incessant long-range fire during the daylight hours of the 19th, but at 0315 the following morning the enemy made a desperate effort to retake Hill 812. After a brief but intense mortar and artillery barrage, North Koreans in at least company strength came pouring around the northern side of the Rock to attack with grenades and burp guns at close range. The left platoon of Easy Company counterattacked but was pushed back by superior numbers to positions on the left flank of the hill.

The enemy immediately took possession of evacuated ground which enabled him to fire into the front lines of Easy Company. At 0500 another Marine counterattack began, with Easy Company making a frontal assault and the 2d Platoon of Fox Company striking the enemy flank. It was the same platoon that had delivered the flank attack resulting in the capture of Hill 812. Again 2/Fox struck the decisive blow with grenades and automatic weapons. The surprise was too

[22] *Cavalry of the Sky*, 162.

much for enemy troops who hastened back to their own side of the Rock, leaving 60 counted dead behind.[23]

This was the last action of a battle that had occupied all three Marine regiments from 11 to 20 September inclusive while the KMC Regiment patrolled aggressively on the Division left flank. Three of the four Division objectives had been secured after savage fights, but Objective CHARLIE (the ridgeline northwest of Hill 1052 in the KMC zone) had yet to be attacked when Division OpnO 26–51 put an abrupt stop to offensive movement.

Not only was the fight west of Hill 812 the last action of the 1st Marine Division's nine-day battle; it was the last action of mobility for Marines in Korea. As time went on, it would become more and more apparent that 20 September 1951 dated a turning point in the Korean conflict. On that day the warfare of movement came to an end, and the warfare of position began.

[23] 1st Marine Division losses of 33 killed and 235 wounded during the three-day attack were incurred for the most part by the 5th Marines in general and 2/5 in particular. Enemy casualties of this period were reported as 972 KIA (265 counted) and 113 prisoners.

CHAPTER X

The New Warfare of Position

Sectors of Major EUSAK *Units—Statement by General Van Fleet—Hill 854 Secured by 3/1—Helicopter Troop Lift to Hill 884—Helicopter Operation* BLACKBIRD*—"To Organize, Construct, and Defend"—Marine Operations of November 1951—The Second Marine Christmas in Korea*

TWO AND A HALF weeks of hard fighting had taken place along the X Corps front when General James A. Van Fleet paid a visit on 16 September 1951. The commanding general of EUSAK wished to inspect the operations and determine the morale of the 1st Marine Division and 2d Infantry Division, both of which had suffered heavy casualties. He found the morale of these X Corps units good and had no adverse criticisms of their operations. While on this tour of inspection, however, he issued the following three directives to X Corps:

> (1) That replacements would be integrated into units only when the battalion or larger-sized unit to which they were assigned was in reserve;
>
> (2) that certain 'choke points' [General Van Fleet pointed out the locations on the map] be interdicted to prevent enemy reinforcements or withdrawals through these points;
>
> (3) *that the Corps Commander firm up his line by 20 September and to plan no further offensives after that date,* as it was unprofitable to continue the bitter operation.[1]

Italics have been added to emphasize the importance of 20 Septem-

[1] EUSAK *Cmd Rpt,* Sep 51, 47. Other sources for this chapter are comments and criticisms by the following officers, all but one of whom are U.S. Marines. Ranks in each instance are those held at the time of interview or correspondence.

General J. A. Van Fleet, USA (Ret.); General G. C. Thomas, Lieutenant General J. T. Selden; Brigadier Generals V. H. Krulak, S. S. Wade, R. G. Weede; Colonels G. P. Groves, B. T. Hemphill, K. L. McCutcheon, J. H. Tinsley, F. B. Nihart, G. D. Gayle, W. P. Mitchell, J. F. Stamm, F. P. Hager, Jr.; Lieutenant Colonels H. W. Edwards, J. G. Kelly; Major R. L. Autry.

634040 O–62–15

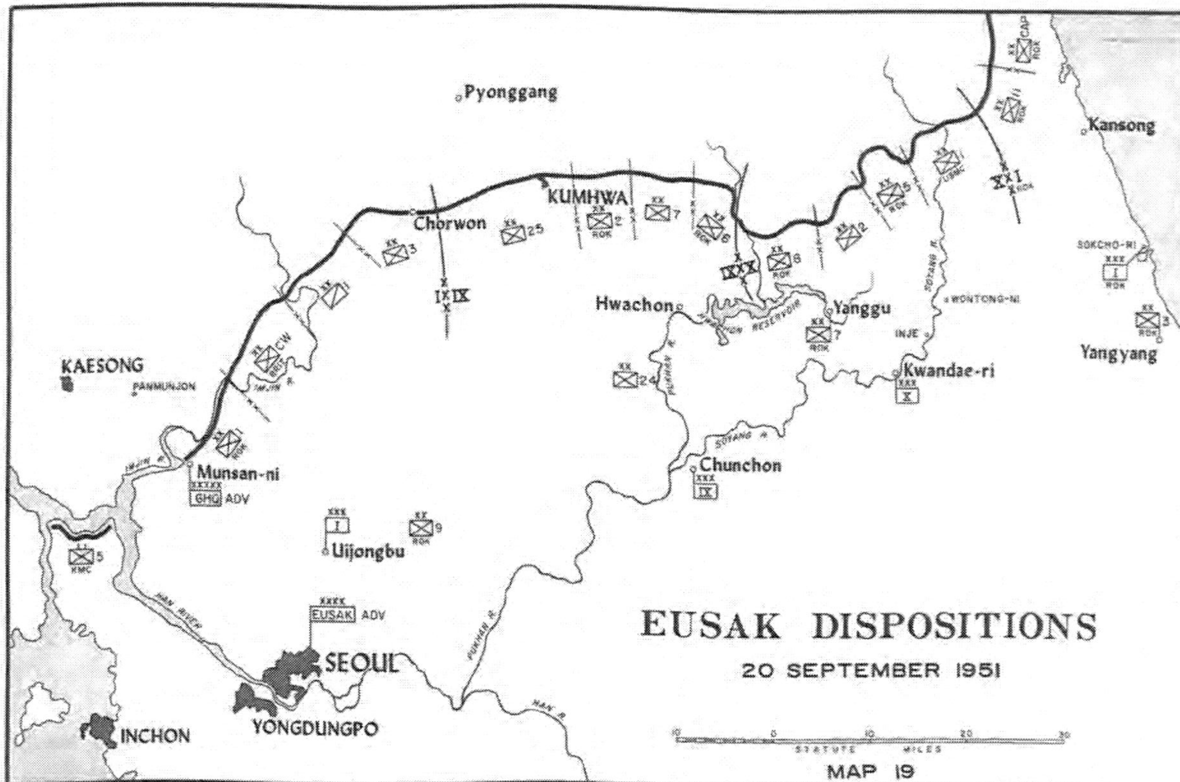

EUSAK DISPOSITIONS
20 SEPTEMBER 1951
MAP 19
Pyonggang
Kansong
KUMHWA
Chorwon
Hwachon
Yanggu
INJE
WONTONG-NI
SOKCHO-RI
Yangyang
Kwandae-ri
Chunchon
KAESONG
PANMUNJON
Munsan-ni
GHQ ADV
Uijongbu
EUSAK ADV
SEOUL
YONGDUNGPO
INCHON
IMJIN R.
HAN RIVER
PUKHAN R.
HAN R.
SOYANG R.
CAP
ROK
USMC
KMC
CW
STATUTE MILES

ber 1951 as the turning point when a warfare of position replaced a warfare of movement throughout the remaining 22 months of the conflict in Korea. There are few dates as important in the entire history of the war.

General Van Fleet reiterated his instructions on the 18th in a confirming directive to the effect that X Corps continue making limited attacks "until 20 September, after which . . . units were to firm up the existing line and to patrol vigorously forward of it." [a]

Sectors of Major EUSAK *Units*

At this turning point the Eighth Army had 14 divisions from four corps committed along a 125-mile front across the peninsula. These units were distributed (Map 19) as follows:

U.S. I CORPS

ROK 1st Division holding the left anchor in the Munsan-ni area and controlling the 5th KMC Battalion on the Kimpo Peninsula;

British 1st Commonwealth Division across the river Imjin to the northeast;

U.S. 1st Cavalry Division (Greek and Thai Battalions attached) still farther to the northeast in the Yonchon area;

U.S. 3rd Infantry Division (Belgian Battalion and Philippine 20th BCT attached) having the responsibility for the vital Chorwon area;

U.S. IX CORPS

U.S. 25th Infantry Division (Turkish Brigade attached) defending the area west of Kumhwa;

ROK 2d Division holding a sector east of Kumhwa;

U.S. 7th Infantry Division (Ethiopian Battalion attached) on the right;

ROK 6th Division with a narrow sector as far east as the Pukhan River, the Corps boundary;

U.S. 24th Infantry Division (Colombian Battalion attached) in Corps reserve south of Hwachon;

U.S. X CORPS

ROK 8th Division on the left flank;

---

[a] EUSAK *Cmd Rpt*, Sep 51, 53.

U.S. 2d Infantry Division (French and Netherlands Battalions attached) in left-central portion of Corps front;
ROK 5th Division occupying a narrow sector to the east;
U.S. 1st Marine Division holding eastern portion of the Corps sector;

ROK I CORPS

ROK 11th Division responsible for left of the Corps front;
ROK Capitol Division holding the line eastward to the Sea of Japan;
ROK 3d Division in reserve at Yangyang for a period of training.[3]

Some rather complicated juggling of units took place on the X Corps front, giving the effect of a game of musical chairs in the tactical sphere. From 18 to 21 September the 1st Marine Division extended its line eastward to relieve the 8th ROK Division on the extreme right of the Corps area. That Division in turn relieved the 5th ROK Division on the extreme left, whereupon the latter leapfrogged the 2d Infantry Division to occupy a new sector on the left of the Marines.

Statement by General Van Fleet

"Theirs not to reason why" could never have been written about American fighting men. From 1775 to the present day, they have always taken a keen interest in the high-level strategic and tactical decisions governing their operations. This applies with particular force to the Marines, who have seldom had a voice in the shaping of operations above the division level.

As if in direct reply to unspoken questions, the commanding general of the Eighth Army made a statement on 30 September explaining the purpose of his strategy. "My basic mission during the past four months," he said, "has been to destroy the enemy, so that the men of Eighth Army will not be destroyed. . . . Each loaded enemy weapon was a definite threat to the Eighth Army. It was imperative that we knock out as many of those weapons as we could find. . . ."

"In prodding the enemy in the deep belly of the peninsula," continued General Van Fleet, "we have taken many casualties. . . . It was mandatory that we control the high ground features, so that we could look down the throat of the enemy and thereby better perform

[3] EUSAK *Cmd Rpt*, Oct 51, 5–6 and Plate 1; 1stMarDiv *HD*, Sep 51, 3.

our task of destruction. . . . In seizing these hills we lost men, but in losing a comparative few we saved other thousands."

Estimated casualties, inflicted on the enemy by UN ground forces alone from 25 May to 25 September, were announced as 188,237 by the EUSAK commander. "As we open our autumn campaign," he added, "the enemy potential along the front line has been sharply reduced by our hill-hopping tactics. The Communist forces in Korea are not liquidated but they are badly crippled." [4]

Even so, EUSAK G–2 summaries credited the enemy on 1 October 1951 with more than 600,000 troops at the front, or in reserve and available as immediate reinforcements. Six CCF armies and one NKPA corps were capable of reinforcing the units on the MLR or participating in an offensive. The enemy also had an estimated 7,000 men in guerrilla forces behind the UN lines.[5]

The maximum strength of UN forces in Korea during October was 607,300. This total included 236,871 U.S. Army troops, 21,020 Fifth Air Force personnel, 30,913 U.S. Marines (including 5,386 officers and men of the 1st Marine Aircraft Wing), 286,000 men in ROK units, and 32,172 Allied troops.[6]

Although it might appear that the opposing forces were about equal, it must be remembered that well over one-fourth of the UN troops were engaged in administrative or maintenance duties behind the front. Thus the Communists had a numerical advantage of at least four to three on the firing line. This was not at all unusual, since they had enjoyed a preponderance in manpower from the beginning.

Hill 854 Secured by 3/1

In accordance with EUSAK instructions, X Corps OI–235 directed the 1st Marine Division to organize and construct defensive positions after relieving the 8th ROK Division on the right and taking over its sector. On the Corps boundary, elements of the 11th ROK Division, I ROK Corps, were to be relieved on Hill 884 (Map 20). This meant the addition of some 9,000 yards to the Marine front, making a total of about 22,800 yards or more than 13 miles.

---

[4] *Ibid.*, 29–30.
[5] *Ibid.*, 7–9 and Plate No. 4.
[6] *Ibid.*, 5–6, and Plate No. 1.

First Marine Division OpnO 27–51, issued on 18 September, relayed the X Corps directions. It also called for such offensive action as might be necessary to complete the securing of Hill 854, in the sector of the 8th ROK Division, if not in friendly hands at the time of the relief.[7]

That the enemy had put up a desperate fight to hold this position is indicated by the EUSAK report for 15–16 September: "The ROK 8th Division, employing all three regiments, attacked against heavy and stubborn resistance to wrest Hill 854 from the three battalions of North Koreans who held the position. The ROK 21st Regiment forced one of these battalions to withdraw and occupied a part of the hill, but at the close of the day were engaged in heavy hand-to-hand fighting to retain the position." [8]

On 20 September, after three weeks of continual combat, the major units of the 1st Marine Division were disposed from left to right (Map 20) as follows:

1st KMC Regiment (Colonel Kim Dae Shik, commanding; Colonel Walter N. Flournoy, senior adviser) occupying the HAYS line on the left flank and patrolling vigorously to the north;

5th Marines (Colonel Richard G. Weede) holding a wide sector in the center, with Hill 812 as the principal terrain feature;

1st Marines (Colonel Thomas A. Wornham) in process of extending eastward to the Corps boundary just beyond Hill 884;

7th Marines (Lieutenant Colonel John J. Wermuth) in Division reserve at Wontong-ni.[9]

Division OpnO 27–51 designated the 1st Marines to relieve the ROKs on Hill 854 and complete the seizure of that terrain feature, if necessary. As a preliminary, the 1st Battalion of the 5th Marines (Lieutenant Colonel William P. Alston) took over the front of the 1st Marines on the HAYS line. This enabled 1/1 and 3/1 to enlarge the Division sector by side-slipping to the east while Lieutenant Colonel Franklin B. Nihart's 2/1 went into immediate reserve just behind the main line of resistance.[10]

The 1st Battalion of the 1st Marines (Lieutenant Colonel John E. Gorman) relieved two battalions of the 10th ROK Regiment in the Hill 854 area. No opposition from the enemy was encountered, but

[7] 1stMarDiv *HD*, Sep 51, 3.
[8] EUSAK *Cmd Rpt*, Sep 51, 47.
[9] 1stMarDiv *HD*, Sep 51, 3–4, 18–22.
[10] The balance of this section is based on the 1stMarDiv *HD*, Sep 51, 18–24, and on 1/1 and 3/1 *HD*, Sep 51.

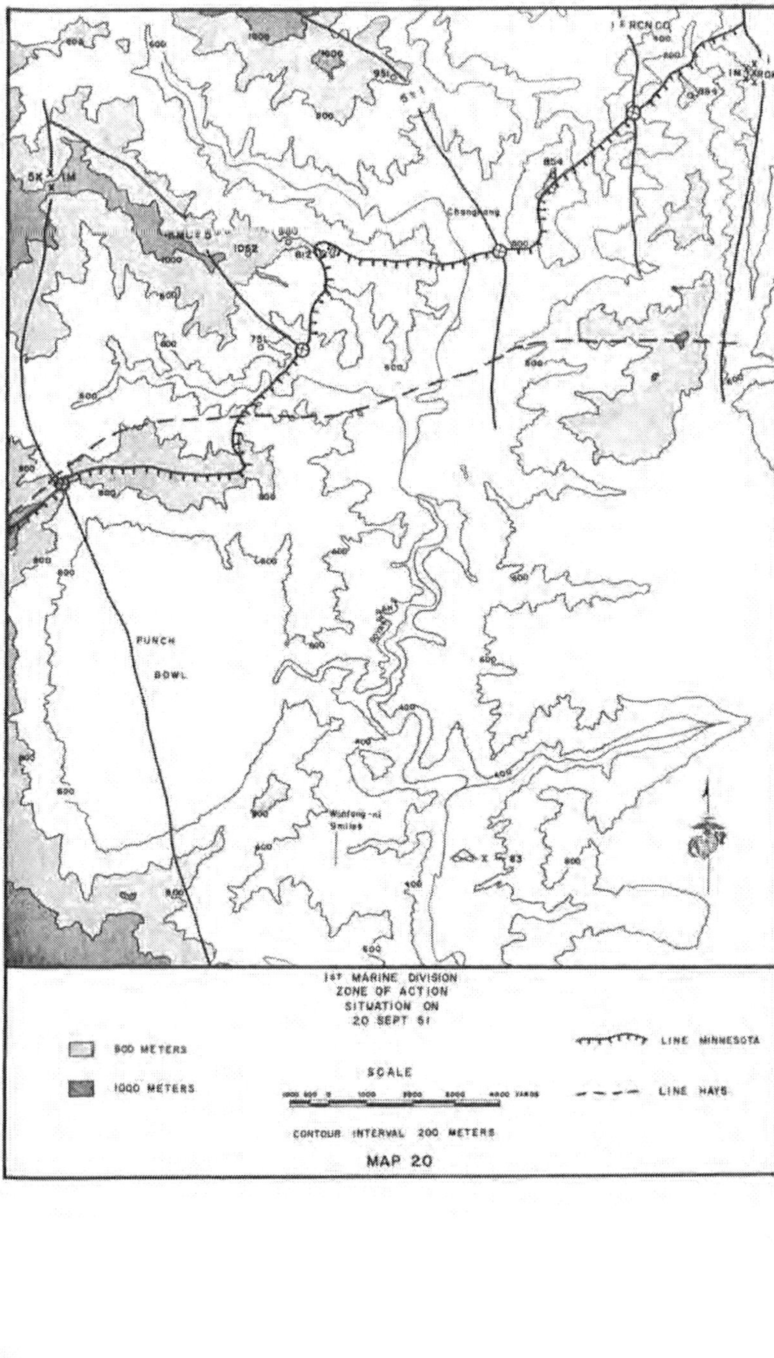
PUNCH
BOWL
Changhong
1ST MARINE DIVISION
ZONE OF ACTION
SITUATION ON
20 SEPT 51
800 METERS
1000 METERS
SCALE
LINE MINNESOTA
LINE HAYS
CONTOUR INTERVAL 200 METERS

MAP 20

the Marines suffered 11 casualties from mines as a consequence of incorrect charts supplied by the ROKs.

By this time it had become an open question whether "friendly" mines did more harm to friend or foe. Certain it was, at any rate, that the prevailing system—or lack of system—resulted in Marine casualties during nearly every offensive operation in zones where the action shifted back and forth.

Lieutenant Colonel Foster C. La Hue's 3/1 relieved two battalions of the 21st ROK Regiment. Although the ROKs had fought their way to the summit of Hill 854, the ridgeline to the southwest remained in the enemy's hands. An attack by 3/1 was planned for 1530 on 20 September, supported by artillery and an air strike. Delays in the arrival of the planes caused a postponement until 1720. How Company jumped off and had advanced 50 yards when a man was killed and another wounded by mines. The attack was called off at dusk so that the ROKs could remove the explosives they had planted.

Air support was requested for 0700 on the morning of the 21st, but it was 1040 before four Air Force F–51s arrived for a strike directed by an observation plane of VMO–6 and a forward air controller. At 1220, following a 10-minute artillery preparation, How Company spearpointed a battalion attack which met stiff resistance. Another air strike was requested but did not materialize. The assault continued with mortar and artillery support until 1745, when How Company reported the ridge line secured.

Casualties of 3/1 for the two days were nine KIA and 55 WIA. Enemy losses totaled 159 counted and 150 estimated KIA, 225 estimated WIA, and 29 prisoners.[11]

"A large number of mines and booby traps were discovered within the battalion sector," the 3/1 report for the 23d concluded, "most of these being U.S. types which were placed by ROK troops, with only a few enemy mines scattered in the central portion of the sector." [12]

Helicopter Troop Lift to Hill 884

Division OpnO 27–51, it may be recalled, had directed the Marines to extend the X Corps boundary eastward by taking over the sector of

[11] Sources for the action on Hill 854 are the 1/1 and 3/1 historical diaries for September 1951.

[12] 3/1 *HD*, Sep 51, 8.

the 11th Regiment, I ROK Corps. Even under ordinary circumstances this would have meant an exhausting 15-hour march for the relieving troops merely to climb Hill 884 (Map 20). The position was accessible only on foot, and supplies had to be brought on the backs of cargadores.

Because of the isolation of this wildly mountainous area, a reconnaissance was deemed essential. Major General Gerald C. Thomas, commanding general of the 1st Marine Division, assigned that mission to the Division Reconnaissance Company after deciding on a troop lift by helicopter.

He was aware, of course, that no such operation had ever been undertaken during the brief history of rotary-wing aircraft. Large-scale helicopter troop lifts were still at the theoretical stage.

Lieutenant Colonel George W. Herring, commanding officer of HMR–161, had but 48 hours for preparation. He and his executive officer, Lieutenant Colonel William P. Mitchell, worked out a tactical and loading plan with the commanding officer of Recon Company, Major Ephraim Kirby-Smith, and the acting Division Embarkation Officer, First Lieutenant Richard C. Higgs.[18]

An air reconnaissance of Hill 884 disclosed only two acceptable locations for landing sites, both approximately 50 feet square with a sheer drop on two sides. About 100 yards apart and some 300 feet below the topographical crest, each could be cleared sufficiently for the landing of a single aircraft.

Major Kirby-Smith decided on the order in which troops of his company and attached units would be landed. The assignment and loading tables were completed on 20 September in time for a rehearsal. All participants were instructed as to their team numbers and embarkation points.

H-Hour of Operation SUMMIT (Map 21) was set for 1000 on 21 September. The plan called for a preliminary landing of a Recon Company rifle squad to provide security. Next, a landing point team from the 1st Shore Party Battalion (Lieutenant Colonel Harry W. Edwards) had the mission of clearing the two sites. These two groups were to disembark from hovering helicopters by means of knotted 30-foot ropes. Strong winds at the 2,900-foot altitude made landing quite hazardous.

[18] Sources for this section, unless otherwise specified, are the following: DivReconCo *HD*, 1stShorePartyBn *HD*, HMR–161 *HD*, Sep 51; Type "C" Spec Rpt, "Employment of Assault Helicopters," 7–13; *Cavalry of the Sky*, 162–165.

The execution was delayed half an hour by the ground fog so prevalent at this time of year. As soon as the two landing sites were cleared (about 40 minutes), word was transmitted by radio for the loading to begin at Field X–83 (Map 21), about 14 miles southwest of Hill 884 by the defiladed route of flight.

Control over the landings and takeoffs on the two Hill 884 sites was exercised by a hovering helicopter. Aircraft landed at 30-second intervals, each carrying five fully equipped men who disembarked in average time of 20 seconds. Two radio nets maintained communications between the landing sites and orbiting aircraft. Voice contact could not be established between the landing point team and X–83, however, and it became necessary for a helicopter to return within sight of the field to restore communications for incoming aircraft.

A total of 224 men, including a heavy machine gun platoon from 2/7, was lifted in flight time of 31.2 hours and over-all time of four hours. In addition, 17,772 pounds of cargo were landed.

Operation SUMMIT ended with the laying of two telephone lines between Recon Company on Hill 884 and the CP of the 1st Marines, about eight miles to the rear. Fifteen minutes were required for dropping each line. The ROKs, following their relief, proceeded on foot to their own Corps area.

From a tactical viewpoint, the importance of Hill 884 lay in its domination of enemy-held terrain. The difficulty of reaching the remote position had been overcome by the helicopter, and Operation SUMMIT was recorded in front page headlines by Stateside newspapers.

Congratulations poured in from all sides. Lieutenant General Lemuel C. Shepherd, Jr., commanding general of FMFPac, complimented HMR–161 on "a bright new chapter in the employment of helicopters by Marines." Major General Clovis E. Byers, commanding X Corps, praised the "organic and attached units of the 1st Marine Division that participated in the first relief of units on the battle position. Your imaginative experiment with this kind of transport is certain to be of lasting value to all the services." [14]

Nobody was more enthusiastic than General Thomas. "Operation SUMMIT, the first helicopter-borne landing of a combat unit in history, was an outstanding success," said his message. "To all who took part, well done!"

[14] Messages of congratulation are quoted from HMR–161 *HD*, Sep 51.

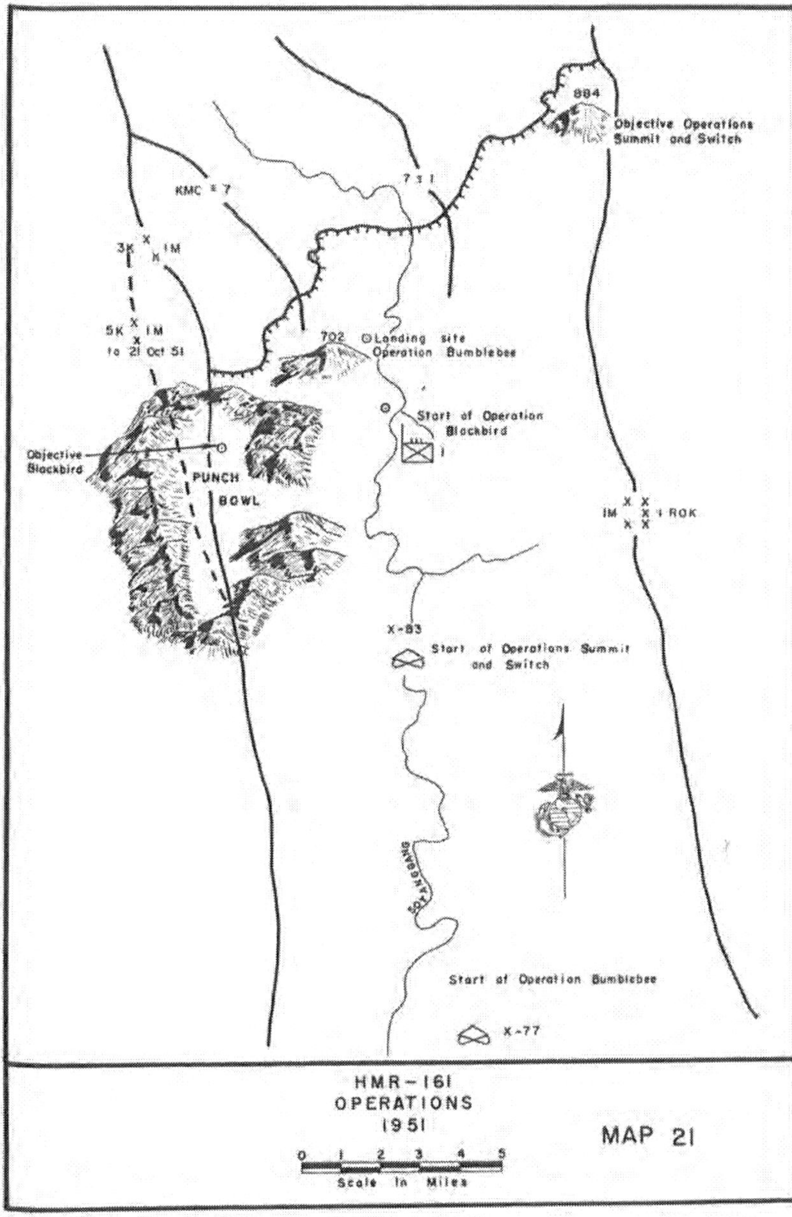
884
Objective Operations Summit and Switch
KMC
7
1
3K
1M
5K
1M
to 21 Oct 51
702
Landing site Operation Bumblebee
Start of Operation Blackbird
Objective Blackbird
PUNCH BOWL
1M
ROK
X-83
Start of Operations Summit and Switch
SOYANGGANG
Start of Operation Bumblebee
X-77
HMR-161
OPERATIONS
1951
0 1 2 3 4 5
Scale In Miles

MAP 21

Helicopter Operation BLACKBIRD

It is not surprising, considering their training, that the Marines found it a difficult transition from offensive to defensive operations after 20 September. As evidence that patrols were conducted with customary aggressiveness, Marine casualties (including the 1st KMC Regiment) for the last 10 days of the month were 59 KIA, 1 MIA, and 331 WIA. Enemy losses for the same period were 505 counted KIA, and 237 prisoners.

1st Marine Division casualties of 2,416 (including 594 reported by the KMCs) for September as a whole were the most severe suffered during any month of the war so far with the exception of December 1950 and June 1951. NKPA losses of the month were 2,799 counted KIA and 557 prisoners.[15]

On the 23d the 1st Marines extended to the eastern boundary of X Corps and relieved the Division Reconnaissance Company on Hill 884. That same day the enemy was treated to a novelty when 100 well aimed 16-inch projectiles, fired from a range of 40,000 yards, roared in like meteors on his positions in the area of Hill 951 (Map 20). Naval gunfire from the USS *New Jersey* was being conducted by Marine spotters in forward OPs, who reported good coverage for the 2,000-pound rounds. Ammunition dumps and artillery pieces were destroyed while NKPA troops in the open suffered heavy personnel casualties, according to observers.

Several more bombardments were contributed by the *New Jersey* at the request of 1/1 and 3/1 during the balance of the month. Marine and attached Army artillery also gave excellent support with fire so accurate as to break up enemy counterattacks before they could be launched. Ammunition restrictions hampered the efforts of the 11th Marines (Colonel Custis Burton, Jr.) but the cannoneers never failed to respond to an emergency. The 90mm rifles of the 1st Tank Battalion (Lieutenant Colonel Holly H. Evans) continued to show good results with direct observed fire on enemy bunkers. Air support in September, concluded the Division report, was "generally inadequate and unsatisfactory." [16]

By the last week of September the Division right (east) flank was well protected, considering the rugged terrain. Not as much could

[15] 1stMarDiv *HD*, Sep 51, 4, 31–32.
[16] *Ibid.*

be said for the other flank, northwest of the Punchbowl, where the sector of the Marines joined that of the 5th ROK Division. Since the Division sector was divided by high, roadless mountains, there was no rapid way of moving reserves other than by helicopter. In short, the 1st Marine Division was hard pressed to man a 22,800-yard MLR while keeping in reserve enough troops to help defend this sensitive area in an emergency.

Plans were completed by General Thomas and the Division staff for the rapid displacement of a company from 2/1, the reserve battalion of the 1st Marines, to meet any such threat. Since a surprise attack was most likely to occur at night, it was decided that a helicopter lift of an element of the Division reserve should be made in the darkness of 27 September after a detailed daytime rehearsal.[17]

In contrast to former Marine helicopters, which had no night-flying aids, the HRS–1 was equipped with few attitude of flight instruments. They were primitive compared to the sophisticated instrumentation of fixed-wing planes, and Lieutenant Colonel Herring sent his pilots on preliminary night indoctrination flights to memorize terrain features.

The route, five air miles in length, amounted to a round trip of 13 miles because of the detours necessary for purposes of concealment. The aircraft were to take off from a dry river bed southeast of Hill 702 (Map 21) and land near the northwestern rim of the Punchbowl, where the troops would march a mile to their final assembly area.

The infantry unit selected for Operation BLACKBIRD was Easy Company of 2/1, commanded by Second Lieutenant William K. Rockey. Lieutenant Colonel Nihart and Major Carl E. Walker, the battalion commander and his executive officer, supervised the daylight rehearsal on the morning of the 27th. Six helicopters lifted 200 men in the overall time of two hours and 10 minutes to a landing site of 50 by 100 feet cleared by a team of the 1st Shore Party Battalion. The troops were proceeding on foot to their assembly area when an antipersonnel mine wounded a man. Nihart called a halt immediately and investigation revealed that the area was filled with mines. Plans were changed to abandon the march, although the landing site remained the same.

---

[17] The remainder of this section is based upon the Type "C" Spec Rpt, "Employment of Assault Helicopters," Part II, 1–9; HMR–161 and 1stShorePartyBn *HD*, Sep and Oct 51; *Cavalry of the Sky*, 165–167.

Operation BLACKBIRD got under way at 1930 on 27 September. The night was dark when the first HRS–1 took off with five combat-equipped men. Three-minute intervals were required between aircraft operating on a shuttle system, so as to avoid the danger of collisions. Different altitudes were assigned to outgoing and incoming helicopters which used running lights only two minutes before entering or leaving the debarkation zone.

A total of 223 troops were landed in over-all time of two hours and 20 minutes instead of the nine hours a movement by foot would have required. Nevertheless, some of the results were not reassuring. Rotor wash blew out many of the flare pots lighting the embarkation area, and the battery-powered beach lanterns on the landing site proved inadequate. Pilots were temporarily blinded by the glare on windshields; and artillery flashes bothered them while making their way through three mountain passes. Fortunately, good radio communications aided pilots who had trouble in locating the landing site in spite of night rehearsals.

Operation BLACKBIRD remained the only night helicopter troop lift during the war in Korea. "Present equipment," said the Marine report, "indicates that under present conditions in Korea these night lifts should be limited to movements within friendly territory." [18]

"To Organize, Construct and Defend"

"The Division continued to organize, construct and defend positions along a 13½-mile front; patrol forward of the MLR and screen rear areas; and maintain one U.S. Marine regiment which could not be committed without authority from X Corps in a reserve area 17 miles behind the lines."

The above quotation, from the opening paragraph of the report of the 1st Marine Division for October 1951, sums up in a nutshell the new trend of operations since 20 September. It is significant that for the first time in 1951 the Division Historical Diary departs from a daily account of events and divides the month into two equal parts for a chronicle of operations. Not enough had happened to justify a day-by-day summary.

This does not mean that the Marines neglected any opportunity to

[18] Type "C" Spec Rpt, "The Employment of Assault Helicopters," Part II, 4.

do the enemy hurt. It means only that the opportunities of defensive warfare were limited as compared to the preceding six months of offensive operations. That the Marines made the best of such opportunities is shown by the fact that the ratio of enemy to friendly casualties increased from the 4-to-1 of September to the 20-to-1 of October, even though the totals of the former month were larger.[19]

As a result of his new defensive policies, the enemy often avoided a fight. Day after day passed during the first two weeks of October without far-ranging Marine patrols being able to make contact.

Line MINNESOTA, the new MLR (Map 20), ran roughly parallel to the HAYS line but included advanced positions taken in the September offensive. During the first 10 days of October the 2d Battalion of the 1st Marines continued to be the Division forward reserve in readiness for a quick shift to any threatened point in the MLR, and the Division Reconnaissance Company had the mission of maintaining daily contact with the 11th ROK Division on the Marines' right flank.

It might seem that the 7th Marines, 17 miles to the rear at Wontong-ni, would be entirely becalmed. Yet this regiment saw as much action on some days as any of the three regiments ranging forward of the MLR. The explanation was that the rear area was infested with elusive North Korean guerrillas who kept the 7th Marines patrols busy.

Early in October the question arose as to how quickly a reserve battalion could be shifted from one point to another. By this time a company-size helicopter lift had become commonplace, having been successfully completed twice by HMR–161 since Operation SUMMIT. It remained to be seen whether a battalion could be transported with comparable celerity, and, on 9 October, Division issued an order warning of 3/7's move.

The 7th Marines was due to exchange places with the 5th Marines on the 11th after relieving that regiment in the center of the Division front. While 1/7 and 2/7 completed a conventional relief of their opposite numbers, 3/7 was selected for a helicopter lift. Lieutenant Colonel Edwards, the new commanding officer, had recently commanded the Shore Party Battalion and helped to train its landing site and loading point teams. He took part in the planning along with Colonel Krulak, Lieutenant Colonels Herring and Mitchell, and the new commanding officer of the Shore Party Battalion, Lieutenant Colonel George G. Pafford.

---

[19] 1stMarDiv *HD*, Oct 51, 1–3.

Planning went on as if for an amphibious operation. Assignment and loading tables were worked out, and each Marine of the six-man embarkation teams had his designated place in the helicopter. On 10 October all officers and men of 3/7 attended a familiarization class at which trial teams were loaded.

Operation BUMBLEBEE began at 1000 on the 11th. Field X–77 (Map 21) had been selected as the loading zone because of its proximity to the assembly area of the 7th Marines. The landing site was just behind the 5th Marines MLR, northeast of Hill 702. A flight path of 15 miles took advantage of the concealment afforded by valleys and defiladed areas.

The two dispatchers in the loading zone were provided with a checkoff flight list containing the names of every team of 3/7. In order to avoid delays, replacements could be summoned from a casual pool to fill understrength teams to plane capacity. Average time for loading was 20 seconds.

Ten to 12 minutes were required for the flight. As the helicopters landed at intervals of a minute, a team could exit and allow the craft to be airborne in an average time of 17 seconds. "Time was saved," according to one Marine report, "when the Shore Party personnel, after opening the door, vigorously assisted the passengers by grasping their arms and starting them away from the craft. The last man out checked to see if any gear had been forgotten. Guides furnished by the battalion directed the passengers toward their respective company assembly areas, thus keeping the landing areas clear at all times." [30]

Twelve helicopters were employed in 156 flights. The flight time was 65.9 hours and over-all time five hours and 50 minutes. A total weight of 229,920 pounds included 958 combat-equipped troops averaging 240 pounds.

These statistics of Operation BUMBLEBEE made it certain that Stateside headlines would proclaim another Marine "first." Only four days later HMR–161 demonstrated its ability to carry out on short notice an emergency resupply and evacuation operation in a combat zone. Help was requested in the IX Corps sector to the west for a completely surrounded ROK unit in need of ammunition and of casualty evacuation. Lieutenant Colonel Mitchell led six HRS–1 aircraft which flew in 19,000 pounds of ammunition. Lieutenant

[30] Type "C" Spec Rpt, "Employment of Assault Helicopters," Part II, 5–9. Other sources for Operation BUMBLEBEE are HMR–161 and 1st ShorePartyBn *HD*, Oct 51, and *Cavalry of the Sky*, 167–170.

DD MC A 132540

Helicopter In The Hills—A large Sikorsky helicopter hovers over the mountainous terrain of Korea. This type of aircraft has been extensively used for many types of transport missions. **B1**

634040 O–62–16

DD MC A 131324

Generals Confer—Above, MajGen Gerald C. Thomas (left) and MajGen Field Harris discuss the situation in April, 1952. Below, MajGen John T. Selden (right) briefs Army MajGen R. D. Palmer. B2

DD MC A 159084

DD MC A 8410

On The Planning Level—Above, BrigGen Whaling, MajGen Thomas, and BrigGen Puller enjoy a bit of humor, while, below, MajGen Thomas, LtGen Shephard, and Col Wade pose for the photographer. B3

DD MC A 157916

DD MC A 161585

Ready For Action—Above, a Marine 105mm howitzer battery preparing to fire a mission. Below, exterior and interior views of a heavy and light machine gun emplacement.

B4

DD MC A 155525

DD MC A 169584

DD MC A 6953

DD MC A 6754

Going Up—Above, left and right, Marines move into position on one of the many Korean hills. Below, a section of the MLR in March, 1952. B5

DD MC A 160250

DD MC A 159304

The Lifeline—Above, a view of the MSR of the 1st Marine Division during January, 1952. Below, during the rainy season it is difficult to move supplies over the poor roads.

B6

DD MC A 164552

DD MC A 162316

Terrain Features—Above, a Marine helicopter flies behind ice-covered slopes to avoid enemy fire. Below, an enemy stronghold nicknamed "Luke The Gook's Castle." B7

DD MC A 160572

USN 439970

USN 441346

Panmunjom "Talkathon"—Above (left) Gen Nam Il starts for the truce talks. Above (right) United Nations' sentries. Below, the Chinese and North Korean Communist negotiators. **B8**

USN 431929

USN 432414

Watchful Waiting—Above, MajGen L. C. Cragie and VAdm C. T. Joy talk to correspondents at Panmunjom. Below, Communist and UN sentries walk posts around the peace talk site. B9

USN 935646

USN 435646

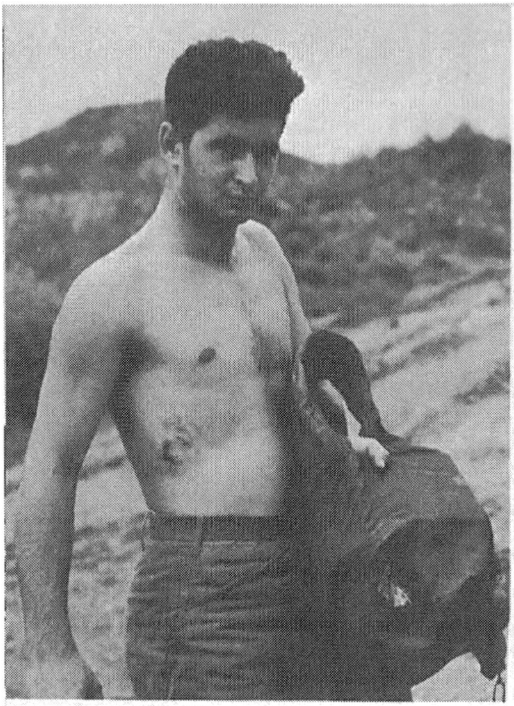

DD MC A 163575

DD MC A 164145

Lifesaver—Above, bruised Marines show the armored vests that saved their lives. Below, a .45 caliber bullet test-fired into cotton contrasted with three removed from an armored vest.

B10

DD MC A 46216

DD MC A 168081

Torso Protection—Above, a Marine rifleman lies prone to exhibit the protective torso armor. Below, a helicopter evacuates a corpsman who was wounded while treating a buddy.

DD MC A 167282

DD MC A 156301

Papa-sans and Pills—Above, a group of Korean patriarchs gaze curiously at a Marine tank. Below, an Army nurse administers medication to a grimacing Marine. B12

SC 358063

DD MC A 159786

A Dog's Life—Above, Marines line up with their pets all packed and ready to go. Below, troops gather for the ever-welcome mail call.

B13

DD MC A 7468

DD MC A 6843

At The Front—Above, Marines advance across a fog-filled valley while supported by machine gun fire. Below, a group of Communist prisoners wait for interrogation by trained experts. B14

DD MC A 157686

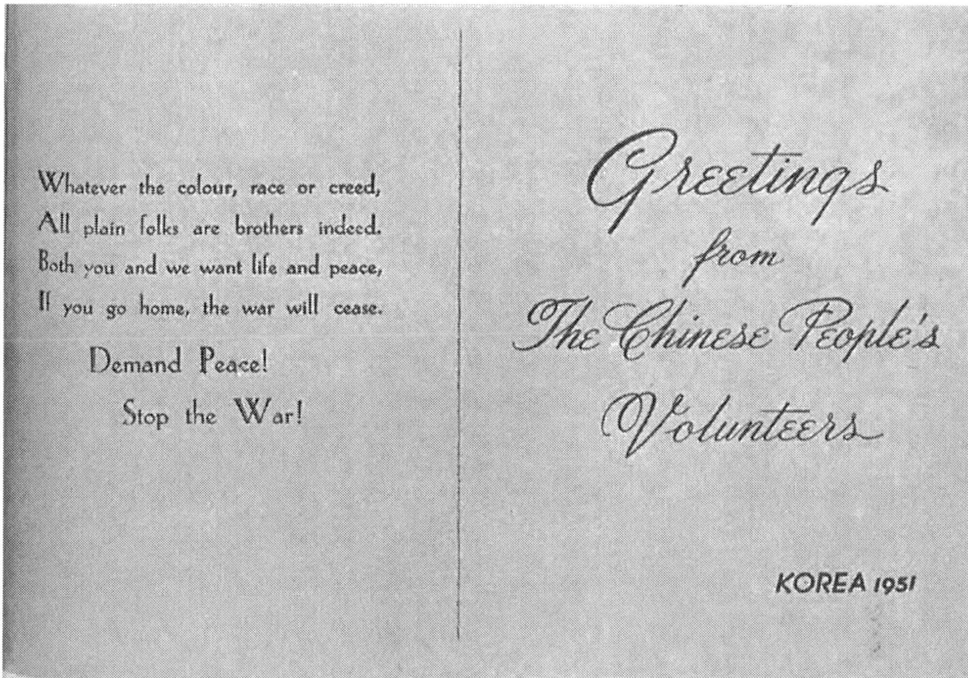

DD MC A 408351

CCF Propaganda—Above, one of the thousands of attractively-colored CCF Christmas cards dropped on the MLR *in December, 1951. Below, a mortar observer crew in action.* B15

DD MC A 6627

DD MC A 162989

Life's Little Problems—PFC Henry A. Friday pauses to rest in a trench and reflect upon the progress of his own particular efforts towards fighting the Communists.

B16

Donald L. Hilian (MC), USN, surgeon of HMR–161, landed to supervise the evacuation of 24 wounded ROKs, several of whom would otherwise have died. Captains James T. Cotton and Albert A. Black made four flights each into the beleaguered area, and all Marine pilots of Operation WEDGE were congratulated in person by Major General Claude F. Ferenbaugh, commanding general of IX Corps.[21]

Seven infantry battalions, with 2/1 in immediate reserve, manned the MLR from 1 to 13 October—three KMC battalions on the left of the Division sector; two 5th Marines battalions (relieved by the 7th Marines on the 11th) in the center; and two 1st Marines battalions on the right. Scout and sniper teams were employed throughout the period, with contacts few and far between. More destruction was inflicted on the enemy by observed artillery, tank, and mortar fire.[22]

A new emphasis was placed on psychological warfare during these defensive operations. Eighty-seven NKPA soldiers surrendered from 1 to 13 October, but whether they responded to leaflets fired by the 11th Marines could not be determined.

Early in October the 1st Marine Division was granted permission by EUSAK to use Sokcho-ri (Map 19) as a port of embarkation and debarkation instead of Pusan. The change proved satisfactory even though troops had to be lightered from ship to shore. A 68-mile truck movement through the I ROK Corps zone replaced the airlift of 200 miles from Pusan to Chunchon, followed by a motor march of 70 miles. It was estimated that the new routing would add from 8,000 to 10,000 man-days a month to the combat potential of the Division.

An improvement in logistics resulted when the Division asked and received permission from EUSAK to use field K–50 near Sokcho-ri for an airhead instead of K–51 at Inje. Although the Marines were limited to five or six sorties a day while sharing K–50 with I ROK Corps, they were able to transfer many airhead activities to the new field.

The mission of the Division remained essentially unchanged from 14 to 31 October. Foot patrols ranged farther into enemy territory, and tank-infantry raids in company strength, supported by air and artillery, were launched at every opportunity.

---

[21] *Cavalry of the Sky*, 171.

[22] The remainder of this section, unless otherwise specified, is based on the 1stMarDiv *HD*, Oct 51, 3–12.

634040 O–62–17

Typical of these operations was the raid staged on 16 October by elements of the 1st Battalion, 7th Marines, (Lieutenant Colonel James G. Kelly) supported by tanks, air, artillery, and engineers. Captain John R. McMahon's Charlie Company was the principal unit involved. The Marine column had as its objective an NKPA strong point overlooking the village of Changhang (Map 2) on the east and the flats on both sides of the Soyang-gang to the south and southwest. Captain McMahon's mission was "to reduce all fortifications and installations . . ." [and] ". . . to seize, occupy and hold ground until the area was thoroughly mined, booby-trapped and infested with trip flares."[23]

A small-scale battle flared up for a few minutes as the enemy put up a stiff resistance with artillery, mortar, and automatic weapons fire. Superior Marine firepower soon prevailed, and at 1540 the attackers reached their objective. During the next hour and 20 minutes enemy installations were destroyed and the strong point rendered untenable by mines and booby traps. The Marines withdrew at 1700 after sustaining casualties of 3 KIA and 18 WIA. Enemy losses were 35 counted KIA.

The next day a reinforced KMC company, supported by tanks, air, artillery, and engineers made a similar raid on enemy positions about 875 yards northwest of Hill 751 and 1,500 yards south of Hill 1052 (Map 20). Twenty-five NKPA bunkers were destroyed with losses to the enemy of 15 counted KIA, 3 prisoners, and 5 captured machine guns.[24]

On 21 October the front of the 1st Marine Division was reduced a mile when elements of the 3d ROK Division relieved the 2d KMC Battalion on the Marine left flank in accordance with instructions of X Corps. Six infantry battalions now manned an MLR of 12¾ miles.

A strong enemy position, menacing the forward elements, had developed to the north of the 1st Battalion, 1st Marines' sector. Three days of reconnaissance and detailed preparation preceded the destructive raid carried out on 30 October. Captain George E. Lawrence's Charlie Company, reinforced with heavy machine guns, was held up by NKPA resistance in estimated company strength. The Marines fought their way up a ridgeline, throwing white phosphorus

---

[23] This account of the raid is derived from the 1/7 *HD*, Oct 51, and the 1stMarDiv *HD*, Oct 51, 7.

[24] 1stMarDiv *HD*, Oct 51, 7–8.

grenades into enemy bunkers. Pinned down momentarily by NKPA mortar and small-arms fire, they reached a defiladed position and withdrew under cover of Marine artillery, air, mortars, and heavy machine guns. At a cost of only one WIA, the raiders inflicted 65 counted KIA casualties on the enemy and destroyed an estimated 40 NKPA bunkers.[25]

All three Marine regiments on Line MINNESOTA were directed by General Thomas to fight the enemy whenever possible with his own weapons in the form of ruses and night ambushes. On 31 October the 3d Battalion of the 1st Marines feigned preparations for an attack even to the extent of a brief artillery barrage. When the firing let up, the Marines sounded an NKPA bugle call as a signal for enemy troops to rush out of bunkers and man open trenches. Thus exposed, they became the victims of intense Marine mortar and artillery fire which inflicted an estimated 47 KIA and 48 WIA casualties.

During the last 2 weeks of October, 11 missions were fired by the battleship USS *New Jersey* and 41 missions by the heavy cruiser USS *Toledo*. Appreciation was expressed in a message to the *Toledo* by General Thomas: "Your accurate and effective fire during period 24–29 October made an important contribution to operations of this division. Many thanks and come again." [26]

Antiguerrilla raids behind the MLR were carried out by Marine ground forces relying upon HMR–161 helicopters for transportation. In Operation BUSHBEATER teams from 1/1 were landed on the Division's east flank to sweep westward toward the Soyang-gang on 22 October while teams from Recon Company patrolled from the opposite direction.

Operations HOUSEBURNER I and II were planned to deprive guerrillas of shelter during the coming winter. As the name implies, helicopter-borne teams set Korean huts afire with flame throwers and incendiary grenades.[27]

Enemy forces facing the Marines at various times in October were believed to comprise the 2d Division, II NKPA Corps, the 1st and 15th Divisions of III Corps, and the 19th Division of VI Corps. NKPA casualties during the month were announced by the 1st Marine Division as 709 counted and 2,377 estimated KIA, 4,927 estimated WIA,

[25] 1/1 *HD*, Oct 51, 16; 1stMarDiv *HD*, Oct 51, 7.
[26] CO 1stMarDiv msg to USS *Toledo*, 1232 30 Oct 51 in G–3 msgs, Oct 51.
[27] *Cavalry of the Sky*, 172–173.

and 571 prisoners. The Marines (including the 1st KMC Regiment) suffered losses of 50 KIA, 2 MIA, and 323 WIA.[28]

Marine Operations of November 1951

On 1 November 1951 the front line strength of the opposing forces was nearly equal—195,000 for the UN, and 208,000 for the enemy. In reserves the Communists held their usual numerical advantage with nine CCF armies totaling 235,000 men plus 138,600 in four NKPA corps. All were readily available either as reinforcements or as assault troops for a great offensive.[29]

Even though the Eighth Army was committed to a warfare of position, General Van Fleet meant to keep the initiative. "If we had stagnated on any one of our many positions since the tide turned in April," he said in a recorded statement of 3 November, "the hydra-headed Communists—who seem to grow two soldiers for each one cut down—would soon have been at our throats. With the enemy's prolific capacity posing an ever-present threat, we had no choice but to destroy the menace before it matured." [30]

Throughout November the 1st Marine Division continued to occupy the eastern portion of the X Corps defense sector in east-central Korea. From left to right the 1st KMC Regiment, 7th Marines, and 1st Marines held the 12¼-mile MLR with two battalions each. The 5th Marines remained in reserve until the 11th, when it relieved the 1st Marines. That regiment went into the new reserve area at Mago-ri (Map 19).[31]

Elements of the 1st, 15th, and 19th Divisions, III NKPA Corps, manned the opposing lines. The Marines continued to organize artillery- and air-supported tank-infantry-engineer task forces in company strength for raids. Squad-size patrols were sent out nightly to ambush the enemy, employing ruses whenever possible.

The howitzers of the 11th Marines and the 90mm rifles of the 1st Tank Battalion were kept busy throughout the month. On 7–8 November, for instance, Marine artillery fired 257 observed missions

[28] 1stMarDiv *HD*, Oct 51, 2.
[29] EUSAK *Cmd Rpt*, Nov 51, 9.
[30] *Ibid.*, 32.
[31] The remainder of this section, unless otherwise specified, is derived from the 1stMarDiv *HD*, Nov 51, 1–20.

in 24 hours—including 34 on enemy artillery positions, 32 on mortar positions, 25 on bunkers, 22 on machine gun positions, 4 in support of friendly patrols, 3 on supply dumps, 2 on trucks, and 1 each on a bridge, a CP, and a 57mm recoilless rifle position.

In spite of such daily pounding, aerial photographs proved that NKPA defenses in depth had become more intricate and formidable in November 1951 than during any previous month.

On the 7th the 14th Replacement Draft added 2,756 officers and men to the 1st Marine Division. Within a few hours 2,066 officers and men of the 10th Rotation Draft were detached. And on the 27th the 11th Rotation Draft represented a further loss of 2,468 Marines whose departure was hastened so that they could be home by Christmas.

A note of grim humor crept into proceedings on 9 November. Division OpnO 50–51 directed that all supporting arms and weapons commemorate the Marine Corps Birthday the next day by firing a TOT on Hill 1052, the key enemy observation point overlooking the friendly sector.[32] While the cruiser USS *Los Angeles* contributed naval gunfire, the Commanding General of 1st MAW, Major General Christian F. Schilt, led an air strike of 83 Marine planes to blast this enemy strong point.

The performance was embellished on the 10th when Marine tanks, mortars, and machine guns added their fire to the grand crescendo of exploding shells and bombs. The Communists were also bombarded with 50,000 leaflets inviting them to the Marine birthday dinner that evening. Twenty Korean Reds actually did surrender, though some doubt remained whether they had responded to the invitation or the TOT. General Van Fleet sent a message to all Marines in his command, congratulating them on "a job well done" in Korea.[33]

On 11 November the 5th Marines carried out its relief of the 1st Marines on Line MINNESOTA. This was the occasion for the largest helicopter troop lift so far, involving the transportation of nearly 2,000 combat-equipped men.

Operation SWITCH began at 0635 on D-Day when three helicopters took off from Field X–83 with Shore Party specialists to signal aircraft into landing sites and supervise the unloading and reloading

---

[32] The initials TOT stand for Time on Target—an artillery order calling for all guns to time their firing so that projectiles will hit the target simultaneously.

[33] EUSAK *Cmd Rpt*, Nov 51, 42.

of troops. Twelve helicopters were employed, each carrying five men and supplies from the 1st Battalion, 5th Marines (Lieutenant Colonel Kirt W. Norton), and returning to Field X–83 with a like load from Lieutenant Colonel Clifford E. Quilici's 2d Battalion, 1st Marines.[84]

Naval gunfire from the USS *New Jersey* helped to keep the enemy quiet during the relief. All told, 950 men were flown to Hill 884—soon to be known unofficially as "Mount Helicopter"—and 952 lifted to Field X–83 in return flights. Total flight time was 95.6 hours and over-all time 10 hours. Once again the Marine Corps had made tactical history.

Ground forces operations throughout November seldom varied from the familiar pattern of squad-size patrols nightly and an occasional daytime raid by a company-size task force with the support of artillery and air. Supporting arms kept enemy strongholds under almost constant fire, and North Korean activity in the construction or improvement of bunkers provided frequent targets of opportunity.

Contacts seemed to be avoided by enemy troops. On the night of 29 November, for instance, 11 Marine ambush patrols ranged from 1,500 to 2,500 yards ahead of the MLR with only a single contact before returning at daybreak. One enemy KIA was inflicted and one prisoner taken at a cost of four Marine WIA casualties.

Total Marine casualties (including the KMCs) during November were 34 KIA and 250 WIA. Enemy losses amounted to 408 counted and 1,728 estimated KIA, 2,235 estimated WIA, and 104 prisoners.

The Second Marine Christmas in Korea

Marine operations in December were shaped in advance by the resumption of armistice negotiations. This time Panmunjom was agreed upon as a conference site instead of Kaesong. Literally a wide place in the road, the tiny hamlet was located just north of the 38th Parallel between Munsan and Kaesong (Map 19). In the lack of houses, tents provided shelter for the UN and Communist delegates who renewed their meetings on 25 October 1951 for the first time since the Reds walked out at Kaesong on 23 August.

Discussions during November were largely devoted to the question of a cease fire based upon a line of demarcation. On the 23d it was

[84] HMR–161 *HD*, Nov 51; *Cavalry of the Sky*, 174.

agreed to accept a line linking up the farthest points of repeated contacts up to 2,000 yards forward of the United Nations MLR. Three days later, representatives of both sides initialed maps to indicate acceptance.[35]

The effect of the so-called cease fire on EUSAK operations was immediate. General Van Fleet sent his corps commanders a letter of instructions warning that active defensive operations were to continue until a full armistice had been concluded. If such an event took place within 30 days after 27 November 1951, the demarcation line would not be altered. But if an agreement had not been reached by that time, the line would be revised in accordance with actual changes.[36]

EUSAK instructions to corps commanders were relayed in a X Corps message of 27 November to the 1st Marine Division:

> *Part I.* The conference at Panmunjom has fixed a military demarcation line as a preliminary step to ending hostilities within a 30-day period.
>
> *Part II.* Every US, UN, and ROK soldier will be informed that hostilities will continue until armistice agreement is signed.
>
> *Part III.* While negotiations continue, X Corps will: (1) Demonstrate its willingness to reach an agreement by reducing operations to those which are essential to insure maintenance of present positions. Counterattacks to regain key terrain lost to enemy assault are authorized, but other clearly offensive actions will be taken only by direction of this Headquarters; patrolling only to that line beyond which contact has been repeatedly established; limiting supporting fires, including air strikes, to destruction of those targets which appear to constitute a major threat, or to improve the enemy's offensive capability. (2) Prepare for offensive action by: Conserving ammunition; maintaining combat effectiveness through intensified training; preparation for and rehearsal of limited-objective attacks, to be launched near the end of the 30-day period in order to improve the MLR.
>
> *Part IV.* Every effort will be made to prevent unnecessary casualties.[37]

In view of these instructions, it is understandable that a lull set in along the X Corps front in December 1951. Most of the cold weather clothing had been issued during the preceding month, and work was largely completed for the "winterizing" of bunkers. It remained only to improve defensive installations as front line elements

[35] References to the Panmunjom decisions are based upon the following sources: William H. Vatcher, Jr., *Panmunjom, The Story of the Korean Military Armistice Negotiations* (New York: F. Praeger, 1958), 72–94, 232–237; Joy, *How Communists Negotiate*, 40–52.

[36] EUSAK *Cmd Rpt*, Nov 51, 58.

[37] X Corps *Cmd Rpt*, Nov 51, 15–16.

continued to send out patrols to maintain pressure against the enemy. And since the Communists were putting similar military policies into effect, both sides kept in contact with relatively small units.[38]

The enemy also busied himself with extending already formidable defenses in depth. And though he did not seek a fight, he showed no hesitation about accepting one.

From 5 to 20 Marine patrols went out nightly during December, some of them manning night outposts called "duck blinds;" [39] occasional raids continued with relatively few contacts. In the rear of the Division area, helicopter patrols continued against guerrillas.

The 13 aircraft of HMR–161 had a busy month with 390 missions and 621 flights. Six thousand pounds of rations, 9,000 pounds of fuel oil in drums, 15,000 pounds of fortification material, and 15,000 pounds of cold weather clothing were among the supplies flown to the front. Personnel to the number of 2,022 were lifted, and cargo to the amount of 149,477 pounds.

The first breakthrough in truce negotiations, at Kaesong, occurred on 18 December, when lists of prisoners held by both sides were exchanged. Prior to this exchange of lists the UN Command could only speculate on the number carried as missing in action who were in reality held as prisoners of war. The Communists had previously reported only a few dozen names, and then only if it suited their propaganda purposes. Radio Peking, in releasing names piecemeal, had broadcast recordings made by UN prisoners under duress. Far Eastern monitors reported these broadcasts were slanted to give the Communist viewpoint.

The 18 December list of 3,198 American POWs revealed only 61 Marines including 2 Navy hospital corpsmen. (Information received from 18 Marines who gained their freedom in May 1951 was sketchy concerning others held at the time and was never accredited as official or authoritative.)[40] Interestingly enough when the Communist negotiators saw the list given them by the UN representative they became irate and tried to withdraw their list. The names of the Chinese and Korean prisoners had been Anglicized and caused considerable difficulty in retranslating the names into oriental characters.

---

[38] The source for the remainder of this section, unless otherwise stated, is the 1stMarDiv *HD*, Dec 51, 1–17.

[39] LtCol Harry W. Edwards, memo to G–3 dtd 3 Feb 1959.

[40] Maj J. Angus MacDonald, "The Problems of Marine POWs," MS available in Historical Archives, G–3, HQMC.

Negotiations hit a snag at this point, and no other list was offered by the Red officials until the first prisoner exchange (Operation LITTLE SWITCH in April 1953). Notwithstanding the protracted and exasperating tactics of the Reds at the truce table, the exchange of prisoner of war lists presaged infinitely better treatment to the UN prisoners than had been accorded them prior to that time. The so-called lenient treatment policy by the Chinese, promulgated in July 1951, was initiated after the exchange of lists.[41]

The lists given by the Communists did not include several Marines captured during the months of October, November, or December of 1951. The families of these men were to sit in anguish waiting for these names until April of 1953. These and other instances of perfidy and treachery at the truce table by the Communist negotiators were to become legion.[42]

On 19 December the 2d Battalion, 5th Marines (Major William E. Baugh) was relieved just behind the MLR by Lieutenant Colonel Norton's 1/5 in helicopter Operation FAREWELL. It was the last flight in Korea for Lieutenant Colonel Herring, who returned to Quantico as commanding officer of Marine Helicopter Experimental Squadron (HMX)-1. His relief as commander of HMR-1 was Colonel Keith B. McCutcheon, and Lieutenant Colonel Mitchell remained as executive officer.[43]

The Marine helicopters of VMO-6 had also been setting records during the last half of 1951 under four commanding officers, Major David W. McFarland (5 April–5 October), Major Allan H. Ringblom (6 October–31 October), Major Edward R. Polgrean (1 November–25 November), and Major Kenneth C. Smedley (26 November–31 January 1952). A total of 1,096 Marine wounded had been flown out during this period, many of whom would otherwise have lost their lives.[44]

The supposed vulnerability of the helicopter was whittled down to a myth by VMO-6 experience. Returning from a front line mission with bullet holes was too commonplace for mention, yet the year

---

[41] Maj G. Fink, interview of 16 Dec 1960; Extract of Interim Historical Report, Korea War Crimes Division, cumulative to 30 Jun 1953, 18.

[42] Joy, *How Communists Negotiate*, 104–105; Maj J. A. MacDonald, "The Problems of Marine POWs," *op. cit.*

[43] HMR-161 *HD*, Dec 51; *Cavalry of the Sky*, 175–176. Two of the original 15 HRS-1 aircraft had been damaged in accidents, but one was later restored to action with parts cannibalized from the other.

[44] VMO-6 *HD*, Jun–Dec 51; *Cavalry of the Sky*, 146, 180–181.

1951 passed without a single helicopter pilot being lost to enemy action, even though several aircraft were shot down. The experience of these 12 months also proved anew the wisdom of combining rotary-wing and fixed-wing aircraft in an observation squadron in fairly equal numbers. When it came to reconnaissance and artillery spotting, the nimble little OYs and OEs (both types are light observation planes) were much better suited than the "choppers."

As for close air support, increased Air Force emphasis on an interdiction campaign beyond artillery ranges added to the limitations imposed on Marine requests. Of the 22 strikes requested in December 1951, only five were approved.

From the 1st to the 10th, units of the Division along the MLR consisted from left to right of the 1st KMC Regiment, 7th Marines, and 5th Marines. The only major change took place on the 11th, when the 1st Marines relieved the 7th and the latter went into Division reserve. Enemy units were believed to be the 1st, 15th, and 19th (soon relieved by the 47th) NKPA Divisions with an estimated strength of 25,750.

Permission was rarely granted by X Corps for Marine raids to cross the EUSAK military limiting line known as Line DUCK, which generally coincided with the line of demarcation. Christmas passed like any other day except for the holiday feast. Nineteen patrols went out on Christmas Eve, two of which had brief fire fights with enemy patrols before returning at dawn. During the day 40 rounds of naval gunfire from the heavy cruiser USS *St. Paul* were credited with destroying seven enemy bunkers.

More than a third of the Marines partaking of Christmas turkey were comparative newcomers who had reached Korea since the warfare of movement ended on 20 September. The 15th Replacement Draft brought 38 officers and 2,278 men early in December, and 127 officers and 1,805 men departed with the 12th Rotation Draft. No Marines who had arrived prior to 1 January 1951 were left among the 1,495 officers and 23,040 men in Korea at the close of the year.

Heavy snow on 26 December impeded foot-patrol activity and increased the danger of mines. Next day, when the 30-day cease-fire agreement ended, it was announced at Panmunjom that the terms had been renewed and that operational restrictions would be extended indefinitely.

Thus December came to an end on a note of troubled uncertainty.

Not a single large-scale combat had been reported, yet 24 Marines were killed (including KMCs) and 139 wounded in patrol actions. That the enemy had sometimes succeeded in the grim quest of both sides for prisoners is shown by the unwonted entry of eight Marines missing in action. NKPA losses for the month consisted of 246 counted KIA, and 56 prisoners.

The year 1951 passed into history at 2400 on 31 December as the 11th Marines saluted 1952 by firing a "toast" at enemy strongholds. The thud of the snow-muffled howitzers was also a fitting farewell to the past year of a war that was not officially a war. Indications were that it would doubtless be concluded by a peace that was not a peace, judging from the attitude of the Communist delegates at Panmunjom. And meanwhile the Marines and other Eighth Army troops would keep on fighting in accordance with the terms of a cease fire was not a cease fire.

CHAPTER XI

Winter Operations in East Korea

Ambush Patrol on New Year's Eve—Marine Raid in Company Strength—Major General John T. Selden Assumes Command—Boot, Combat, Rubber, Insulated—500 Armored Vests Flown to Korea—Helicopter Operations MULETRAIN *and* CHANGIE-CHANGIE—*The Five Days of Operation* CLAM-UP

AS THE NEW YEAR began, the 1st Marine Division occupied practically the same front it had held along Line MINNESOTA for the last three months (Map 20) and would continue to hold for the next two and a half. The major units were disposed from left to right on 1 January 1952 as follows:

1st KMC Regiment (Colonel Kim Dong Ha commanding, LtCol Alfred H. Marks, senior advisor);
1st Marines (Colonel Sidney S. Wade);
5th Marines (Colonel Frank P. Hager, Jr.);
11th Marines (Colonel Bruce T. Hemphill) in artillery support.

The 7th Marines (Colonel John J. Wermuth) was in reserve until 10 January, when it relieved the 5th Marines on line. That regiment then went into reserve and could not be committed to action without the approval of X Corps.[1]

Tactical units not organic to the 1st Marine Division but attached at this time were, in addition to the 1st KMC Regiment, the 1st Korean Artillery Battalion, the 1st Platoon, 92d U.S. Army Searchlight Company, and Battery C, 1st 4.5" Rocket Battalion.

The new Korean artillery battalion consisted of two medium (155mm) and two light (105mm) howitzer batteries. Major General

[1] 1stMarDiv *HD*, Jan 52, 1–2.

Gerald C. Thomas, commanding general of the 1st Marine Division, approved a plan for placing this unit in the Punchbowl on 9 January to reinforce Lieutenant Colonel Sherman W. Parry's 1st Battalion, 11th Marines.

Enemy units opposing the 1st Marine Division up to 23 January 1952 were the 1st, 15th, and 47th NKPA Divisions with an estimated combined strength of 25,750 men. On the 23d the 15th Division was relieved by the 45th.

The enemy, according to the Division report, showed "greater caution than he had in previous months, and friendly outposts and ambuscades noted fewer contacts. His harassing mortar and artillery fires increased in volume through the month. Meanwhile, extensive efforts to improve his defenses continued with particular attention being given to reverse slope installations." [2]

Ambush Patrol on New Year's Eve

The new year was but a few minutes old when the first Marine action took place. Captain Charles W. McDonald's Baker Company had been directed by Lieutenant Colonel Kirt W. Norton, commanding the 1st Battalion, 5th Marines, to send out an ambush patrol on New Year's Eve.

A rifle squad, a light machine gun squad, an interpreter, and a corpsman composed the little column wearing white camouflage clothing which made the men all but invisible against a background of snow. After getting into position, the patrol settled down for the usual long wait. Darkness was the enemy's element, and Marine ambushers ran the risk of being ambushed themselves. This time, however, a six-man North Korean patrol came within five yards before the Marines let the enemy have it with machine gun and rifle fire which inflicted one KIA and four estimated WIA casualties. Efforts to take a prisoner were frustrated as the NKPA survivors melted away into the darkness. The Baker Company patrol returned without casualties at 0400.[3]

Marine operations were still limited by the EUSAK "cease fire" directive which went into effect for a month on 27 November 1951 in accordance with a decision reached during the armistice negotiations

[2] *Ibid.*, 1, 6, 7.
[3] 1/5 *HD*, Dec 51, 31; 1stMarDiv *HD*, Jan 52, 3.

at Panmunjom. UN and Communist delegates agreed on a line of demarcation, known to the Eighth Army as Line DUCK. It linked up points of repeated EUSAK patrol contacts, not to exceed 2,000 yards beyond the MLR. Operations past this line, running generally parallel with Line MINNESOTA, could not be launched without permission from corps commanders.

When the agreement expired on 27 December, it was renewed indefinitely. Actually, it brought about few changes in the warfare of position which had replaced a warfare of movement on 20 September 1951. Each Marine infantry regiment on the MLR continued to send out several squad-size patrols nightly for such purposes as ambush, reconnaissance, and taking prisoners. Raids were employed for special missions where formidable enemy resistance might be expected. These forces usually ranged from a platoon to a company in strength, reinforced by supporting weapons. Operations of this sort were planned with meticulous thoroughness and carried out with minimal risks.

Marine Raid in Company Strength

The first company-size raid of the new year was conducted by units of the 3d Battalion of the 1st Marines (Lieutenant Colonel Spencer H. Pratt) in the darkness of 1–2 January 1952. Captain James B. Ord, Jr.'s How Company was alerted on 30 December to prepare for a night raid with a mission of reconnaissance and capturing or destroying any enemy that might be encountered. On the afternoon of the 30th, Ord made a preliminary reconnaissance with Second Lieutenants Milo J. See and John E. Watson, commanding the 2d and 3d Platoons respectively. That evening the company commander held a briefing at his OP (observation post) which was attended by the sergeants and squad leaders of the two platoons selected for the raid.[a]

This command group carried out a second reconnaissance forward of the MLR on 31 December, proceeding until they ran into enemy sniper fire. Captain Ord requested aerial reconnaissance and three missions were flown by observation planes of Major Kenneth C. Smedley's VMO–6.

[a] Sources for this account of the raid, unless otherwise specified, are Maj J. B. Ord, Jr., intervs of 3 Sep and 24 Oct 58; and Appendix VI, 1stMarDiv *HD*, Jan 52, a five-page special action report of the operation.

Line DUCK and the assigned battalion sector limited the objective area. On a basis of these restrictions as well as reconnaissance reports, Ord recommended an operational area containing three objectives, each of which represented a point where the enemy was not likely to be encountered. These objectives were approved by Lieutenant Colonel Pratt and formed the basis of the battalion order.

The task organization for the raid included two attached How Company units, the machine gun platoon (–), and 60mm mortar section, commanded by Second Lieutenants John D. Koutsandreas and James J. Hughes respectively. Another infantry unit, the 1st Platoon of Item Company, 3/1 (Second Lieutenant William E. Harper), was also attached.

First Lieutenant Francis E. White, How Company executive officer, remained at the OP with the tactical air-control party, which had an observation plane on strip alert in case the raiders ran into artillery or mortar fire. A forward air controller with radioman accompanied the raiding party as well as artillery, 4.2″, and 81mm mortar forward observers. An interpreter, the assistant battalion surgeon, and a corpsman were included, and wiremen had the assignment of laying a line.

Hill 812 (Map 20) was the jumping-off place for the column of files in ghostly white snow suits with hoods. Boots were dark in contrast but the snow was deep enough to hide them. The drifts slowed up the wiremen and an infantry fire team protected them at their work.

The first objective consisted of bunkers and suspected mortar positions which had been reported by tactical air observers as recently occupied by the enemy. They were empty when the raiding party reached them, and the Marine column proceeded toward Objective 2, an ambush site overlooking and commanding a crossing of the Soyang-gang.

The selected area for the support group was located nearby, and there the machine gun section and riflemen took positions on a nose with the wiremen, radiomen, and corpsman in the center. While these elements peeled off, the raiding party continued toward the ambush site, where it was planned to lie in wait two hours for the enemy. A suspected mine field had to be crossed and Captain Ord directed his men to advance in single file, stepping carefully in the footprints ahead. Twelve Marines had passed safely when the 13th became the victim of a mine explosion. The corpsman found broken bones but none of the usual torn flesh and hemorrhaging, thanks to the new thermal boots issued during the winter of 1951–1952.[a]

The temperature was zero with a sharp wind blowing. Some of the Marines had to shed clothing to keep the casualty warm during the forced immobility, and the raiding party commander broke radio silence by requesting permission of Captain Ord, in the support group area, to pull back to that position and set up the ambush.

Permission was granted by Ord after radio consultation with the battalion commander on the How Company OP. The raiding party remained in ambush formation on Objective 2 for two hours without seeing or hearing an enemy. By that time the condition of the mine casualty had deteriorated to such an extent that Lieutenant Colonel Pratt gave permission for a return to the MLR without proceeding to Objective 3.

He directed that the raiders split and take two routes in the hope of capturing a prisoner, since a light enemy probing attack on the MLR had just been reported by Item Company of 3/1. This proved to be a fortunate decision, for two NKPA soldiers were seized. The main object of the raid had thus been fulfilled, even though little action was seen during the five-hour operation.

Raids of this sort may seem anticlimactic when compared to the fights in the same area during the first three weeks of September. But the Marines were showing adaptability in conforming to a warfare of position that was contrary to all their offensive training. Careful reconnaissance, detailed planning, and minimal risks—these were the elements of defensive tactics in which large forces had to content themselves with small gains.

Major General John T. Selden Assumes Command

On 11 January 1952 the 1st Marine Division had its second change of command in Korea when Major General John T. Selden relieved General Thomas. The new commanding general was born at Richmond, Virginia, and educated there at McGuire's University School. Before the United States entered World War I, he tried to join the Canadian Army but was warned that he would lose his American citizenship. In January 1915, at the age of 21, he enlisted as a private

[a] Later in the chapter this innovation will be described.

634040 O–62–18

in the Marine Corps and saw two years of active duty on jungle patrols in Haiti. Commissioned as a second lieutenant in 1918, he served in ocean convoys during World War I.

Sea duty, China duty, and more Haiti duty occupied him during the postwar years. The outbreak of World War II found him a Scouting Force Marine Officer aboard the *Indianapolis*. After that he had three main assignments: personnel and intelligence officer of I Marine Amphibious Corps; commanding officer of the 5th Marines in the New Britain operation; and chief of staff of the 1st Marine Division at Peleliu.

Brigadier General William J. Whaling remained on duty as Assistant Division Commander. The new staff officers were Colonel Richard G. Weede, Chief of Staff; Colonel Walter N. Flournoy, G–1; Lieutenant Colonel James H. Tinsley, G–2; Lieutenant Colonel Gordon D. Gayle, G–3; and Colonel Custis Burton, Jr., G–4.

A change of FMFPac command had taken place on 1 January. Lieutenant General Franklin H. Hart relieved General Shepherd, who became Commandant of the Marine Corps as General Cates finished his four-year term. General Hart paid his first visit to the 1st Marine Division late in January.

The new FMFPac commander found the Marines occupying essentially the same positions they had defended since late September. About two-thirds of the 12¼-mile MLR on Line MINNESOTA (Map 20) was good defensive ground. It had been strengthened by an elaborate system of trenches and bunkers behind miles of barbed wire.[8]

In the left-central portion of the Marine sector, the enemy held the dominating terrain. This was particularly true of the rugged area just west of Hill 812, where the opposing trenches were only 50 to 150 yards apart. There a fire-raked landmark, known to the Marines as Luke the Gook's Castle, had been made into a strong point by the enemy. Its base was a maze of trenches and bunkers, and the 20-foot granite knob could have been taken only at an excessive cost in casualties. Although this bastion was hit repeatedly by almost every type of supporting ordnance, it was never completely destroyed nor denied to the enemy.

Operations of trench warfare had inevitably shaken down into a daily routine of sniping by day and patrols or raids by night. Marine

---

[8] Sources for this section, unless otherwise indicated, are the 1stMarDiv *HD*, Jan, Feb, and Mar 52, and *PacFlt Interim Rpt* No. 4, IX.

artillery, mortars, and stationary tank fire, occasionally reinforced by naval guns, played an increasingly important part in the coordinated destruction of NKPA defenses. As a result the enemy was limited for the most part to well camouflaged reverse slope positions.

Because of the 1st Marine Division's defensive mission and the constant rotation of the more experienced personnel back to the United States, it was considered that men assigned to infantry elements, in particular, needed additional training in small unit leadership and offensive tactics. Consequently the regiments were rotated at monthly intervals to the reserve area near Wontong-ni, where Camp Tripoli had been established for training. An average of 84 NCOs a week completed a 168-hour special course of instruction over a four-week period. The program for the rank and file was so intensive, according to one report, that "it was considered a relief by some Marines to cease training and return to the relatively quiet life on the front lines." [7]

The truce talks at Panmunjom continued to influence operations at the front. A demilitarized zone having been proposed in anticipation of an armistice, preparations were begun by the 1st Marine Division to develop the defenses along Line ICELAND, generally conforming to the Line KANSAS of Marine fights early in September. It was to be used as a new line of defense if the UN and Communist delegates reached an agreement.

Perhaps because other offensive tactics were so curtailed, psychological warfare had its heyday in the winter months of 1952. Propaganda leaflets were dropped from planes or fired by 105mm howitzers. At vantage points along the front, loud speakers bombarded the Communists with surrender appeals in their own language. The effects could not be evaluated with any degree of certainty, but it was hoped that the enemy did not respond with the amused indifference shown by the Marines toward Red propaganda.

Boot, Combat, Rubber, Insulated

The average low temperature for January 1952, was 11 degrees Fahrenheit. This was mild weather as compared to the subzero read-

[7] *PacFlt Interim Rpt* No. 4, IX, 9–11.

ings of the previous winter. Only 10 slight frostbite cases were reported for the month in contrast to the 3,083 nonbattle casualties, nearly all frostbite cases, incurred during the two weeks (27 November to 10 December 1950) of the Chosin Reservoir breakout.

The improvement in January 1952 could not be credited entirely to more clement weather. It was due in greater measure to one of the most noteworthy innovations of the Korean war—the insulated rubber combat boot, which proved much superior to the shoe pac of the past winter.

U.S. Army experiments dated back to 1944. They were dropped three years later after efforts to perfect a boot with sealed insulation failed to meet the test of long marches. The Navy had more promising results with the boot during the winter of 1948–1949 when Arctic clothing tests were conducted at Point Barrow, Alaska. Army and Navy tests at Mt. Washington, New Hampshire, the following winter were inconclusive. Marine Corps tests were held during the first four months of 1951 at the following places: MCEB, Quantico; Fort Churchill, Manitoba; Big Delta, Alaska; Pickel Meadows, California; and the Naval Medical Field Research Laboratory (NMFRL), Camp Lejeune.

"In addition to engineering tests," states the Marine report, "the insulated rubber boots have been worn by test subjects selected from a variety of backgrounds; under conditions of activity varying from strenuous marching for 20 miles to complete immobility; in ambient temperatures from 58° to —42° F.; over terrain ranging from soft snow [to] hard snow, ice, sand, rocky ground, mud, gravel, water, and iced river banks; for periods of time corresponding to a normal working day and more than 72 hours. As now constructed, the insulated rubber boot, employing the vapor barrier principle, meets the requirements outlined previously and is satisfactory for use by Marine Corps ground troops in cold climate areas, supplanting the shoe-pac combination. . . ."[8]

The distinguishing feature of the "thermal boot," as it came to be popularly known, is an air space between the inner and outer layers

[8] LtCol G. W. Hardwick, "Summary of Marine Corps Experience with IRB [Insulated Rubber Boot], Rpt of 8 May 1951." Other sources for the development of the boot, also found in G–4 files, Headquarters Marine Corps, are as follows: G. E. Folk, Abstract of Bowdoin College Rpt, Jun 1951, "The Penetration of Water into the Human Foot;" G–4 Rpt, "Resume of Activity re Insulated Rubber Boot," 7 Feb 1952; G–4 Rpt, "Boot, Rubber, Insulated, Cold Weather," 28 Nov 51; G–4 Rpt, "Fact Data Sheet, Boot, Insulated, Rubber," n.d.; MajGen J. T. Selden memo to CMC, 26 Apr 52.

of wool pile insulation, both of which are completely sealed off by latex from any contact with moisture. This air space, under pressure, produces a vapor barrier such that heat cannot readily escape when it is emitted from the foot. Thus the wearer of the boot supplies his own warmth, which is retained as long as he is active, regardless of prevailing temperatures. If, however, the walls of the air space are punctured and the insulation becomes wet, the moisture collected within the boot freezes at low temperatures if the wearer remains inactive. In such cases, severe frostbite may result.

Some of the tests were spectacular. One subject poured water containing pieces of ice into his boots and donned frozen socks before putting on the footgear. After 10 minutes of walking, the ice in the boots had turned to warm water, and there was no harmful effect on the man.

Another subject waded across a knee-deep creek at a temperature of zero. Before he had marched a mile in the snow, his feet had warmed the water in the boots, although his pants were frozen so stiff that he could scarcely walk.

Seldom has a military innovation been tested so thoroughly and scientifically in such a short time. Colonels Ion M. Bethel and John F. Stamm of Marine Corps Headquarters took a leading part in the development and procurement phases along with Lieutenant Colonel Gordon A. Hardwick. Major Vernon D. Boyd and Captain David R. McGrew, Jr. were active in the troop acceptance tests.

A good many "bugs" had to be eliminated before the boot met with complete Marine approval. The manufacturer's modifications were effected with minimal delay.

It is perhaps needless to add that the thermal boot was not foolproof. Protection continued in subzero weather for at least an hour after the termination of activity, but it was inviting frostbite to remain motionless much longer. Socks had to be changed every 12 hours, and foot cleanliness and hygiene could not be neglected.

If a few such simple rules were observed, a man had virtually perfect frostbite protection in the coldest weather. In fact, it was seriously proposed that a Marine casualty of this sort should be charged with misconduct if he acquired his frostbite while provided with thermal boots and a change of socks.

In view of the tests and negotiations with the manufacturers, it was a marvel of promptness when the first shipment of boots reached the 1st Marine Division in August 1951, long before the advent of cold weather.

Distribution to the Division was completed by 15 November. Throughout the winter the experience of all units concerned was reported to Division headquarters. And in a memorandum of 26 August 1952 to the Commandant, General Selden expressed his approval: "The boot, rubber, insulated, is considered an excellent item of cold weather equipment. It is far superior to the shoe pac."

The acceptance by the rank and file went so far that the "Mickey Mouse boot," as it was sometimes dubbed, acquired a reputation for protecting the wearer against antipersonnel mines. Some wounds apparently were reduced in severity by this protection, but it could not be claimed that the boot qualified as armor.

Production by the manufacturer kept pace with Division and Air Wing requirements in Korea. By 14 December 1951 about 90,000 pairs of boots and 2,000 patching kits had been received at San Francisco—more than enough to take care of the 6,500 pairs needed monthly for resupply under combat conditions.

The thermal boot was here to stay.

500 Armored Vests Flown to Korea

Marine body armor was just then about to meet its first large-scale test in the field. It had cleared its preliminary hurdle during the tests from 14 June to 13 October 1951 (see Chapter VIII) when a joint Army-Navy Medical Commission endorsed 40 vests worn in action by troops of the 5th Marines and two Army infantry regiments.

On 9 November, at Marine Corps Headquarters, Marine officers were briefed on the successful results in Korea by the two Navy officers who helped supervise the tests, Commander John S. Cowan (MC) USN, and Lieutenant Commander Frederick J. Lewis (MSC) USN.

That same day the commanding general of FMFPac stated an operational requirement for 500 armored vests to be sent to the 1st Marine Division. And on 16 November the Commandant approved the standardization and procurement of vests to be designed by the Naval Medical Field Research Laboratory at Camp Lejeune and air-shipped to Korea not later than 31 January 1952.[8]

---

[8] Sources for this section, except when otherwise specified, are the following: ACofS,

So many problems remained to be solved that it was nip and tuck whether Lieutenant Commander Lewis and his NMFRL colleagues would make the deadline. On 11 December 1951 another body armor meeting was held at Marine Corps Headquarters, attended by Marine representatives. Lieutenant Commander Lewis and Mr. John F. Quinlan, reporting for the NMFRL, explained that as a consequence of changes in design to speed up manufacture, samples submitted to them weighed as much as 10 pounds.

Under no circumstances, said Lewis, would he approve a vest weighing more than eight pounds, since its success depended so much on troop acceptance. Despite the fact that only a few weeks remained before the deadline, Lewis exhibited a vest that he and Quinlan had redesigned by working around the clock until the armor came within the weight limit without any sacrifice in protection. This vest was immediately put into production as the M–1951.

A plastic fibre manufacturer agreed to supply 70,000 Doron plates, and a Philadelphia sportswear company contracted to manufacture the first 500 vests, plus an additional 2,500 to be delivered by 30 March 1952. The M–1951 was described in Marine reports as "a zippered, vest-type, sleeveless jacket constructed of water-resistant nylon incorporating two types of armor. One, a flexible pad of basket-weave nylon, covers the upper chest and shoulder girdle; the other, overlapping curved Doron plates, covers the lower chest, back and abdomen. These Doron plates consist of several layers of fibre glass cloth, bonded or laminated together with a resin. . . . Although the ballistic properties of the flexible pads of basket-weave nylon and the Doron plates are virtually the same, by using the rigid plates where flexibility is not mandatory the problem of protrusion and the resultant wounds under the armor is reduced."[10]

Marine wearers of the M–1951 were warned that it would not stop rifle or machine gun bullets unless they had lost much of their velocity at long ranges. The vest was protection against most grenade, mortar, and artillery fragments, as well as .45 caliber pistol and burp gun slugs of less than 1,000 feet per second initial muzzle velocity. Wearers did not escape entirely unscathed, for the impact of the fragment or slug left painful bruises.

G–4, Rpts of 2 Jan, 29 Feb, and 15 May 52 (in G–4 files, Headquarters Marine Corps); Rpt of Test (Project 671) by MCEB, Quantico, Va., 3 Jan 1952; LtCol G. A. Hardwick, ltr of 30 Jun 1954; LtCdr F. J. Lewis (MSC) USN, ltr of 21 Jun 1954.

[10] ACofS, G–4, "Instructional Information, Vest, Armored, M–1951," 5–6.

It was a close squeak but the first 500 vests reached Korea with only a few days to spare. Captain David R. McGrew, Jr. accompanied the shipment as project officer with a mission of supervising and observing the use made of the M–1951 in action. His first letter to Headquarters Marine Corps, dated 4 February 1952, commented that "up to tonight we have had nine men hit while wearing the vest. One was killed outright as a 120mm mortar round landed right in his lap. However, the other eight showed excellent results. All of the eight were wounded in other places not covered by the vest—but they are all WIA instead of KIA." [11]

Captain McGrew cited the instance of a Pfc of the 2d Battalion, 7th Marines, wounded by the explosion of an 82mm mortar shell only 15 feet in front of him. He received several fragments in the face and his leg was fractured. But there were some 45 holes in his vest, without any penetrations. Fifteen of the fragments had been large enough to inflict mortal chest or abdomen wounds.

The 500 vests were issued only to troops in particularly hazardous situations, such as patrols to the enemy lines. Upon returning from a patrol or raid, the wearers turned in their armor to be worn by other Marines under fire.

"The reaction of the user to the vest," reported McGrew, "is closely related to the amount of enemy activity. In sectors of the OPLR and MLR [outpost and main lines of resistance] where heavy incoming mortar and artillery fire was received, there were no complaints regarding the weight or restrictive features of the vest. In other sectors where there was little or no enemy activity, approximately 15 percent of the personnel complained that the vest was heavy and restricted movement to some degree. Approximately 2 percent of the wearers in these sectors thought the vest was not worth the trouble and would wear it only when ordered to do so." [12]

The project officer believed that a "significant reduction" in KIA casualties could be credited to the M–1951, but that WIA figures were only slightly lessened. That was because so many wearers were wounded who would have been killed save for the armor. Captain McGrew listed the following case histories, confirmed by medical officers:

> Men who would have been killed instead of wounded if they had lacked armor protection—23:

[11] Capt D. W. McGrew, Jr. to LtCol G. W. Hardwick, ltr of 4 Feb 52.
[12] ACofS, G–4, "Report of Field Test of Armored Vest, M–1951," 15 May 51.

Men who had potentially severe wounds reduced to superficial wounds —29;

Men who had superficial wounds prevented altogether—31.

The project officer had no opportunity to compare the casualties of vest wearers with those of an equal number of unprotected Marines taking part in the same action. It was his conclusion, based on observation, that "use of the vest by all personnel who are habitually forward of battalion command posts may result in as much as a 30 percent reduction in battle casualties. Because many WIA cases are the result of wounds of the extremities and/or multiple wounds, there probably will not be a large reduction of casualties in this category. It is believed that the largest reduction will occur in the KIA category and that this reduction will be substantial." [13]

The introduction of body armor was not heralded in the press by page one headlines such as had announced the first transport helicopter operations in Korea. Occasionally a photograph on page eight showed a Marine grinning triumphantly while pointing to a hole in his armored vest and holding aloft the jagged mortar fragment that might otherwise have killed him. But it is safe to say that a majority of Stateside newspaper readers and radio listeners in 1951 were unaware of the Marine revival of armor adapted to 20th-century warfare.

Press correspondents in Korea did not appear to grasp the tactical significance of an innovation which they regarded entirely as a humanitarian achievement. From a strictly military viewpoint, however, it was apparent that if the M–1951 could reduce casualties by 30 percent, as Captain McGrew estimated (and his estimate was later regarded as conservative), it would mean that a like reduction had been effected in the destructive potential of the enemy's best antipersonnel weapons. It was as if the Marines were able to slip behind the enemy's lines and silence 3 out of 10 of his howitzers, mortars, burp guns, and grenades.

This was of particular importance in overcoming the numerical superiority of the Communists. Not only did each American wound casualty reduce the effectiveness of a unit, but four or more comrades were often neutralized as stretcher bearers in Korean mountain terrain. If body armor could prevent 3 casualties out of 10, therefore, it

[13] *Ibid.*

would be a significant addition to a unit's numerical strength as well as combat morale.

Any doubts about Marine troop acceptance of the M–1951 were laid to rest by the approval of the 500 vests issued early in February 1952. An additional 2,500 arrived early in March and on the 13th of that month the Division ordered 25,000 more. The armored vest, like the thermal boot, had needed only a thorough trial to become standard equipment.

Helicopter Operations MULETRAIN *and* CHANGIE-CHANGIE

The combat helicopter, oldest of the three Marine tactical innovations in Korea, had already managed to make routine performances out of operations that once claimed headlines. Battalion troop lifts were no longer a novelty, and supplying a front-line company by air was taken for granted. But nothing quite as ambitious as Operation MULETRAIN had ever been attempted—the mission of completely supplying a battalion on the MLR for a week with a daily average of four helicopters.

Hill 884 was again the objective. Colonel Keith B. McCutcheon's HMR–161 was given the task of flying tentage, stoves, rations, and ammunition from supply dumps to the 1st Battalion of the 1st Marines, commanded by Lieutenant Colonel John E. Gorman.

It was the first opportunity for HMR–161 to try out improvements in helicopter "flying crane" techniques credited to Major Charles E. Cornwell. He had adapted the underslung nets, controlled manually from the cabin, which did a better job than the pallet, or portable platform, for many types of cargo.

An average altitude of 2,300 feet for the five landing places made it necessary to reduce the payload to 850 pounds. Yet HMR–161 handled the assignment during the first week of 1952 with about one-third of its aircraft while the remainder went about routine chores. So well did four helicopters keep ahead of schedule that sometimes they flew in more cargo than could be immediately unloaded at the objectives. Following are the statistics of the seven days:

> Pounds lifted, 150,730; Hours of flight time, 91.7; Loads lifted, 219; Average of miles flown, 9.6

Three days later, Operation CHANGIE-CHANGIE began on 10 January 1952. Like Operation BUMBLEBEE three months earlier, this was a battalion relief lift. Yet it differed from its predecessors in that troops were to be flown from Field X–83 to sites on the company instead of battalion level, the former being only 200 yards behind the front line.[14]

In December the loading zone and landing site duties formerly assigned to a platoon of the 1st Shore Party Battalion, were taken over by the 1st Air Delivery Platoon, Service Command, FMFPac. First Lieutenant William A. Reavis and 35 enlisted men had a mission "to prepare and deliver supplies by air, whether by parachute, air freight, or helicopter." These specialists were in charge during Operation CHANGIE-CHANGIE when the 2d Battalion, 7th Marines (Lieutenant Colonel Edward G. Kurdziel) relieved Lieutenant Colonel Norton's 1st Battalion, 5th Marines. The operation was conducted smoothly by helicopters flying in defilade throughout the approach, landing, and return phases.

Operation MOUSETRAP, from 14 to 17 January, was planned primarily as a test of the ability of HMR–161 to launch an antiguerrilla attack on short notice. Colonel McCutcheon and Lieutenant Colonel Mitchell were alerted at 0100 in regard to a two-company lift scheduled for 1000 that same morning. With "only minor difficulties" they transported 500 Marines to a landing site cleared by the Air Delivery Platoon. Three similar troop movements were completed by HMR–161 during the next three days.

If ever a bronze plaque is awarded in commemoration of the first history-making helicopter troop and supply lifts, it would be fitting to install it on Hill 884. That bleak and roadless height had its fifth large-scale operation on 24 February when Lieutenant Colonel Harold C. Howard's 1st Battalion, 7th Marines, relieved the 2d Battalion, of that same regiment on "Mount Helicopter." Operation ROTATE was completed without incident as further evidence that battalion reliefs by helicopter were now routine.

In spite of the demands made upon HMR–161 helicopters in cold weather and mountainous terrain, it is noteworthy that no serious

[14] Sources for the helicopter operations described in this section are the following: HMR–161, *HD*, Jan and Feb 51; *Cavalry of the Sky*, 176–175. Veterans of the Korean conflict will recall that "changie-changie" meant "swap" in the pidgin English serving as a conversational medium between Americans and Orientals. Hence it was applicable to a relief operation.

mechanical defects had developed. This six-month record came to an end on 24 February 1952 when Captain John R. Irwin was returning from Seoul to X–83. Warned by alarming vibrations, he landed to discover that the broken remnants of the tail assembly had dropped behind him in the snow.

Four days later, while flying a load of logs for bunkers, Captain Calvin G. Alston's aircraft was so shaken by vibrations that he suspected damage from enemy artillery fragments. He made a forced landing in the snow only to discover another instance of a tail assembly breakdown.

Colonel McCutcheon grounded all HMR–161 aircraft until the trouble could be corrected. Not until 14 March, after 16 modified tail assemblies had been flown to Korea did the Marine transport helicopter squadron take to the air again.

The Five Days of Operation CLAM-UP

Ground operations continued with little change during February and the first two weeks of March. The only departure from the well-worn tactical norm came on 10 February, when EUSAK put Operation CLAM-UP into effect across the entire UN front.

The purpose was to feign a withdrawal and lure the enemy into sending out patrols which would yield prisoners to Eighth Army units. A EUSAK letter of instruction, dated 4 February 1952, asserted that "a policy of aggressive patrolling has led the enemy to rely upon our patrols for the maintenance of contact. This situation enables him to maintain contact without subjecting his troops to the hazard of capture or casualty." [15]

All corps were directed to ". . . attempt to decoy the enemy into dispatching patrols against our lines and ambush and capture such patrols."

First Marine Division orders called for an elaborate series of deceptions. Immediately prior to CLAM-UP, on 9–10 February, the 11th Marines fired 471 harrassing and interdiction missions, as if to cover a large-scale withdrawal. Over 12,000 artillery rounds were expended.[16] Then CLAM-UP commenced, and the three regiments on the

[15] This section, unless otherwise specified, is based upon the 1stMarDiv *HD*, Feb 52, 1–12; and *PacFlt Interim Rpt* No. 4, 9–11 to 9–14.

[16] 11thMar *HD*, Feb 52, 13; Col B. T. Hemphill comments, 20 Jan 59.

MLR—from left to right, the KMCs, 1st Marines, and 7th Marines—did their part to hoodwink the enemy. Reserve battalions executed daylight marches on foot to the rear and returned after dark by means of motor lifts. The 5th Marines, in Division reserve at Camp Tripoli, executed similar feigned withdrawals.

After the Marine cannoneers completed their supposed covering fires, the front was plunged into an eerie silence. It did not take long, of course, for the enemy's curiosity to be aroused. NKPA patrols reconnoitred the Marine lines on the night of 10–11 February without being fired upon. The following night a patrol attempted to draw Marine fire in the Hill 812 area by advertising its presence with loud talk. The enemy's fire was not returned until the patrol attacked a Marine position with white phosphorous grenades. In sheer self-defense the Marines retaliated, and the North Koreans made a hurried exit, leaving behind 10 dead and 2 wounded men who became prisoners.

At first light on the 12th another enemy patrol tried to penetrate the wire in front of a 1st Marines position and paid the penalty with nine men killed and three wounded in a 15-minute fire fight.

On 13 February the Marines were pounded with the month's heaviest concentration of NKPA fire—344 artillery and 1,469 mortar rounds. Thus did the enemy serve notice of his realization that Marine positions on the MLR were being held in strength. NKPA patrol actions on the nights of the 13th and 14th were launched at Marine trenches on Hills 812 and 854 at the estimated cost of heavy casualties.

When Operation CLAM-UP came to an end on 15 February, it had admittedly fallen short of EUSAK expectations. Although NKPA patrol losses had been considerable, they were offset by fewer casualties in rear areas enjoying a five-day immunity from UN artillery fire. Worse yet, the enemy was enabled during this period of grace to bring up ammunition and other supplies without interference. As a final disillusionment, it was reckoned that across the whole Eighth Army front the Communists had lost fewer prisoners than during the preceding five-day period.

In the Marine combat zone a gain was recorded in enemy casualties. General Selden congratulated the Division on "the fire discipline practiced by MLR troops and by platoon and company commanders. As a consequence of the fire discipline, the line companies were able to kill 56 enemy and wound 54." These totals, it was pointed out, were larger than the losses normally inflicted on the enemy in a five-day period.[17]

On the other hand, five deserters from the mortar company of the 1st Battalion, 91st Regiment, 45th NKPA Division revealed that advantage had been taken of Operation CLAM-UP by detailing mortar personnel and men from the rifle companies to carry ammunition. During the five-day lull, according to the prisoners, 2,600 rounds were brought up for the company's nine mortars.[18]

After the brief flurry of Operation CLAM-UP the front quickly settled down to its old routine of patrols. An average of eight Marine night ambush patrols and five daylight reconnaissance patrols forward of the MLR was maintained. The results left much to be desired. Of the last 110 ambuscades and 75 reconnaissance patrols reported in February, only 1 of the former and 6 of the latter claimed contacts. All but one of the contacts had negligible results.

The Marine fire attack did the enemy more damage. Artillery fired 679 observed missions during the month—211 on troops, 175 on bunkers, 121 on mortars, 96 on artillery, and 75 on such miscellaneous targets as OPs, vehicles, machine guns, and supply points. This total was recorded in spite of an ammunition shortage which would ultimately become the subject of debate in Congress.

Even with supplies of ammunition limited by X Corps orders, Marine artillery drove the enemy from untenable forward-slope positions to underground fortifications on the reverse slope.

Naval gunfire was limited by the extreme range to the Division zone of action.[19] Only large targets forward and to the right of center could be taken under fire. Even so, the *Wisconsin* and the *St. Paul* scored some devastating hits in February on enemy reverse slope positions.

On one occasion, the *Wisconsin* erroneously calculated its deflection. Two 16-inch rounds landed between the front line and the 3/7 mortar positions before the fire could be stopped. Fortunately, no one was injured. The *Wisconsin* Marine officer happened to be visiting the Division CP that day, and on hearing the news he came up to 3/7 and

---

[17] 1stMarDiv *HD*, Feb 52, 3.

[18] 1stMarDiv *PIR* No. 486, Feb 52.

[19] The battleship *Wisconsin* had a main battery of 16-inch guns with a maximum range of about 23 miles. The heavy cruiser *St. Paul* had a main battery of 8-inch guns with a maximum range of 16 miles.

collected a large shell fragment. He stated that he intended to mount the jagged piece of steel in the ship's CIC room as a reminder to future gunners to make no errors in plot.

Observed direct fire by the 90mm rifles of the 1st Tank Battalion (Major Walter E. Reynolds, Jr.) continued to be effective against NKPA bunkers and gun emplacements. Utilizing the high ground along the MLR, particularly on Hills 812 and 854, tanks sniped at the enemy both by day and night.

This was made possible by the powerful lights of a platoon from the 92d U.S. Army Searchlight Company, attached to the 11th Marines. The mountainous terrain in East Korea was not particularly suited to "artificial moonlight"—the indirect illumination of a large area which results from "bouncing" the rays of searchlights off low-lying clouds. But direct illumination permitted aimed 90mm fire in the darkness and had the further advantage of blinding the enemy to the tanks themselves as well as to troop movements behind them. Not a single light was shot out during the winter in spite of persistent NKPA attempts.

The lessons taught by battlefield illumination in Korea were to be incorporated into two instructive bulletins after the war. "The enemy does *not* have any better night vision than we do," asserted USMC Landing Force Bulletin No. 6. "No racial or national group of people has any inherent physical advantage over another as to capability for seeing in darkness. . . .[20] The apparent advantage which the enemy sometimes displays in night operations is due only to a difference in training. In the case of the Oriental soldier, or the Eskimo, for example, training usually begins early in life, where he does not have the convenience of artificial light to the degree we have, and has been forced to make maximum use of his natural night vision in many of his normal activities.

"U.S. Forces have conducted many successful night operations after adequate training. Some units have reported that after intensive night training, personnel have become so proficient that they sometimes prefer night operations to daylight operations."

In support of this conclusion, records for the winter of 1951–1952 reveal that the Marines held their own very well in the night combats

[20] U.S. Marine Corps Landing Force Bulletin No. 6, "Night Vision and Night Combat," 5 Dec 53. See also Bulletin No. 18, "Battlefield Illumination," 4 Jun 56.

of no man's land, where the outcome depended upon immediate decisions based upon seeing in the dark.

Marine casualties for February, the last full month in East Korea, were 23 KIA, 102 WIA, and 1 MIA, including the KMC Regiment. Enemy losses were reported as 174 counted and 381 estimated KIA, 606 estimated WIA, and 63 prisoners.[21]

After a winter of positional warfare, the Marines could recall with better understanding the tales their fathers had told them about France in World War I. For history was staging one of its repetitions; and, allowing for improvements in weapons, the trenches of Korea in 1951–1952 differed but slightly from the trenches of the Western Front in 1917–1918.

[21] 1stMarDiv *HD*, Feb 51, App No. 5. Other sources for this chapter are comments and criticism by the following officers: (Ranks listed are those held at time of interview or comment.) Gen G. C. Thomas; LtGen J. T. Selden; BrigGen S. S. Wade; BrigGen C. R. Allen; Col J. H. Tinsley; Col F. B. Nihart; Col J. F. Stamm; Col B. T. Hemphill.

CHAPTER XII

The Move to West Korea

Truce Talks—Tactical Innovations—The Marines in Operation MIXMASTER—*Operations of Fifteen Months in Retrospect*

NO CHRONICLE of activities in Korea would be complete without a discussion of the truce talks which began in the summer of 1951. When the Communists proposed these meetings early in June, their motives were transparent; they were hurt, staggering, and badly in need of a breathing spell. Pretending a sudden interest in peace, the hard-pressed enemy requested talks at Kaesong for the purposes of recuperation.

The enemy would never admit the real damage he suffered. A typical excuse for the smashing CCF defeat was given in a book by Wilford G. Burchett, an Australian Communist who was a press correspondent behind the Chinese lines.

"Immediately prior to the beginning of the talks," he explained, "the Korean-Chinese troops had withdrawn extensively along the East Coast, hoping to entice the Americans as deep as possible into a trap which would be sprung and would cut them off by an encircling move. The Americans were seriously nibbling at the bait when the proposal for cease-fire talks was made. The line was immediately frozen and Korean-Chinese troops started to dig in." [1]

This beginning of static warfare was unquestionably the great turning point of a war whose course from that time on was to be decided at the conference table of Kaesong and later Panmunjom. Any doubts as to the actual motives of the Communists might have been dispelled

[1] Wilford G. Burchett: *This Monstrous War* (Melbourne, 1953): J. Waters, 121–122. Burchett was a Communist free lance correspondent for left-wing newspapers. He wrote several books and articles lauding the Communist cause in the Korean War.

634040 O–62–19

upon reading in Burchett's book this naive boast of the advantage taken of the truce talks by the Reds:

> Digging in is an understatement of the way the Korean-Chinese troops literally burrowed into the mountains, constructed two and three story dwellings underground, linked mountains and hills by underground tunnels and carved deep communication trenches linking flank with flank and front with rear. They raked the insides out of mountains as you would rake ashes out of a furnace. Each hill, mountain or ridge was connected with its neighbors by deep, zig-zagged inter-communication trenches, at least two yards below ground level and with yard-high antiblast walls. In emergency, troops could be switched from hill-top to hill-top with the enemy never knowing. Similar trenches extended well to the rear, so that supplies could be brought up and withdrawals if necessary made in comparative safety. . . . Everything was deep underground with many yards of rock and earth between them and shells and bombs, atomic or otherwise. Back of the front line positions, similar scooped-out mountain ridges stretched all the way back to Pyongyang and further. It was against these positions that Van Fleet began hurling his troops in August, 1951.[2]

The breathing spell provided by preliminary truce talk discussions gave the Communists an opportunity they had not previously enjoyed. Not only did they have time to prepare sturdy and effective entrenchments, but they were able to bring up additional mortars and artillery to equal those of the Allied forces. As a further advantage, while "free from the compulsion of impending military disaster," [3] they made use of the interlude to reorganize and train NKPA divisions to a new and increased level of effectiveness.

Communists are never embarrassed in the least to deny an agreement already reached, and once having accomplished their intermediate goal, the Red delegates broke off the Kaesong talks for a while. Once the pressure on them was reduced, the enemy was in a position to try to obtain the most favorable terms for armistice talks, even if it meant prolonging the fighting.

The change in tactics soon became apparent. "Since the opening of the Kaesong conference," commented a FECom G–2 report, "the enemy has deviated from his usual tactics of 'flexible defense' which he so skilfully employed during the buildup period prior to all his past offensives—to that of a more orthodox 'fixed defense.' Where the enemy in the past has defended key terrain features with relatively

[2] *Ibid.* General Van Fleet did not "hurl" his troops against anything. He began limited offensives for the purpose of improving Eighth Army morale and maintaining offensive spirit. See Gen James A. Van Fleet, ltr of 28 Feb 59.

[3] C. Turner Joy, *How Communists Negotiate*, 28.

small groups to delay friendly forces, he has now changed over to tactics of a fixed line of defense to be defended at all costs." [4]

"The most extended delay imposed upon the Korean Armistice Conference by the Communists was in connection with the exchange of prisoners of war," [5] which subject will be discussed in Volume V of this series. The United Nations contended that all prisoners should be "screened" to determine whether they wished to return to their side of origin. No prisoner was to be returned against his wishes. The Communists claimed this treatment consisted of a reign of terror in which CCF prisoners were held at gunpoint.

Some prisoners held in UN camps rioted and injuries and deaths resulted. This provided the Communists with excellent propaganda on which to denounce our principles of no forced repatriation.[6] In the end, after a delay of more than 14 months of war, the Communists finally did accept this principle, and an armistice was achieved.

The Communist delaying tactics were not entirely without benefits to the Allied forces, for the major part of the 1st Marine Division had the opportunity to go into reserve and engage in several weeks' intensive training. While the 1st Marine Aircraft Wing was busily participating in the interdiction activities of Operation STRANGLE, General Van Fleet and his ground commanders felt frustrated over their orders to "sit tight" rather than attack and prevent further enemy buildup.

An agreement to resume cease-fire talks, this time at Panmunjom, led to a EUSAK order which committed the 1st Marine Division and other major units to a defensive stand behind a fixed line of demarcation on 20 September 1951 (Map 19). Further negotiations resulted in a month's lull which was brought about by the fact that the delegates could not agree on where the lines would remain if the fighting stopped. The United States delegates pressed for a settlement within a 30-day period. The Communists continued to stall. The United States then consented to accept the present (then current) demarcation line if the Communists agreed within the 30-day period.[7]

The significance of these dates was to become more and more plain as the conflict dragged on into 1952 with both sides on the defensive, limiting themselves to the raids and patrols of positional warfare

[4] FECom G–2 Intelligence Summary, 18 Sep 51.
[5] Joy, *How Communists Negotiate*, 53.
[6] *Ibid.*
[7] Col J. C. Murray, Comments, Jan 59.

while the appointed representatives haggled for a truce. Although the Marines did not realize it, the war had already turned into a contest of watchful waiting and fierce local fights.

This line of demarcation left the Eighth Army holding a MLR across one of the narrowest parts of the peninsula (Map 22). Just behind the Communist MLR the peninsula bulged to the west. This meant that the enemy had to devote much of his effort to mining the waters and defense of many beaches against a surprise amphibious attack, and it necessitated keeping in operation long and vulnerable supply lines.

It is probable that a UN breakthrough or successful amphibious operation could have been mounted at this time,[8] for several high ranking officers expressed such opinions. All the necessary ingredients were available, yet the high level decision for such an operation was not made.

Tactical Innovations

Until World War II, it had been a deserved reproach throughout the brief history of our country that Americans were never prepared at the outset of a war. A welcome departure from this tenet came in 1942 when the Marine Corps and Navy introduced the new amphibious tactics they had developed during the 1930s. Victory in the Pacific War was due in large measure to the techniques, landing craft, and vehicles of the Navy-Marine Corps ship-to-shore attack.

As a result, North Africa, Europe, and the Japanese-occupied islands of the Pacific were opened to invasion without a single major reverse. In contrast, Hitler's *Wehrmacht* lacked both the techniques and equipment to launch a cross-channel attack on England in 1940, and Operation SEA LION was of necessity abandoned by an army that dominated the rest of Europe as a result of victories in land warfare.

Again, in Korea, the Marines demonstrated their foresightedness by taking a prominent part in the development of such important innovations as combat helicopters, body armor, and thermal footwear.[9] By the first month in 1952 the combat helicopter had proved to be of immeasurable assistance in modern warfare. In the beginning of the Korean War the "chopper" was initially used for command and liaison flights and reconnaissance missions. Evacuation of casualties

[8] BGen V. H. Krulak, Comments, Jan 59.

[9] Previous chapters discuss the background and development of these innovations.

and rescue missions also became routine duties, and within a short time the helicopter became the favorite "workhorse" for a variety of tasks. In September of 1951 tactical troop movements began. These operations made newspaper headlines everywhere.

Of greater tactical importance, at least in the opinion of the front-line rifleman, was the physical protection provided him. The armored vest and the new thermal boots were first tested by Marines late in 1951 and soon came to be highly desired items of equipment.

The fighting men in Korea would not disagree with Benjamin Franklin's statement that "there never was a good war," but modern inventions certainly improved conditions by providing for the safety and comfort of the fighting men. Marine transport helicopters and body armor were of particular importance because they added to the human resources of UN forces opposed by an enemy with a contempt for life, based on seemingly endless reserves of manpower. UN commanders in their fight against the Communist forces could not recklessly expend lives as did the enemy; therefore, the Allies had need of tactical innovations and life-saving devices in order to compensate for a lack of numbers.

The Marines in Operation MIXMASTER

In the spring of 1952, when the UN and Communist forces were facing each other from static positions and fighting local engagements, Operation MIXMASTER took place. MIXMASTER was a complicated rearrangement of UN divisions across the entire Korean front during March, and involved the shuffling of about 200,000 men and their equipment over distances from 25 to 180 miles. It was a severe test of Eighth Army mobility.[10]

General Van Fleet visited the 1st Marine Division CP on 12 March 1952, and announced an important command decision. After six months of defensive warfare in the same sector along Line MINNESOTA (20 September 1951 to 16 March 1952) the Division was to move across the peninsula to West Korea.

The Marines had orders to relieve the 1st ROK Division and take over a sector at the extreme left of the Eighth Army line under the operational control of I Corps (Map 22). There they would have the

[10] Col B. T. Hemphill, Comments, 30 Jan 59.

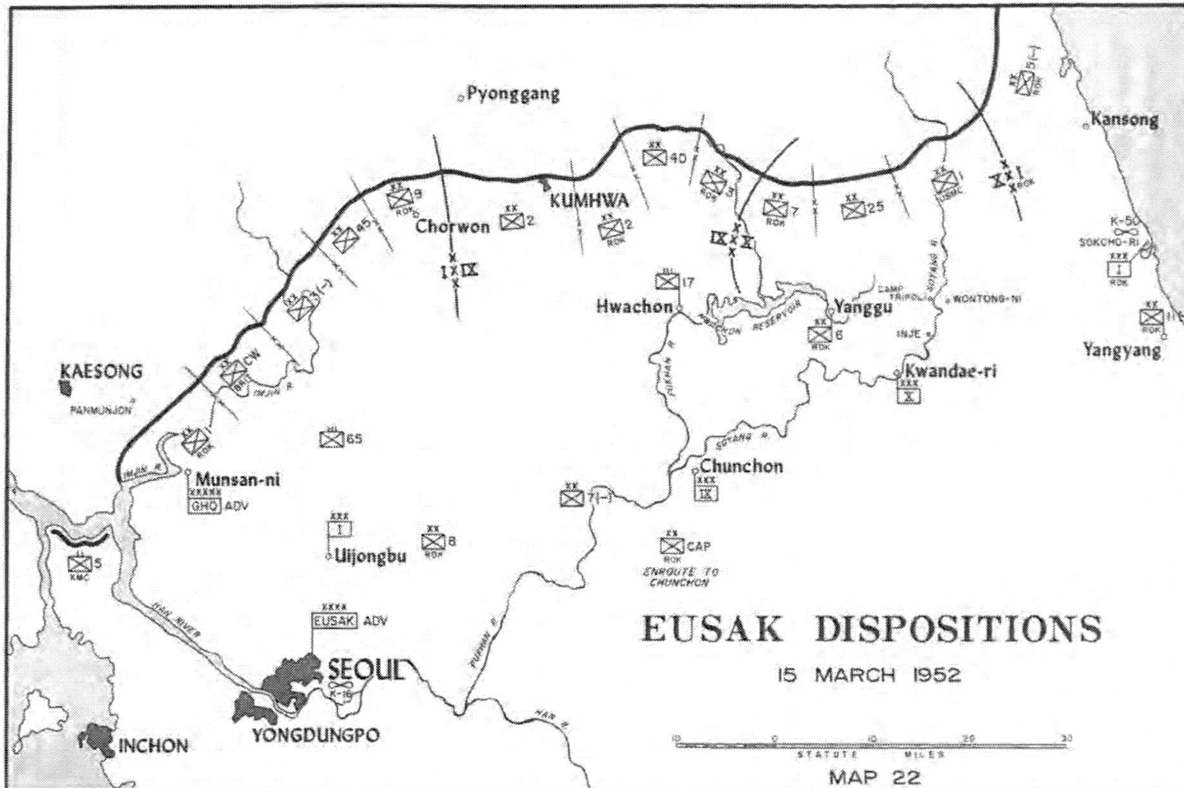

EUSAK DISPOSITIONS
15 MARCH 1952
STATUTE MILES
MAP 22
Pyonggang
Kansong
KUMHWA
Chorwon
Hwachon
Yanggu
Kwandae-ri
Yangyang
KAESONG
PANMUNJOM
Munsan-ni
GHQ ADV
Uijongbu
EUSAK ADV
SEOUL
YONGDUNGPO
INCHON
Chunchon
ENROUTE TO CHUNCHON
SOKCHO-RI
K-50
WONTONG-NI
CAMP TRIPOLI
INJE
HAN RIVER
IMJIN R.
PUKHAN R.
SOYANG R.
HAN R.

responsibility for blocking Korea's historic invasion route to Seoul. The reasons behind this EUSAK decision were summarized in the 1st Marine Division report as follows:

> (1) The abandonment of plans to carry out an amphibious envelopment somewhere on the east coast;
> (2) Concern over weaknesses in the Kimpo area defenses;
> (3) The overall situation would not permit loss of ground on the EUSAK left (South Korea) as this would endanger the capital at Seoul; that if retraction of lines was necessary, territory could better be sacrificed on the right (North Korea) where the country was mountainous and had little economic or strategic value.[11]

Up to this time the four corps of the Eighth Army had defended a 125-mile front across the peninsula (Map 22) with the following units in line from left to right on 15 March 1952.

I CORPS—ROK 1st Division; British Commonwealth Division; U.S. 3d Infantry Division (–); U.S. 45th Infantry Division (Oklahoma National Guard); ROK 9th Division. In reserve were the ROK 8th Division and RCT–65 of the U.S. 3d Infantry Division.

IX CORPS—U.S. 2d Infantry Division; ROK 2d Division; U.S. 40th Infantry Division (California National Guard); ROK 3d Division. In reserve were the U.S. 7th Infantry Division (–), RCT–17 of that Division, and the ROK Capitol Division.

X CORPS—ROK 7th Division; U.S. 25th Infantry Division; U.S. 1st Marine Division (including 1st KMC Regiment). In reserve was the ROK 6th Division (–).

I ROK CORPS—ROK 5th Division (–). In reserve was the ROK 11th Division (–).[12]

Allowing for a few changes, these were the positions held by major EUSAK units through the winter of 1951–1952.

The Marine move was launched by Division Operation Plan 2–52 and provided that the 1st Marine Division would be relieved by the 8th ROK Division as a preliminary to movement overland and by sea to the relief of the 1st ROK Division and defense of Line JAMESTOWN in the I Corps sector in the west. According to verbal orders later confirmed by EUSAK OI 272, transportation by truck and ship was specified, and the move was to be completed prior to 1 April.[13]

Obviously such a transplacement—moving entire divisions great distances from one sector of the MLR to another—necessitated careful

[11] 1stMarDiv *HD*, Mar 52, 1–2.
[12] EUSAK *Cmd Rpt*, Mar 52, 13–14.
[13] Sources for this section are 1stMarDiv *HD*, Mar 52, 9–10; 1st MT Bn *HD*, Mar 52; 7th MT Bn *HD*, Mar 52.

timing and close coordination, but the planners involved were equal to the task. In referring to detailed plans by the Division G–3 Section (Lieutenant Colonel Gordon D. Gayle) and the G–4 Section (Colonel Robert A. McGill), several unit commanders expressed the opinion that "the move from east to west was a masterpiece of logistical efficiency with no unnecessary paper work and no undue harrassment." [14]

In addition to transporting the Division, the arrival of replacements and departure of personnel to be rotated to the United States were smoothly coordinated into the over-all plan. The transport *General W. H. Gordon* anchored at Sokcho-ri on 16 March with 174 officers and 1,135 enlisted men of the 18th Replacement Draft. The newly arrived Marines scarcely had time to drop their seabags before they joined the motor march to West Korea. The *Gordon* departed with 103 officers and 1,135 Marines homeward bound, and the 2d Logistical Command (Army) received a 1st Marine Division request to route the 19th Replacement Draft, due in April, to Inchon instead of Sokcho-ri.

At K–50, near Sokcho-ri on the east coast, air freight and passenger service was discontinued and diverted to the new Division airhead, K–16, at Seoul. The Division railhead was changed to Munsan-ni (Map 22).

The first Marine unit to depart for West Korea was the KMC Regiment with its organic battalion of artillery. Since the artillery had to be moved and repositioned all across the front with as little interruption as possible in overall support available at any one time, the 11th Marines CO planned to move his battalions directly into their new firing positions. This was preceded by an initial detailed reconnaissance.

Elements of the U.S. 25th Infantry sideslipped to the right and assumed responsibility for the Marine sector on the 17th (Map 22), and the KMCs and the 1st Battalion, 11th Marines moved into their new positions on 18 March. The other artillery battalions followed at two-day intervals, all battalions firing from their new positions by 24 March.

The movement of the 1st Armored Amphibian Battalion (less Company A), commanded by Lieutenant Colonel John T. O'Neill, was an unforgettable experience. Embarking on LSTs manned by a skeleton Japanese crew, the vessels headed for the Kimpo Peninsula.

[14] Col T. A. Culhane, Jr., Comments, 4 Mar 59, and others.

The weather was squally and foggy throughout, and the ships were completely blacked out at night with no facilities for emergency transmission of messages. There were many navigational hazards, but in spite of this, and the lack of adequate navigational equipment, the LSTs arrived at their destination without incident.

Two days later, on 20 September, the 1st Tank Battalion and the antitank companies of the three infantry regiments also took the sea route to the new Division area in the west.

Division Operation Order 8–52, dated 18 March, directed the 1st Marines to proceed by motor march from the Division reserve area at Camp TRIPOLI to the new Division area east of Munsan-ni, and there to move into front line positions. The 7th Marines, after being relieved on the 20th by elements of the 8th ROK Division, assembled at Camp TRIPOLI and moved by truck to West Korea. Colonel Austin R. Brunelli, who had replaced Colonel Custis Burton, Jr., as chief of staff, moved the forward CP personnel and prepared the new Division command post.

After being relieved by the ROKs on the 23d, the 5th Marines departed their east coast area. Two days later the regiment arrived in the Munsan-ni area behind the 7th Marines and the remaining elements of the artillery regiment.

The 5th Marines had originally been scheduled to occupy reserve positions on the Kimpo Peninsula, but plans were changed en route. The commanding general and his G–3 were appalled at the Division sector's width, and after General Selden had a chance to inspect the areas to be defended and talk over the situation with the commanders of the 1st and 7th Marines (Col Sidney S. Wade and Col Russell E. Honsowetz), he decided that the 5th Marines should go into the line.[15]

A few hours after the 5th Marines convoy left the east coast on their 140-mile trans-Korea move, helicopters picked up the regimental and battalion commanders from their respective vehicles in the convoy and took them to the new Division CP. There they were assigned new defensive sectors and immediately reconnoitered the ground while awaiting the arrival of their units. By the time the regiment arrived, all preparations were made for them to move into positions and relieve a portion of the thinly stretched line of the 1st Marines.

It had been a busy week for the 1st and 7th Motor Transport Battalions, commanded respectively by Lieutenant Colonel Howard E.

---

[15] *Ibid.*

Wertman and Major Herbert E. Pierce. Two hundred Division trucks and a like number of U.S. Army vehicles made up the long columns that shuttled back and forth across the peninsula. The plan provided for moving an infantry regiment every third day. For the drivers this meant a 140-mile trip, a return trip the following day, and a one-day layover for maintenance before commencing the new cycle. The artillery battalions, by order of X Corps, were retained until the latest possible date.

The statistics of Operation MIXMASTER are impressive. It took 5,716 truck loads and 80 DUKW loads to move most of the Division personnel, gear, and supplies. Sixty-three lowboys (flat-bed trailers) and 83 railroad cars were also utilized in addition to hundreds of jeeps and jeep trailers. Three LSDs and 11 LSTs sailed from Sokcho-ri to Inchon with the heaviest equipment.

During the previous winter a sizable number of prefabricated shelters had been set up for supporting and headquarters units. Since timber, logs, and salvage materials were in short supply, the 1st Marine Division moved large quantites of these materials to the west coast in order to live as comfortably as possible under static warfare conditions.

The operations of the 1st Marine Division in defense of the western sector of Line JAMESTOWN do not come within the scope of Volume IV. The account of Marine activities in the new sector, under the operational control of I Corps, will be discussed in the fifth and final volume of this series.

Operations of Fifteen Months in Retrospect

During 1951 the Korean War became a most unpopular military venture among Americans. As a consequence, letters and newspapers from home caused a certain amount of anxiety among citizen-soldiers in Korea. To counter any spirit of doubt which may have arisen, military leaders issued frank and honest replies to inquiring politicians.

The *esprit de corps* of Marines was high, and they were well aware of their purpose in Korea. One noted author, on spending a couple of days among front-line Marines during January of 1952, told a group of officers at the Division CP that he "was impressed with the morale of the Marines on the MLR." He stated that he "had been prepared to find that they didn't know what they were fighting for or why they were there." However, he was encouraged to find that they knew exactly their purpose in the Korean fighting.[16]

The period of nearly 15 months covered by Volume IV was at that time the longest stretch of land warfare ever experienced by a major Marine unit. Even during the numerous island-hopping campaigns of World War II, the periods of combat were relatively brief for each.

Glancing back over the year 1951 with the benefit of hindsight, it is evident that Marine "uncommon valor" during this period was supplemented by such outstanding innovations as helicopter-borne assaults and lightweight body armor, concepts brought to fruition by the pressure of combat.

It is also apparent that Marine training, both for officers and enlisted men, paid off handsomely under the demands of practically every type of land warfare. The Division chalked up a commendable record of service fighting on the east-central front. Since the UN commander desired to have EUSAK's only amphibious trained and equipped division near a coast offering a suitable selection of landing beaches, the Division was originally positioned in the east. Not since the Inchon landing, however, had the Marines been employed in their specialty, amphibious assault.

Subsequent to the unprecedented Chosin Reservoir campaign of late 1950 the Division reorganized and refitted in South Korea near Masan. Then in January and February of 1951 came the prolonged guerrilla-hunting campaign (Map 5) some 60 air miles north of Masan. Division operations in this area covered more than 1,000 square miles.[17]

The mountainous terrain offered cover and concealment for the clandestine operations of far too many enemy groups. A solution to this problem was found in "rice paddy patrols"—groups ranging from a fire team to a squad in size which penetrated the mountain areas on foot to flush out small enemy bands. In retrospect, had one squadron of helicopters been available at that time, and its quick lift capabilities utilized, the increased mobility and surveillance would have made quite a difference in the conduct of the action.

Although land-based Marine air power had been under operational control of the Fifth Air Force during the Chosin Reservoir fighting, a verbal agreement allowed the 1st MAW commander to provide

[16] Col F. B. Nihart, Comments regarding author James Michener's visit to 1stMarDiv, ltr of 23 Mar 59.

[17] Gen O. P. Smith, USMC (Ret.), ltr of 28 Jan 59.

directly necessary support to the 1st Marine Division. At the same time, carrier-based Marine planes were flying on the west coast along with other Allied planes harrassing enemy traffic.

During the guerrilla hunt VMO–6 planes provided air support to the 1st Marine Division while Marine attack aircraft were busy elsewhere along the Eighth Army front. Marine pilots, operating under JOC control, felt frustrated because they were unable to provide the timely close air support desired by the infantry. The Marine viewpoint held that too many links in the Air Force system of control caused an excessive delay in bringing air power over the target. This system continued for the remainder of the year.

As an operation, the guerrilla hunt was merely a series of minor engagements, but it accomplished its purpose of clearing out most of the North Korean irregulars who had been a constant threat in the Eighth Army's rear. In addition, the numerous small patrols provided excellent training for the newly arrived replacements.

The Eighth Army seemed to gain new vitality under General Ridgway. On the 18th of February, when the general learned that the enemy was withdrawing, he ordered a limited offensive. Operation KILLER began three days later, and was followed by Operation RIPPER on 7 March. The purpose of these operations was twofold: (1) General Ridgway wanted to restore his army's fighting spirit after its two defeats during the 1950–1951 winter; and (2) he wished to keep the Chinese Reds off balance while they prepared for another Communist offensive.

For the Marines these two operations were an experience with a strictly limited offensive. The advance was "buttoned up" as major units paid close attention to lateral contact. As the advance continued in March and April, mud proved to be an adversary second only to a formidable enemy using delaying tactics, and the Division as a whole had a thorough workout in the logistics of the offensive under adverse conditions.

In early April the Division, as part of the Eighth Army, crossed the 38th parallel and continued the attack to the north, the purpose being to threaten the suspected enemy buildup for an offensive. EUSAK forces rolled onward while the enemy, using his roving defensive tactics, fought vigorously and withdrew.

The long-expected enemy counterblow fell on the night of 22 April and resulted in the 1st Marine Division bearing the brunt of a 48-hour attack (Map 10). This opening CCF assault in the IX Corps area of east-central Korea was intended to throw the Eighth Army off balance as a preliminary to aiming the main blow at I Corps in west Korea.

The CCF attack opened a hole in the MLR large enough for a major breakthrough, and the Communists apparently expected to exploit this success to the fullest. However, the Allied line pulled back, consolidated, and held, as the Division's reserve regiment was thrown in to stem the tide. As the Marine flank was refused, the units on the left found themselves facing to the west while stopping the enemy thrust. Slowly, trading space for time, the Marines contained the enemy attack while the entire Eighth Army line organized new positions.

The enemy effort ground to a halt in the east-central sector, and the Chinese Reds were contravened in their attempt to take Seoul by May Day. Surprise and impetus were lost on the western front when they struck several days later, only to be stopped with frightful losses after a few gains on regimental fronts. The Allied line now held firm.

The Division's war of maneuver had worked well in halting this round of the CCF offensive, but the Communists were far from finished. As 17 enemy divisions were still available to attack, the Marine division was shifted to the east on 1 May in preparation for an expected battle.

On the 16th of May the Chinese offensive again opened, with the enemy hitting more to the east than had been expected, and making a deep but narrow penetration near the coast. The Marines moved eastward, established blocking positions, and engaged fringe units of the drive. This allowed the right flank Army division to move farther east and brake the enemy's rush.

The enemy was dangerously overextended when the UN counterstroke hit him late in May. For a month the Eighth Army attacked and advanced, the Marines slugging ahead day after day in the X Corps zone of action. CCF casualties mounted high, and Marine veterans of only a few months of Korean service saw scores of enemy corpses left behind on the battlefield as the enemy withdrew northward.

This great UN counteroffensive netted prisoners all along the EUSAK front as remnants of CCF platoons and even companies threw down their arms. Marines captured their share. Upwards of 10,000 Chinese surrendered to the Allies in a 10-day period—more prisoners than had been taken up to this time.

As the Chinese withdrew northward they left determined NKPA troops behind. The 1st Marine Division moved slowly forward, fighting for every inch of ground. So fierce was the enemy's resistance that at times during June the division commander was forced to commit all four regiments (the KMCs included) in the attack at the same time in order to seize designated objectives. This was a modification of accepted tactical doctrine, necessitated by the situation.

Throughout March, April, and part of May, Marine pilots continued to provide close air support not only for the 1st Marine Division, but also for other Allied units as directed by JOC. From the beginning of Operation STRANGLE on 20 May this interdiction effort had first priority, and close air support to all infantry units was secondary. Difficulties in air-ground communication continued as radio frequencies were heavily burdened with traffic. Although the 1st Marine Division received a proportionate share of the few air support missions flown, the frustrating time lag between requests for air support and the arrival of planes on target continued into the next year.

Some planes were always available for front line support, although rarely ever enough according to infantrymen's opinion. When they had the chance, 1st MAW pilots viciously attacked the fleeing enemy to ease the way for advancing ground troops. During June the unrelenting pressure of combined air-ground attacks sometimes caused large groups of enemy to surrender. Marines also captured thousands of rounds of enemy ammunition and other equipment.

By the last week in June the Marines had entrenched themselves along the Division's assigned portion of the MLR and "caught their breath" after two months of hard fighting. In driving from the Hwachon Reservoir area to the Punchbowl, they had employed practically every weapon and tactic that could be used in an all-out offensive. The Division then settled down to stable positions for a while, and some units had the opportunity to go into reserve and train.

It was a recharged 1st Marine Division (the 5th and 11th Marines did not go into reserve during this period) which moved back into the lines at the end of August. The offensive which opened northeast of the Punchbowl on the 30th and lasted with few and brief interludes until 20 September was the equal of the June fighting in sustained ferocity. All four infantry regiments (including the KMCs) went up against seemingly impregnable opposition.

The enemy's "stubborn defense of strong positions and many well-placed log and earth bunkers was similar to the tenacious tactics of the Japanese in World War II," according to a Navy report. "His artillery and mortar fires were effective, his minefields continued to be hazardous for many weeks, and his ability to dig in and fortify his positions [was] always impressive." [18]

After the 20th of September the EUSAK commander ordered that no further offensives be launched and that the MLR be stabilized. This was a period of aggressive patrolling, local attacks for more advantageous pieces of terrain, and watchful waiting to determine the outcome of truce negotiations. In spite of Operation STRANGLE, enemy vehicular movements increased at the end of the year, but 1st MAW pilots continually attempted to provide more support for all the infantry divisions.

The mission of the 1st Marine Division at this time was to organize, construct, and defend its sector of the MLR, a front of more than 13 miles. Although there were heavy local skirmishes, during the latter months of 1951 and the first 3 months of 1952, no great offensive drives were launched. Essentially, the Marines were engaged in an aggressive defense of their positions until they moved to West Korea.

While all Marines were hoping that the conflict would soon end, there was no slackening of the customary vigilance. All hands remembered General Ridgway's words of the previous year, that it was ". . . a fight for our own freedom, our own survival . . . ," [19] and this was their creed.

These lines would have made a fitting epitaph for Marines who gave their lives in Korea. They had as worthy a cause as any fighting men of our history, for it had become increasingly plain since World War II that a stand must eventually be made against Communist encroachments. By going halfway around the world to fight the enemy on his own doorstep, Americans may well have spared themselves a more bloody and costly future struggle nearer to their own homeland if not actually on their own soil. The designs of Red China and Soviet Russia were unmasked in Korea, and the people of the United States awakened to their peril after neglecting the Nation's defenses since 1945. To that extent, therefore, the operations in Korea were a defeat for Communism.

[18] *PacFlt Interim Rpt* No. 3, 15–25.
[19] See Ridgway's Declaration of Faith, Chapter 1.

APPENDIX A

Glossary of Technical Terms and Abbreviations

ADC—Assistant Division Commander
AdmO—Administrative Order
AD—Douglas "Skyraider" single engine attack plane
AF—Air Force
AH—Hospital Ship
AirDelPlat—Air Delivery Platoon
AirO—Air Officer
AirSptSec—Air Support Section
AmphTracBn—Amphibian Tractor Battalion
AmphTrkBn—Amphibian Truck Battalion
ANGLICO—Air and Naval Gunfire Liaison Company
ArmdAmphBn—Armored Amphibian Battalion
AT—Antitank
AutoMaintCo—Automotive Maintenance Company
AutoSupCo—Automotive Supply Company
BB—Battleship
BLT—Battalion Landing Team
Bn—Battalion
Btry—Battery
BuMed—Bureau of Medicine and Surgery
C-47—Douglas Transport used by Air Force (same as R4D)
CA—Heavy Cruiser
CCF—Chinese Communist Forces
CG—Commanding General
CIC—Counter Intelligence Corps, USA
CinCFE—Commander in Chief, Far East
CinCPacFlt—Commander in Chief, Pacific Fleet
CinCUNC—Commander in Chief, United Nations Command
CL—Light Cruiser
CO—Commanding Officer
Co—Company
ComFltAirWing—Commander Fleet Air Wing
ComNavFe—Commander Naval Forces Far East
ComPacFlt—Commander Pacific Fleet
ComPhibGruOne—Commander Amphibious Group One
ComSeventhFlt—Commander Seventh Fleet
ComUNBlockandCortFor—Commander United Nations Blockade and Escort Force
CP—Command Post
CR—Command Report
C/S—Chief of Staff
CSG—Combat Service Group
CSUSA—Chief of Staff, U. S. Army
CTF—Commander Task Force
CTG—Commander Task Group
CVE—Escort Aircraft Carrier
CVL—Light Aircraft Carrier
DD—Destroyer
DE—Destroyer Escort
Det—Detachment
DOW—Died of Wounds
EmbO—Embarkation Order/Officer
EngrBn—Engineer Battalion
EUSAK—Eighth U.S. Army in Korea
FABn—Field Artillery Battalion (USA)

634040 O-62-20

FAC—Forward Air Controller
FAF—Fifth Air Force
FEAF—Far East Air Force
FECOM—Far East Command
F4U—Chance-Vought "Corsair" Single-Engine Fighter-Bomber
F4U–5N—Chance-Vought "Corsair" Single-Engine Night Fighter
F7F–3N—Grumman "Tigercat" Twin-Engine Night Fighter
FMFPac—Fleet Marine Force, Pacific
FO—Forward Observer
FragOrder—Fragmentary Order
Fum&BathPlat—Fumigation and Bath Platoon
GHQ—General Headquarters
Gru—Group
H&SCo—Headquarters and Service Company
HD—Historical Diary
Hedron—Headquarters Squadron
HO3S—Sikorsky Helicopter
HqBn—Headquarters Battalion
HQMC—Headquarters, U.S. Marine Corps
InfDiv—Infantry Division (USA)
Interv—Interview
ISUM—Intelligence Summary
JANIS—Joint Army-Navy Intelligence Studies
JCS—Joint Chiefs of Staff
JMS—Japanese Minesweeper
JSPOG—Joint Strategic Planning and Operations Group
JTF—Joint Task Force
KIA—Killed in Action
KMC—Korean Marine Corps
Ln—Liaison
LSD—Landing Ship, Dock
LSM—Landing Ship, Medium
LSMR—Landing Ship, Medium-Rocket
LST—Landing Ship, Tank
LSTH—Landing Ship, Tank-Casualty Evacuation
LSU—Landing Ship, Utility
Ltr—Letter
LVT—Landing Vehicle, Tracked
MAG—Marine Aircraft Group
MAW—Marine Aircraft Wing
MS—Manuscript
MedBn—Medical Battalion
MedAmbCo—Medical Ambulance Company (USA)
MIA—Missing in Action
MISD—Military Intelligence Service Detachment (USA)
MLR—Main Line of Resistance, the main front line
Mosquito—North American AT–6 "Texan" Trainer; Single Engine Plane used as Airborne FAC and Target Spotting
MP—Military Police
MRO—Movement Report Office
Msg—Message
MSR—Main Supply Route
MSTS—Military Sea Transport Service
MTACS—Marine Tactical Air Control Squadron
MTBn—Motor Transport Battalion
NavBchGru—Naval Beach Group
NavFE—Naval Forces Far East
NCO—Noncommissioned Officer
NK—North Korea(n)
NKPA—North Korean People's Army
N.d.—Date not given
N.t.—Time not given
O—Officer; Order
OCMH—Office of the Chief of Military History (USA)
OI—Operation Instruction
OpnO—Operation Order
OpnPlan—Operation Plan
OrdBn—Ordnance Battalion
OY—Consolidated-Vultee Single-Engine Light Observation Plane
PhibGru—Amphibious Group
PIR—Periodic Intelligence Report
PLA—People's Liberation Army
Plat—Platoon
POL—Petroleum, Oil, Lubricants
POR—Periodic Operation Report

POW—Prisoner of War
QMSubsistSupCo—Quartermaster Subsistence Supply Company (USA)
R4D—Douglas Twin-Engine Transport (Navy and Marine designation of C–47)
R5D—Douglas Four-Engine Transport
RCT—Regimental Combat Team
Recon—Reconnaissance
Reinf—Reinforced
RktBn—Rocket Battalion
RM—Royal Marines
ROK—Republic of Korea
R&O File—Records and Orders File
ROKA—Republic of Korea Army
ROKN—Republic of Korea Navy
Rpt—Report
SAC—Supporting Arms Coordinator
SAR—Special Action Report
Sec—Section
SecDef—Secretary of Defense
ServBn—Service Battalion
SigBn—Signal Battalion
SigRepCo—Signal Repair Company
SitRpt—Situation Report
SP—Shore Party
SMC—Marine Supply Squadron
TAC—Tactical Air Coordinator; Tactical Air Commander
TACP—Tactical Air Control Party
Tacron—Tactical Air Control Squadron
TADC—Tactical Air Direction Center
T–AP—Transport operated by MSTS
TBM—General Motors "Avenger" Single-Engine Torpedo Bomber. Also used for Utility Purposes.
TE—Task Element
T/E—Table of Equipment
Tel—Telephone Message
TF—Task Force
TG—Task Group
TkBn—Tank Battalion
Trk—Truck
T/O—Table of Organization
TU—Task Unit
UDT—Underwater Demolition Team
U/F—Unit of Fire
UN—United Nations
UNC—United Nations Command
URpt—Unit Report
USA—United States Army
USAR—United States Army Reserve
USAF—United States Air Force
USMC—United States Marine Corps
USMCR—United States Marine Corps Reserve
USN—United States Navy
USNR—United States Navy Reserve
VMF—Marine Fighter Squadron
VMF(N)—Marine All-Weather Fighter Squadron
VMO—Marine Observation Squadron
VMR—Marine Transport Squadron
WD—War Diary
WD Sum—War Diary Summary
WIA—Wounded in Action

APPENDIX B

Effective Strength of 1st Marine Division

Listed below are selected dates and figures which represent the effective strength of the 1st Marine Division throughout the period 1951–1952.

Date	Organic USMC and USN	Attached U.S. Army	Attached KMC	Total
30 Mar 51	25,831	236	3,128	29,195
30 May 51	25,820	302	3,266	29,388
30 Sep 51	24,160	54	3,035	27,249
30 Mar 52	26,140	59	4,378	30,577

APPENDIX C

Command and Staff List December 1950—March 1952 1st Marine Division

Commanding General	MajGen Oliver P. Smith (to 23 Feb 1951)
	BrigGen Lewis B. Puller (from 24 Feb)
	MajGen Oliver P. Smith (from 5 Mar)
	MajGen Gerald C. Thomas (from 25 Apr)
	MajGen John T. Selden (from 11 Jan 1952)
Asst Division Commander	BrigGen Edward A. Craig (to 20 Jan 1951)
	MajGen Edward A. Craig (from 21 Jan)
	BrigGen Lewis B. Puller (from 2 Feb)
	BrigGen William J. Whaling (from 20 May)
Chief of Staff	Col Gregon A. Williams (to 22 Jan 1951)
	BrigGen Gregon A. Williams (from 23 Jan)
	Col Edward W. Snedeker (from 27 Jan)
	Col Francis M. McAlister (from 23 May)
	Col Richard G. Weede (from 10 Jun)
	Col Victor H. Krulak (from 29 Jun)
	Col Richard G. Weede (from 26 Nov)
	Col Custis Burton, Jr. (from 15 Feb 1952)
	Col Austin R. Brunelli (from 23 Mar)
G-1	LtCol Bryghte D. Godbold (to 13 Feb 1951)
	Col Bryghte D. Godbold (from 14 Feb)
	Col Wesley M. Platt (from 31 May)
	Col Gould P. Groves (from 27 Sep)
	Col Walter N. Flournoy (from 20 Nov)
G-2	Col Bankson T. Holcomb, Jr. (to 5 Feb 1951)
	LtCol Ellsworth G. Van Orman (from 6 Feb)
	LtCol Joseph P. Sayers (from 8 Mar)
	LtCol James H. Tinsley (from 13 Aug)
G-3	Col Alpha L. Bowser, Jr., (to 7 May 1951)
	Col Richard G. Weede (from 8 May)
	Col Bruce T. Hemphill (from 30 Jul)
	LtCol Gordon D. Gayle (from 14 Nov)
G-4	Col Francis M. McAlister (to 25 Jan 1951)
	LtCol Charles L. Banks (from 26 Jan)
	Col Charles L. Banks (from 14 Feb)
	Col Frank P. Hager (from 24 May)
	Col Custis Burton, Jr. (from 19 Nov)
	Col Robert A. McGill (from 9 Feb 1952)

Special Staff

Adjutant .... Maj Philip J. Costello (to 18 Feb 1951)
LtCol Foster C. LaHue (from 19 Feb)
LtCol Homer E. Hire (from 19 Jun)
Maj James K. Young (from 15 Oct)

Air Officer .... Maj James N. Cupp (to 20 Apr 1951)
LtCol Edward V. Finn (from 21 Apr)

Amphibian Tractor Officer. . LtCol Erwin F. Wann, Jr. (to 26 Sep 1951)
LtCol Michiel Dobervich (from 27 Sep)

Anti-Tank Officer .... Maj John H. Blue (to 27 Apr 1951)
Maj William L. Bates (from 28 Apr)
Maj Robert E. Baldwin (from 3 Sep)
Maj Franklin J. Harte (from 9 Nov)
Maj John P. Lanigan (from 31 Dec)
Maj Harold C. Howard (from 2 Mar 1952)

Armored Amphibian Officer. LtCol Francis H. Cooper (to 15 Jun 1951)
Maj George M. Warnke (from 16 Jun)
LtCol John T. O'Neill (from 2 Oct)

Artillery Officer .... LtCol Carl A. Youngdale (to 5 Mar 1951)
Col Joseph L. Winecoff (from 6 Mar)
LtCol Custis Burton, Jr. (from 5 Aug)
LtCol George B. Thomas (from 8 Nov)
LtCol Dale H. Heely (from 1 Jan 1952)
Col Bruce T. Hemphill (from 11 Jan)
Col Frederick P. Henderson (from 27 Mar)

Chaplain .... Cmdr Robert M. Schwyhart, USN (to 17 Feb 1951)
Cmdr Francis W. Kelly, USN (from 18 Feb)
Cmdr Walter S. Peck, Jr., USN (from 8 Oct)

Chemical Warfare and Radiological Defense Officer .... Maj John H. Blue (to 15 Jul 1951)
Maj Robert E. Baldwin (from 3 Sep)
Maj Luther H. Hake (from 21 Nov)
Maj John P. Lanigan (from 31 Dec)
Maj Harold C. Howard (from 29 Feb 1952)

Dental Officer .... Capt Mack Meradith, USN (to 20 May 1951)
Cmdr James L. Bradley, USN (from 21 May)
Capt Francis C. Snyder, USN (from 15 Jul)

Embarkation Officer .... Maj Jules M. Rouse (to 9 Mar 1951)
LtCol Louis C. Griffin (from 10 Mar)
LtCol Clifford E. Quilici (from 11 Aug)
LtCol Corbin L. West (from 26 Oct)
LtCol John H. Papurca (from 6 Dec)

Engineer Officer .... LtCol John H. Partridge (to 10 Jun 1951)
LtCol John V. Kelsey (from 11 Jun)
LtCol August L. Vogt (from 19 Sep)

Exchange Officer . . . . Capt Wilbur C. Conley (to 16 May 1951)
1stLt Frank C. Trumble (from 17 May)
1stLt George W. Krahn (from 29 Aug)
Capt Robert W. Schmidt (from 26 Oct)
Capt Robert J. McKay (from 6 Mar 1952)
Capt Benjamin Reed (from 26 Mar)

Food Director . . . . LtCol Norman R. Nickerson (to 6 May 1951)
LtCol George G. Pafford (from 7 May)
1stLt Herbert E. McNabb (from 16 Aug)

Historical Officer . . . . 1stLt John M. Patrick (to 26 Jun 1951)
1stLt Theodore L. Richardson (from 27 Jun)
2dLt Francis X. Goss (from 8 Jan 1952)

Inspector . . . . Col John A. White (to 26 Apr 1951)
Col Gould P. Groves (from 27 Apr)
LtCol Charles W. Harrison (from 21 Jun)
Col Russell N. Jordahl (from 30 Jun)
LtCol Alfred H. Marks (from 1 Oct)
Col William K. Davenport, Jr. (from 19 Nov)

Legal Officer . . . . LtCol Albert H. Schierman (to 8 May 1951)
LtCol Randolph S. D. Lockwood (from 9 May)
Cmdr Geoffrey E. Carlisle, USN (from 28 Oct)
LtCdr Arnold W. Eggen, USN (from 6 Mar 1952)

Motor Transport Officer . . . . LtCol Henry W. Seeley, Jr. (to 26 Jun 1951)
LtCol Howard E. Wertman (from 27 Jun)
Maj Herbert E. Pierce (from 17 Aug)
Maj Walter R. O'Quinn (from 3 Jan 1952)

Naval Gunfire Officer . . . . . . LtCol Loren S. Fraser (to 12 Aug 1951)
Maj Charles A. Lipot (from 13 Aug)
Maj John V. Downes (from 23 Mar 1952)

Ordnance Officer . . . . Capt Donald L. Shenaut (to 9 Jul 1951)
Maj Frank W. Keith (from 10 Jul)
Maj James M. Rogers (from 1 Nov)
Maj Harold G. Borth (from 11 Jan 1952)

Postal Officer . . . . Maj Frederick Bove (to 13 May 1951)
1stLt Robert P. Sanders (from 14 May)
1stLt Robert W. Blum (from 26 Jul)
1stLt Edward D. Gelzer, Jr. (from 10 Aug)
CWO George C. Hunter (from 9 Feb 1952)

Provost Marshall . . . . Capt John H. Griffin (to 20 Apr 1951)
Capt Donald D. Pomerleau (from 21 Apr)
Maj Raymond L. Luckel (from 6 Aug)
LtCol William F. Pulver (from 18 Oct)

Public Information Officer . . . Capt Michael C. Capraro (to 14 Apr 1951)
1stLt Jeremiah A. O'Leary, Jr. (from 15 Apr)
1stLt Robert S. Gray (from 27 Dec)

Shore Party Officer........LtCol Henry P. Crowe (to 10 May 1951)
LtCol Horace S. Figuers (from 11 May)
LtCol Harry W. Edwards (from 7 Jul)
LtCol George G. Pafford (from 29 Sep)
LtCol Franklin B. Nihart (from 20 Dec)
LtCol Warren S. Sivertsen (from 9 Mar 1952)

Signal Officer ....LtCol Robert L. Schreier (to 7 Jun 1951)
LtCol Jino J. D'Alessandro (from 8 Jun)

Special Services Officer......LtCol John M. Bathum (to 10 Sep 1951)
Maj Paul H. Bratten, Jr. (from 11 Sep)
LtCol Franklin B. Nihart (from 28 Oct)
1stLt Joseph H. McDannold (from 20 Dec)
Capt John W. Algeo (from 16 Feb 1952)
LtCol John E. Gorman (from 9 Mar)

Supply Officer ....Col Gordon E. Hendricks (to 29 Jun 1951)
Col Chester R. Allen (from 30 Jun)

Surgeon ....Capt Eugene R. Hering, USN (to 24 Jan 1951)
Cmdr Howard A. Johnson, USN (from 25 Jan 1951)
Capt Louis R. Kirkpatrick, USN (from 10 Jul 1951)

Tank Officer ....LtCol Harry T. Milne (to 22 Apr 1951)
LtCol Holly H. Evans (from 23 Apr)
Maj Walter E. Reynolds (from 9 Feb 1952)

Commanding Officer, Division Rear Echelon Headquarters ....Col Harvey S. Walseth (to 23 Jul 1951)
Col Wilburt S. Brown (from 24 Jul to 19 Nov)

Headquarters Battalion

Commanding Officer ......LtCol Marvin T. Starr (to 23 Apr 1951)
LtCol William P. Alston (from 24 Apr)
Col Gould P. Groves (from 11 May)
LtCol Charles W. Harrison (from 29 Jun)
LtCol Alfred H. Marks (from 29 Aug)
Col William K. Davenport, Jr. (from 19 Nov)
Maj Corbin L. West (from 15 Jan 1952)
Col Robert T. Stivers (from 18 Feb)

Executive Officer ....Maj Frederick Simpson (to 15 Aug 1951)
Maj William O. Cain, Jr. (from 16 Aug)
Maj Corbin L. West (from 10 Dec)
Capt "J" E. Hancey (from 22 Jan 1952)
Maj Corbin L. West (from 18 Feb 1952)

Commanding Officer,
Headquarters Company . . . Maj Frederick Simpson (to 15 Aug 1951)
Maj William O. Cain, Jr. (from 16 Aug)
Maj Corbin L. West (from 10 Dec)
Capt "J" E. Hancey (from 21 Jan 1952)

Commanding Officer,
Military Police Company . . Capt John H. Griffin (to 20 Apr 1951)
Capt Donald D. Pomerleau (from 21 Apr)
Maj Raymond L. Luckel (from 19 Sep)
LtCol William F. Pulver (from 18 October)

Commanding Officer,
Reconnaissance Company . . Maj Walter Gall (to 26 Mar 1951)
Capt Robert L. Autry (from 27 Mar)
Maj Ephraim Kirby-Smith (from 10 Sep)

1st Marines

Commanding Officer ...... Col Lewis B. Puller (to 24 Jan 1951)
Col Francis M. McAlister (from 25 Jan)
Col Wilburt S. Brown (from 19 May)
Col Thomas A. Wornham (from 18 Jul)
Col Sidney S. Wade (from 13 Oct)

Executive Officer .... LtCol Robert W. Rickert (to 7 Jan 1951)
LtCol Alan Sutter (from 8 Jan)
LtCol Robert W. Rickert (from 16 Jan)
LtCol Alan Sutter (from 12 Feb)
LtCol Donald M. Schmuck (from 31 May)
LtCol John A. McAlister (from 3 Sep)
LtCol Clifford F. Quilici (from 7 Jan 1952)

S–1 .... Capt William G. Reeves (to 8 Jan 1951)
Capt David M. Cox (from 9 Jan)
Capt John S. Court (from 5 Sep)
Maj Elizia M. Cable (from 21 Oct)
Capt Thomas C. Palmer (from 12 Feb 1952)
Capt Leroy V. Corbett (from 28 Feb)

S–2 .... Capt Stone W. Quillian (to 10 May 1951)
Capt Glenn F. Miller (from 11 May)
Capt Robert G. Cadwallader (from 2 Oct)
Capt Fred K. Cottrell (from 15 Dec)
Capt Edwin H. Heim (from 4 Mar 1952)

S–3 .... Maj Robert E. Lorigan (to 20 Jul 1951)
Maj Ralph "C" Rosacker (from 21 Jul)
Maj John P. Lanigan (from 4 Mar 1952)

S–4 .... Maj Thomas T. Grady (to 27 Apr 1951)
Capt Augustine B. Reynolds, Jr. (from 28 Apr)
Maj Thomas A. Burns (from 5 Jul)
Maj John L. Kelly (from 5 Oct)
Maj Fletcher R. Wycoff (from 27 Dec)

Commanding Officer, Headquarters and Service Company .... Maj Robert K. McClelland (to 11 Mar 1951)
Maj Carl E. Walker (from 12 Mar)
Capt George E. Petro (from 11 May)
1stLt Roscoe L. Barrett, Jr. (from 15 Aug)
1stLt James L. Burnett (from 3 Oct)
Capt James P. Egan (from 23 Feb 1952)

Commanding Officer, Anti-Tank Company ..... Capt George E. Petro (to 10 May 1951)
1stLt John A. Dudrey (from 11 May)
1stLt Magness W. Marshall (from 2 Oct)
Capt Frederick A. Hale (from 27 Nov)

Commanding Officer, 4.2 Inch Mortar Company .... Capt Frank J. Faureck (to 8 Feb 1951)
1stLt Edward E. Kauffer (from 9 Feb)
Capt Otis R. Waldrop (from 5 Mar)
Capt Edward E. Kauffer (from 4 Jun)
1stLt Robert W. Jorn (from 9 Aug)
1stLt Thomas J. Holt (from 2 Oct)
Capt Robert G. Cadwallader (from 23 Dec)
Capt George E. Lawrence (from 18 Mar 1952)

1st Battalion, 1st Marines

Commanding Officer ...... LtCol Donald M. Schmuck (to 27 Feb 1951)
LtCol Robley E. West (from 28 Feb)
Maj Thomas T. Grady (from 15 Jun)
LtCol Horace E. Knapp, Jr. (from 7 Jul)
Maj Edgar F. Carney, Jr. (from 14 Sep)
LtCol John E. Gorman (from 16 Sep)
LtCol John H. Papurca (from 7 Mar 1952)

Executive Officer .... Maj Robley E. West (to 27 Feb 1951)
Maj David W. Bridges (from 28 Feb)
Maj Thomas T. Grady (from 10 Jun)
Maj Wesley C. Noren (from 15 Jun)
Maj Edgar F. Carney, Jr. (from 20 Jul)
Maj Leo V. Gross (from 18 Dec)
Maj Ralph "C" Rosacker (from 4 Mar 1952)

Commanding Officer, Headquarters and Service Company .... Capt William B. Hopkins (to 30 Jan 1951)
1stLt Bruce E. Geisert (from 31 Jan)
1stLt Norman W. Hicks (from 1 Jul)
1stLt John B. Franklin (from 18 Aug)
1stLt Stuart P. Barr, Jr. (from 22 Oct)
1stLt Nicholas J. Sheppard (from 28 Nov)

1stLt Harry A. Spaight (from 26 Dec)
Capt Edwin H. Heim (from 20 Feb 1952)
2ndLt Vinton L. Spencer (from 4 Mar)

Commanding Officer, Company A .... Capt Robert H. Barrow (to 30 Jan 1951)
Capt Thomas J. Bohannon (from 31 Jan)
1stLt Calvin R. Baker (from 1 Jul)
Capt Edwin H. Heim (from 20 Oct)
1stLt Clifton M. Grubbs (from 20 Feb 1952)
Capt Anthony Novak (from 17 Mar)
1stLt Morace M. Dritley (from 26 Mar)

Commanding Officer, Company B .... Capt Wesley C. Noren (to 12 Mar 1951)
Capt John F. Coffey (from 13 Mar)
1stLt James H. Cowan, Jr. (from 8 Jun)
1stLt Robert G. Work (from 1 Aug)
1stLt Richard S. Kitchen (from 18 Aug)
Capt Roy J. Wride (from 16 Dec)

Commanding Officer, Company C .... Capt Robert P. Wray (to 9 May 1951)
1stLt William A. Craven (from 10 May)
1stLt William F. Koehnlein (from 12 Jun)
Capt Michael D. Harvath (from 21 Jul)
Capt George E. Lawrence (from 10 Oct)
Capt Kenneth F. Swiger (from 7 Jan 1952)

Commanding Officer, Weapons Company ...... Maj William L. Bates (to 28 Feb 1951)
1stLt William F. Koehnlein (from 1 Mar)
Capt Wesley C. Noren (from 13 Mar)
Maj John F. Coffey (from 8 Jun)
Capt Benjamin W. Muntz (from 5 Jul)
Maj William O. Cain, Jr. (from 14 Jul)
Maj John F. Morris (from 14 Aug)
Maj Fletcher B. Wycoff (from 9 Sep)
Capt James P. Egan (from 27 Dec)
Capt George E. Lawrence (from 21 Feb 1952)
1stLt Joseph E. Lee (from 18 Mar)
Maj Stanley N. McLeod (from 27 Mar)

2d Battalion, 1st Marines

Commanding Officer ...... LtCol Allan Sutter (to 7 Jan 1951)
Maj Clarence J. Mabry (from 8 Jan)
LtCol Allan Sutter (from 15 Jan)
Maj Clarence J. Mabry (from 13 Feb)
LtCol Robert K. McClelland (from 15 Mar)
Maj Clarence J. Mabry (from 5 Jun)
LtCol Robert K. McClelland (from 20 Jun)

	LtCol Franklin B. Nihart (from 14 Aug)
	LtCol Clifford F. Quilici (from 28 Oct)
	LtCol Theil H. Fisher (from 3 Jan 1952)
Executive Officer	Maj Clarence J. Mabry (to 7 Jan 1951)
	Maj Whitman S. Bartley (from 8 Jan)
	Maj Clarence J. Mabry (from 15 Jan)
	Maj Whitman S. Bartley (from 13 Feb)
	Maj Clarence J. Mabry (from 15 Mar)
	Maj Jules M. Rouse (from 10 Jun)
	Maj John P. Lanigan (from 6 Aug)
	Maj Franklin J. Harte (from 26 Dec)
Commanding Officer, Headquarters and Service Company	Capt Raymond DeWees, Jr. (to 9 Sep 1951)
	2dLt Robert A. Arning (from 10 Sep)
	1stLt George H. Benskin, Jr. (from 30 Oct)
	1stLt Frank E. Guthrie (from 3 Dec)
Commanding Officer, Company D	Capt Welby W. Cronk (to 4 Mar 1951)
	1stLt Theodore Culpepper (from 5 Mar)
	1stLt Alexander L. Michaux, Jr. (from 19 Apr)
	1stLt Jay "J" Thomas (from 11 Jun)
	1stLt George H. Benskin, Jr. (from 9 Aug)
	1stLt Robert E. Lundberg (from 15 Sep)
	2dLt Arthur H. Woodruff (from 25 Sep)
	1stLt Richard A. Bonifas (from 5 Oct)
	1stLt George H. Benskin, Jr. (from 16 Oct)
	Capt Richard A. Bonifas (from 30 Oct)
	1stLt Robert J. Lahr (from 3 Nov)
	Capt Robert N. Kreider (from 13 Nov)
	Capt John H. Lauck (from 26 Jan 1952)
Commanding Officer, Company E	Capt Jack A. Smith (to 9 Mar 1951)
	1stLt Johnny L. Carter (from 10 Mar)
	1stLt Donald L. Evans, Jr. (from 9 Aug)
	Capt Ralph V. Harper (from 14 Aug)
	1stLt Robert J. Lahr (from 14 Sep)
	2dLt William K. Rockey (from 25 Sep)
	1stLt Kenneth E. Will (from 5 Oct)
	Capt James H. Reeder (from 16 Oct)
	Capt Charles J. Irwin, Jr. (from 21 Feb 1952)
	Capt Jack H. Hagler (from 17 Mar)
Commanding Officer, Company F	Capt Goodwin C. Groff (to 9 Jun 1951)
	1stLt Patrick McGrotty (from 10 Jun)
	Capt Frederick A. Hale, Jr. (from 4 Sep)
	Capt Neville G. Hall, Jr. (from 21 Nov)

	1stLt John A. Barry (from 29 Dec)
	1stLt Robert J. Lahr (from 11 Mar 1952)
	Capt Victor A. Kleber, Jr. (from 18 Mar)
Commanding Officer, Weapons Company	Capt William A. Kerr (to 28 Feb 1951)
	1stLt Russell A. Davidson (from 1 Mar)
	Maj Carl E. Walker (from 12 May)
	Capt Russell A. Davidson (from 2 Jul)
	Maj John I. Kelly (from 22 Jul)
	Maj William S. Witt (from 5 Oct)
	Capt John W. Algeo (from 20 Nov)
	Maj William S. Witt (from 20 Jan 1952)
	Capt John W. Algeo (from 3 Feb)
	1stLt Clarence G. Moody, Jr. (from 17 Feb)
	Capt Charles J. Irwin, Jr. (from 18 Mar)

3d Battalion, 1st Marines

Commanding Officer	LtCol Thomas L. Ridge (to 15 Feb 1951)
	LtCol Virgil W. Banning (from 16 Feb)
	Maj Joseph D. Trompeter (from 25 Apr)
	Maj Edwin H. Simmons (from 8 May)
	LtCol Homer E. Hire (from 15 May)
	LtCol Foster C. LaHue (from 19 Jul)
	LtCol Spencer H. Pratt (from 13 Nov)
Executive Officer	Maj Reginald R. Myers (to 25 Apr)
	Maj Edwin H. Simmons (from 26 Apr)
	Maj Joseph D. Trompeter (from 15 May)
	Maj Ralph "C" Rosacker (from 7 Jun)
	Maj Rodney V. Reighard (from 22 July)
	Maj Thell H. Fisher (from 3 Oct)
	Maj Robert V. Perkins (from 4 Jan 52)
Commanding Officer, Headquarters and Service Company	Capt Roy N. Courington (to 16 Feb 1951)
	1stLt Edgar A. Crum (from 17 Feb)
	1stLt Daniel R. Evans (from 3 Mar)
	Capt Clarence E. Corley, Jr. (from 20 Mar)
	1stLt Thomas J. Holt (from 9 Aug)
	Capt Earle E. Carr (from 1 Sep)
	2dLt Joseph D. Reed (from 3 Oct)
	2dLt Robert C. Morton (from 4 Jan 1952)
	Capt Harold R. Connolly (from 22 Feb)
	Capt Donald C. Mack (from 15 Mar)
Commanding Officer, Company G	Capt Carl L. Sitter (to 13 Feb 1951)
	1stLt Horace L. Johnson (from 14 Feb)
	1stLt Thomas J. Holt (from 26 May)

	1stLt Fred G. Redmon (from 1 Jun)
	Capt Varge G. Frisbie (from 5 Jun)
	1stLt Harold R. Connolly (from 20 Jul)
	Capt Fred A. Kraus (from 8 Nov)
	1stLt Richard A. Krajnyak (from 19 Feb 1952)
	Capt Wilford L. Stone (from 17 Mar)
Commanding Officer, Company H	Capt Clarence E. Corley, Jr. (to 19 Mar 1951)
	1stLt William J. Allert (from 20 Mar)
	1stLt Daniel R. Evans (from 8 May)
	1stLt James L. Burnett (from 8 Jun)
	1stLt Herbert M. Anderson (from 15 Jun)
	1stLt James L. Burnett (from 21 Sep)
	Capt Earle E. Carr (from 3 Oct)
	Capt James B. Ord, Jr. (from 17 Dec)
Commanding Officer, Company I	1stLt Joseph R. Fisher (to 7 Apr 1951)
	1stLt William Swanson (from 8 Apr)
	Capt Stone W. Quillian (from 15 May)
	1stLt Norbert D. Carlson (from 5 Aug)
	Capt Leroy V. Corbett (from 7 Sep)
	Capt Donald C. Mack (from 19 Jan 1952)
	Capt Richard B. Smith (from 22 Feb)
Commanding Officer, Weapons Company	Maj Edwin H. Simmons (to 25 Apr 1951)
	1stLt James F. Williams (from 26 Apr)
	Capt Otis R. Waldrop (from 6 Jun)
	Maj Henry Brzezinski (from 19 Jun)
	Capt Varge G. Frisbie (from 6 Aug)
	Maj Thell H. Fisher (from 31 Aug)
	1stLt Thomas C. Holleman (from 2 Oct)
	Maj Robert V. Perkins (from 15 Nov)
	Capt Earle E. Carr (from 4 Jan 1952)
	1stLt Hugh P. Murphy (from 25 Jan)

5th Marines

Commanding Officer	LtCol Raymond L. Murray (to 23 Jan 1951)
	Col Raymond L. Murray (from 24 Jan)
	Col Richard W. Hayward (from 14 Mar)
	Col Richard G. Weede (from 7 Aug)
	Col Frank P. Hager, Jr. (from 19 Nov)
	Col Thomas A. Culhane, Jr. (from 23 Feb 1952)
Executive Officer	LtCol Joseph L. Stewart (to 13 Feb 1951)
	LtCol John W. Stevens, II (from 14 Feb)
	LtCol Joseph L. Stewart (from 14 Mar)
	LtCol Donald R. Kennedy (from 4 Apr)
	LtCol Francis H. Cooper (from 17 Jun)

LtCol Virgil W. Banning (from 22 Sep)
LtCol John T. Rooney (from 13 Dec)
LtCol John A. Saxten (from 19 Mar 1952)

S–1 .... Capt Alton C. Weed (to 1 Mar 1951)
Capt Jack E. Hawthorn (from 2 Mar)
Capt George A. Rheman, Jr. (from 17 Mar)
Capt Harley L. Grant (from 25 Aug)

S–2 .... 1stLt Richard M. Woodard (to 3 Feb 1951)
Capt Eugene F. Langan (from 4 Feb)
Maj Nicholas G. W. Thorne (from 9 Aug)
Maj Paul H. Bratten, Jr. (from 17 Nov)
Maj John C. Lundrigan (from 31 Jan 1952)

S–3 .... Maj Lawrence W. Smith, Jr. (to 8 Mar 1951)
Maj Robert E. Baldwin (from 9 Mar)
LtCol Glen E. Martin (from 24 Jun)
Maj Merwin H. Silverthorn, Jr. (from 11 Jul)
Maj Gerald P. Averill (from 10 Oct)
Maj David A. Brewster, Sr. (from 15 Dec)

S–4 .... Maj Harold Wallace (to 9 Mar 1951)
Maj William E. Baugh (from 10 Mar)
Maj Robert S. Hudson (from 11 Aug)
Maj Warren F. Lloyd (from 22 Dec)

Commanding Officer, Headquarters and Service Company .... Capt Jack E. Hanthorn (to 1 Mar 1951)
1stLt Richard M. Woodard (from 2 Mar)
1stLt Lee J. Cary (from 22 Jun)
Capt Howard H. Dismeier (from 12 Sep)
1stLt George "T" Capatanos (from 1 Dec)

Commanding Officer, Antitank Company ...... 1stLt Almarion S. Bailey (to 8 Apr 1951)
1stLt Jo M. Van Meter (from 9 Apr)
1stLt William E. Kerrigan (from 23 Jul)
Capt Edgar F. Moore, Jr. (from 15 Aug)

Commanding Officer, 4.2 Inch Mortar Company .... 1stLt Robert M. Lucy (to 25 Feb 1951)
1stLt Robert H. Uskurait (from 26 Feb)
1stLt John A. Buchanan (from 11 Sep)
Capt Yale B. Davis (from 29 Dec)

1st Battalion, 5th Marines

Commanding Officer ...... LtCol John W. Stevens, II (to 20 Feb 1951)
LtCol John W. Hopkins (from 21 Feb)
LtCol William P. Alston (from 21 Jun)
Maj Kirt W. Norton (from 9 Nov)
Maj Lowell T. Keagy (from 25 Nov)

6340 O–62–21

LtCol Kirt W. Norton (from 2 Dec)
LtCol Louis N. King (from 13 Jan 1952)
LtCol Franklin B. Nihart (from 12 Feb)

Executive Officer .... Maj Merlin R. Olson (to 8 Apr 1951)
Maj Donald J. Kendall, Jr. (from 9 Apr)
Maj Kirt W. Norton (from 9 Aug)
Maj Robert L. Autry (from 9 Nov)
Maj Lowell T. Keagy (from 2 Dec)
Maj Hildeburn R. Martin (from 31 Dec)

Commanding Officer, Headquarters and Service Company .... Capt George A. Rheman, Jr. (to 11 Mar 1951)
2dLt Robert H. Corbet (from 12 Mar)
1stLt Andrew V. Marusak (from 29 Mar)
1stLt Frank J. Meers (from 12 Jul)
2dLt Vincent B. Murphy, Jr. (from 3 Oct)
1stLt Parks H. Simpson (from 25 Oct)
1stLt Thomas J. Hermes (from 13 Nov)

Commanding Officer, Weapons Company ...... Capt Almond H. Sollom (to 5 Mar 1951)
1stLt Poul F. Pedersen (from 6 Mar)
Capt Donald D. Pomerleau (from 6 Apr)
Maj Albert Hartman (from 13 Apr)
Capt Raymond H. Spuhler (from 8 May)
1stLt Frank J. Meers (from 4 Jun)
Capt Lucian F. May (from 12 Jul)
Maj David A. Brewster, Sr. (from 1 Sep)
Capt Harry A. Mathew (from 9 Nov)
Capt Nicholas G. W. Thorne (from 17 Nov)
Maj Lowell T. Keagy (from 31 Dec)

Commanding Officer, Company A .... 1stLt Loren R. Smith (to 16 Feb 1951)
Capt Walter E. G. Godenius (from 17 Feb)
Capt John L. Kelly (from 9 Apr)
Capt Richard M. Woodard (from 1 Jul)
Capt Eugene F. Langan (from 12 Aug)
Capt Frederick B. Clunie (from 5 Nov)
1stLt Merrill Waide, Jr. (from 24 Jan 1952)
1stLt Ernest S. Lee (from 18 Feb)

Commanding Officer, Company B .... 1stLt John R. Hancock (to 7 Feb 1951)
1stLt Michael V. Palatas (from 8 Feb)
1stLt James T. Cronin (from 17 Feb)
1stLt William E. Kerrigan (from 8 Jun)
1stLt Stuart H. Wright (from 30 Jun)

	1stLt John A. Hayes (from 12 Jul)
	Capt Louis R. Daze (from 21 Jul)
	Capt Charles M. MacDonald, Jr. (from 21 Nov)
Commanding Officer, Company C	Capt Jack R. Jones (to 8 May 1951)
	1stLt Richard J. Schening (from 9 May)
	1stLt Robert E. Warner (from 29 May)
	Capt Lucian F. May (from 4 Sep)
	Capt Harry A. Mathew (from 22 Jan 1952)

2d Battalion, 5th Marines

Commanding Officer	LtCol Harold S. Roise (to 19 Feb 1951)
	LtCol Glen E. Martin (from 20 Feb)
	Maj Merwin H. Silverthorn, Jr. (from 24 Jun)
	LtCol Houston Stiff (from 8 Jul)
	Maj William E. Baugh (from 3 Dec)
	LtCol George G. Pafford (from 27 Dec)
	LtCol William P. Cushing (from 14 Mar 1952)
Executive Officer	Maj John L. Hopkins (to 20 Feb 1951)
	Maj Theodore F. Spiker (from 21 Feb)
	Maj Merwin H. Silverthorn, Jr. (from 9 Apr)
	Maj Robert E. Baldwin (from 25 Jun)
	Maj Gerald P. Averill (from 3 Sep)
	Maj Robert W. Rynerson (from 9 Sep)
	Maj Warren F. Lloyd (from 26 Sep)
	Maj William L. Sims (from 9 Dec)
	Maj Robert S. Hudson (from 27 Dec)
	Maj William P. Cushing (from 21 Feb 1952)
	Maj Robert S. Hudson (from 14 Mar)
Commanding Officer, Headquarters and Service Company	Capt Franklin B. Mayer (to 9 Jan 1951)
	1stLt Charles "H" Dalton (from 10 Jan)
	Capt William O. Cain, Jr. (from 21 Feb)
	1stLt John R. Hinds (from 2 Jul)
	1stLt Richard T. Hauar (from 12 Jul)
	1stLt Harold L. Michael (from 8 Aug)
	1stLt Dexter H. Kimball (from 25 Sep)
	1stLt Otis "Z" McConnell, Jr. (from 23 Dec)
	1stLt Emmett T. Hill, Jr. (from 15 Mar 1952)
Commanding Officer, Company D	Capt Samuel S. Smith (to 11 Jun 1951)
	1stLt John P. Cooney (from 12 Jun)
	Capt Ray N. Joens (from 28 Jun)
	Capt Victor Sawina (from 26 Sep)
	1stLt Tom G. Fagles (from 7 Oct)

	Capt Philip A. Davis (from 23 Dec)
	1stLt Emmitt T. Hill (from 13 Feb 1952)
	Capt William A. Harper (from 25 Feb)
Commanding Officer, Company E	1stLt James F. Roberts (to 9 Jan 1951)
	Capt Franklin B. Mayer (from 10 Jan)
	Capt William E. Melby (from 9 Apr)
	1stLt Bernard W. Christofferson (from 20 Apr)
	1stLt Warren H. Allen (from 12 Jun)
	Capt William E. Melby (from 18 Jun)
	1stLt Warren H. Allen (from 9 Jul)
	Capt William L. Wallace (from 3 Aug)
	Capt Warren H. Allen (from 3 Oct)
	1stLt Jo M. Van Meter (from 18 Oct)
	Capt Charles C. Matthews (from 4 Jan 1952)
Commanding Officer, Company F	1stLt Charles "H" Dalton (to 8 Jan 1951)
	1stLt George Janiszewski (from 9 Jan)
	Capt William O. Cain, Jr. (from 20 Jan)
	1stLt George Janiszewski (from 20 Feb)
	1stLt James H. Honeycutt, Jr. (from 9 Apr)
	1stLt Harold L. Michael (from 23 Jul)
	Capt William E. Melby (from 11 Aug)
	Capt Arvil B. Hendrickson (from 4 Nov)
	Capt Harold C. Fuson (from 14 Mar 1952)
Commanding Officer, Weapons Company	Maj Glen E. Martin (to 19 Feb 1951)
	Capt John Stepanovich (from 20 Feb)
	Capt Elliot B. Lima (from 6 Apr)
	1stLt Arvil B. Hendrickson (from 17 Aug)
	Maj Warren F. Lloyd (from 15 Sep)
	Capt Arvil B. Hendrickson (from 25 Sep)
	Maj William L. Sims (from 4 Nov)
	Capt William A. Harper (from 23 Dec)
	Capt Harold C. Fuson (from 25 Feb 1952)
	Capt Russell L. Silverthorn (from 16 Mar)

3d Battalion, 5th Marines

Commanding Officer	LtCol Robert D. Taplett (to 13 Feb 1951)
	LtCol Joseph L. Stewart (from 14 Feb)
	LtCol Donald R. Kennedy (from 14 Mar)
	Maj Morse "L" Holladay (from 4 Apr)
	LtCol Donald R. Kennedy (from 16 Jun)
	Maj William E. Baugh (from 23 Sep)
	LCol Bernard W. McLean (from 13 Oct)
	LtCol William S. McLaughlin (from 25 Feb 1952)

Executive Officer .... Maj Harold E. Swain (to 7 May 1951)
Maj Albert Hartman (from 8 May)
Maj William E. Baugh (from 11 Aug)
Maj Donald D. Pomerleau (from 27 Sep)
Maj William E. Baugh (from 13 Oct)
Maj Donald D. Pomerleau (from 30 Nov)
Maj Paul H. Bratten (from 4 Feb 1952)

Commanding Officer, Headquarters and Service Company .... 1stLt Harold D. Fredericks (to 13 Feb 1951)
1stLt Duncan McRae (from 14 Feb)
1stLt Carlisle G. Kohl, Jr. (from 25 Mar)
1stLt Herbert Preston (from 27 Jun)
Capt Robert J. McKay (from 25 Aug)
Capt Charles W. Marker, Jr. (from 23 Dec)

Commanding Officer, Company G .... 1stLt Charles D. Mize (to 5 Mar 1951)
1stLt August L. Camarata (from 6 Mar)
1stLt William G. Robinson (from 18 Jul)
Capt John M. Fallon (from 10 Sep)
Capt James Irving, Jr. (from 5 Nov)
1stLt Wilson L. Cook (from 28 Feb 1952)

Commanding Officer, Company H .... Capt Harold I. Williamson (to 1 Apr 1951)
1stLt Herbert Preston, Jr. (from 2 Apr)
Capt Clarence H. Pritchett (from 1 May)
1stLt Bruce F. Meyers (from 5 Aug)
Capt Raymond J. McGlynn (from 4 Nov)
Capt Matthew A. Clary, Jr. (from 21 Feb 1952)

Commanding Officer, Company I .... 1stLt Donald E. Watterson (to 5 Mar 1951)
Capt Raymond H. Spuhler (from 6 Mar)
Capt John A. Pearson (from 1 Apr)
1stLt Raymond J. McGlynn (from 10 Aug)
1stLt Lawrence W. Payne (from 29 Aug)
Capt Neil Dimond (from 5 Oct)

Commanding Officer, Weapons Company ...... Capt Raymond H. Spuhler (to 31 Jan 1951)
Maj Thomas A. Durham (from 1 Feb)
Maj Ilo J. Scatena (from 26 Jul)
Maj Donald D. Pomerleau (from 20 Sep)
Maj Ilo J. Scatena (from 27 Sep)
Maj James H. Pope (from 13 Oct)
Capt Charles W. Marker, Jr. (from 3 Dec)
Capt Robert J. McKay (from 23 Dec)
1stLt Anthony R. Kurowski (from 6 Mar 1952)
Capt Robert W. Lowe (from 17 Mar)

7th Marines

Commanding Officer ...... Col Homer L. Litzenberg (to 15 Apr 1951)
Col Herman Nickerson, Jr. (from 16 Apr)
LtCol John J. Wermuth (from 20 Sep)
Col John J. Wermuth (from 13 Dec)
Col Russell E. Honsowetz (from 11 Mar 1952)

Executive Officer .... LtCol Raymond G. Davis (to 3 Jun 1951)
LtCol Woodrow M. Kessler (from 4 Jun)
LtCol John J. Wermuth (from 30 Jun)
LtCol Gordon D. Gayle (from 20 Sep)
LtCol James G. Kelly (from 3 Nov)
LtCol Noel C. Gregory (from 2 Dec)
LtCol John D. Wiggins (from 23 Feb 1952)

S–1 .... Capt John R. Grove (to 15 Apr 1951)
Capt Hugh E. McNeely (from 16 Apr)
Maj Robert R. Sedgwick (from 5 Sep)
Capt William K. Dormady (from 5 Jan 1952)

S–2 .... Capt John D. Bradbeer (to 4 Jul 1951)
Capt Walter E. Lange (from 5 Jul)
Capt Clifford E. McCollam (from 29 Jul)
Maj Henry V. Joslin (from 25 Aug)
1stLt George W. Barnes (from 8 Nov)
Capt Donald E. Euchert (from 19 Dec)
Capt Harry E. Leland, Jr. (from 17 Mar 1952)

S–3 .... Maj Henry J. Woessner, II (to 8 Jan 1951)
Maj Joseph L. Abel (from 9 Jun)
Maj George Codrea (from 22 Sep)

S–4 .... Maj Maurice E. Roach (to 8 Jan 1951)
Maj William E. Voorhies (from 9 Jan)
Maj John D. Bradbeer (from 5 Jul)
Maj Franklin C. Bacon (from 5 Oct)
Maj Robert B. Prescott (from 3 Jan 1952)
Maj James K. Linnan (from 19 Jan)

Commanding Officer, Headquarters and Service Company .... 2dLt Arthur R. Mooney (to 17 Feb 1951)
1stLt Harrol Kiser (from 18 Feb)
1stLt John C. Beauparlant (from 6 Mar)
1stLt Welton R. Abell (from 14 Mar)
Capt James J. Bott (from 19 Mar)
Capt Thomas A. Robesky (from 9 May)
Capt Walter R. Anderson (from 18 Jun)
Capt Hugh E. McNeely (from 5 Sep)
Capt Donald S. McClellan (from 20 Sep)
Capt David A. McKay (from 28 Nov)
Capt Robert C. Hendrickson (from 17 Mar 1952)

Commanding Officer, Antitank Company ...... 1stLt Earl R. DeLong (to 5 May 1951)
1stLt Raymond J. Eldridge (from 6 Mar)
Capt Thomas Santamaria (from 19 Apr)
1stLt Francis W. Tief (from 13 May)
1stLt William F. Dyroff (from 10 Aug)

Commanding Officer, 4.2 Inch Mortar Company .... Maj Rodney V. Reighard (to 1 Jul 1951)
1stLt Samuel E. Piercy (from 2 Jul)
Capt Alvin F. Mackin (from 24 Sep)
Capt Dean F. Johnson (from 28 Nov)
Capt John F. McMahon, Jr. (from 28 Dec)

1st Battalion, 7th Marines

Commanding Officer ...... Maj Webb D. Sawyer (to 25 Apr 1951)
LtCol John T. Rooney (from 26 Apr)
LtCol James G. Kelly (from 23 Aug)
Maj Harold C. Howard (from 8 Nov)
LtCol George W. E. Daughtry (from 28 Feb 1952)

Executive Officer .... Maj Raymond V. Fridrich (to 20 Feb 1951)
Maj Thomas B. Tighe (from 21 Feb)
Maj Raymond V. Fridrich (from 24 Mar)
Maj Thomas B. Tighe (from 26 May)
Maj Robert J. Polson (from 5 Jul)
Maj George Codrea (from 4 Aug)
Maj Harold C. Howard (from 15 Sep)
Maj Henry V. Joslin (from 8 Nov)

Commanding Officer, Headquarters and Service Company .... 1stLt Wilbert R. Gaul (to 19 Jan 1951)
Capt John C. Johnson (from 20 Jan)
Capt Nathan R. Smith (from 18 Mar)
1stLt Eugenous M. Hovatter (from 28 Mar)
Capt Donald F. J. Field (from 11 May)
Capt Wilburt R. Gaul (from 7 Jun)
1stLt Robert C. Taylor (from 9 Aug)
Capt Orville E. Brauss (from 24 Nov)
1stLt Guy R. Cassell (from 14 Dec)
1stLt Edward L. Nadeau (from 1 Jan 1952)
Capt Seneker Woll (from 18 Jan)
2dLt Henry D. Bruns (from 10 Feb)
2d Lt Lawrence P. Flynn (from 9 Mar)

Commanding Officer, Company A .... 1stLt Eugenous M. Hovatter (to 27 Mar 1951)
Capt Nathan R. Smith (from 28 Mar)
1stLt Van D. Bell (from 3 Jun)

Capt Everett Hampton (from 2 Sep)
2dLt Carl F. Ullrich (from 2 Jan 1952)
Capt Earl W. Thompson (from 27 Mar)

Commanding Officer, Company B .... Capt James J. Bott (to 5 Mar 1951)
Capt John C. Johnston (from 6 Mar)
1stLt Orville W. Brauss (from 22 Jul)
1stLt Dean F. Johnson (from 23 Aug)
1stLt James W. Sweeney (from 14 Sep)
Capt Henry A. Glockner (from 29 Sep)
1stLt Donald L. Smith (from 14 Dec)
1stLt "J" Alan Myers (from 1 Jan 1952)
1stLt Donald M. Russ (from 14 Feb)
Capt Lyle S. Whitmore, Jr. (from 28 Feb)

Commanding Officer, Company C .... Capt John F. Morris (to 17 Jan 1951)
Capt Eugene H. Haffey (from 18 Jan)
Capt Daniel F. J. Field (from 8 Jun)
1stLt Donald E. Euckert (from 23 Jul)
Capt John F. McMahon (from 10 Aug)
Capt Robert W. Hughes, Jr. (from 21 Nov)
Capt Seneker Woll (from 7 Jan 1952)
Capt Robert W. Hughes, Jr. (from 18 Jan)
Capt Roger L. Johnson (from 3 Mar)

Commanding Officer, Weapons Company ...... Maj William E. Voorhies (to 5 Jan 1951)
Capt Robert J. Polson (from 6 Jan)
Maj Joseph L. Abel (from 12 Jan)
Maj Robert J. Polson (from 15 May)
Capt Alonzo C. Thorson (from 5 Jul)
Capt John C. Johnston (from 5 Aug)
Capt Dean F. Johnson (from 5 Nov)
Capt John R. McMahon (from 22 Nov)
1stLt Guy R. Cassell (from 31 Dec)
Capt Robert W. Hughes, Jr. (from 4 Jan 1952)
1stLt Frank P. Shannon (from 18 Jan)
1stLt Carlton R. Appleby (from 16 Feb)

2d Battalion, 7th Marines

Commanding Officer ...... LtCol Robert L. Bayer (to 15 Feb 1951)
Maj James I. Glendinning (from 16 Feb)
LtCol Wilbur F. Meyerhoff (from 21 Mar)
LtCol Louis C. Griffin (from 21 Jul)
LtCol Noel C. Gregory (from 11 Nov)
Maj Edward G. Kurdziel (from 1 Dec)
LtCol Noel C. Gregory (from 27 Feb 1952)

Executive Officer .... Maj James F. Lawrence, Jr. (to 2 Jan 1951)
Maj James I. Glendinning, Jr. (from 3 Jan)
Maj James F. Lawrence, Jr. (from 20 May)
Maj Edward G. Kurdziel (from 4 Jul)
Maj Edwin Madsen (from 2 Dec)

Commanding Officer, Headquarters and Service Company .... 1stLt Kent D. Thorup (to 19 Jan 1951)
Capt Jerome D. Gordon (from 20 Jan)
1stLt Kent D. Thorup (from 6 Feb)
Capt Thomas "A" Robesky (from 15 Mar)
1stLt Joseph R. Walsh (from 8 May)
1stLt George G. Flood (from 8 Jun)
1st Lt John J. Robinson, Jr (from 1 Sep)
Capt Charles P. Logan, Jr. (from 5 Nov)
1stLt Donald D. MacLachlan (from 16 Dec)
1stLt Edward R. Hannon (from 27 Feb 1952)

Commanding Officer, Company D .... 1stLt James D. Hammond, Jr. (to 1 Jan 1951)
Capt Patsy Algieri (from 2 Jan)
Capt Jerome D. Gordon (from 8 Feb)
Capt Alvin F. Mackin (from 7 Apr)
1stLt Thomas W. Burke (from 21 Jul)
Capt John H. Chafee (from 15 Sep)
Capt Charles P. Logan, Jr. (from 15 Dec)

Commanding Officer, Company E .... 1stLt David H. Vanderwart (to 21 Jan 1951)
1stLt Robert T. Bey (from 22 Jan)
Capt Walter R. Anderson, Jr. (from 8 Feb)
Capt Merlin T. Matthews (from 17 Feb)
1stLt Robert W. Schmidt (from 14 Jun)
1stLt Charles P. Logan, Jr. (from 18 Sep)
Capt Embree W. Maxson (from 5 Oct)
Capt Donald McGuire (from 21 Mar 1952)

Commanding Officer, Company F .... 1stLt Ronald J. Rice (to 1 Mar 1951)
1stLt Ross R. Minor (from 2 Mar)
Capt Raymond N. Bowman (from 6 Mar)
1stLt Ross R. Minor (from 1 May)
Capt Donald S. McClellan (from 23 Jun)
1stLt Don G. Phelan (from 24 Aug)
Capt Harry E. Leland, Jr. (from 14 Oct)
1stLt Rex C. Wells (from 17 Jan 1952)

Commanding Officer, Weapons Company ...... Maj Joseph L. Abel (to 7 Jan 1951)
Maj James P. Metzler (from 8 Jan)
Capt John R. Grove (from 19 Apr)

Capt Harry L. Givens (from 20 May)
Capt Alvin F. Mackin (from 8 Aug)
Capt David A. McKay (from 24 Sep)
Capt Walter Oberg (from 26 Nov)
1stLt Elmer R. Phillips (from 17 Feb 1952)
Maj Dennis D. Nicholson (from 16 Mar)
Capt Owen G. Jackson, Jr. (from 30 Mar)

3d Battalion, 7th Marines

Commanding Officer ...... Maj Maurice E. Roach, Jr. (to 13 Jan 1951)
LtCol Wilbur F. Meyerhoff (from 14 Jan)
Maj Maurice E. Roach, Jr. (from 16 Feb)
LtCol Bernard T. Kelly (from 8 May)
LtCol Harry W. Edwards (from 4 Oct)
LtCol Houston Stiff (from 12 Mar 1952)

Executive Officer .... Maj Warren Morris (to 8 Jan 1951)
Maj Maurice E. Roach, Jr. (from 9 Jan)
Maj Warren Morris (from 16 Feb)
Maj James J. Bott (from 4 Jul)
Capt Howard L. Mabie (from 4 Aug)
Maj Robert B. Prescott (from 6 Aug)
Maj Franklin G. Bacon (from 3 Jan 1952)

Commanding Officer, Headquarters and Service Company .... 1stLt Samuel B. Abston (to 7 Jan 1951)
Capt John DeCloud (from 8 Jan)
1stLt Samuel D. Miller (from 5 Mar)
1stLt Frank N. Winfrey (from 15 May)
1stLt Robert H. Starek (from 25 May)
1stLt William R. Bennett (from 21 Jul)
1stLt Dennis E. Youngblood (from 6 Oct)
1stLt Raymond B. McGill (from 28 Nov)
Capt Clayton A. Lodoen (from 2 Mar 1952)

Commanding Officer, Company G .... 1stLt George R. Earnest (to 31 Dec 1950)
Capt Walter E. Lange (from 1 Jan 1951)
1stLt George R. Earnest (from 11 Mar)
1stLt Frank N. Winfrey (from 22 Mar)
Capt William C. Airheart (from 28 Mar)
1stLt Edward J. Sullivan (from 22 Jul)
Capt Robert C. Hendrickson (from 12 Aug)
Capt Thomas D. Smith, Jr. (from 14 Dec)
1stLt Harry H. Saltzman (from 11 Feb 1952)
Capt Thomas P. O'Callaghan (from 23 Feb)

Commanding Officer, Company H .... 1stLt William C. Airheart (to 19 Jan 1951)
Capt James A. Hoey, Jr. (from 20 Jan)
Capt Reed T. King (from 5 Jun)

1stLt Dwight A. Young (from 4 Aug)
Capt Clayton A. Lodoen (from 9 Nov)
1stLt William B. Stengle (from 22 Feb 1952)
Capt William B. Cosgrove (from 17 Mar)

Commanding Officer, Company I .... Capt Howard L. Mabie (to 15 Feb 1951)
1stLt Alfred I. Thomas (from 16 Feb)
1stLt Victor Stoyanow (from 29 Mar)
1stLt Frank N. Winfrey (from 5 Jun)
1stLt Thomas N. Preston (from 20 Jun)
1stLt Richard L. Shell (from 23 Jul)
Maj Hildeburn R. Martin (from 5 Sep)
Capt Clifford G. Moore (from 14 Sep)
1stLt Charles H. Hammett (from 27 Dec)
1stLt Hubert McEntyre (from 2 Mar 1952)
Capt Gifford S. Horton (from 9 Mar)

Commanding Officer, Weapons Company ...... Maj Jefferson D. Smith, Jr. (to 16 Feb 1951)
Capt Howard L. Mabie (from 17 Feb)
1stLt Frederick Van Brunt (from 8 Apr)
Capt Howard L. Mabie (from 19 Apr)
Maj James J. Bott (from 4 Jun)
1stLt Alfred I. Thomas (from 4 Jul)
Capt Claudie "M" Hollingsworth (from 8 Jul)
Capt William C. Airheart (from 12 Aug)
Capt Theodore E. Metzger (from 4 Nov)
Capt Thomas P. O'Callaghan (from 27 Dec)
1stLt Louis A. Mann (from 22 Feb 1952)

11th Marines

Commanding Officer ...... LtCol Carl A. Youngdale (to 5 Mar 1951)
Col Joseph L. Winecoff (from 6 Mar)
Col Custis Burton, Jr. (from 5 Aug)
Col Bruce T. Hemphill (from 17 Nov)
Col Frederick P. Henderson (from 27 Mar 1952)

Executive Officer .... LtCol Douglas A. Reeve (to 5 Mar 1951)
LtCol Carl A. Youngdale (from 6 Mar)
LtCol Douglas A. Reeve (from 7 May)
LtCol Merritt Adelman (from 13 Jun)
LtCol Albert H. Potter (from 15 Aug)
LtCol Lewis A. Jones (from 23 Nov)

S-1 .... Maj Floyd M. McCorkle (to 10 Jun 1951)
Capt Arthur L. Jackson (from 11 Jun)
1stLt Jessie R. Collins (from 2 Oct)

S-2 .... Capt William T. Phillips (to 26 Aug 1951)
Capt Vernon K. Ausherman (from 27 Aug)

	Capt Phillip A. Schloss, Jr. (from 17 Dec)
	Capt Marshall R. Hunter, Jr. (from 20 Feb 1952)
S–3	LtCol James O. Appleyard (to 19 Jul 1951)
	LtCol William H. Gilliam (from 20 Jul)
	LtCol William F. Pala (from 18 Nov)
S–4	Maj Donald V. Anderson (to 5 Feb 1951)
	Maj Thomas M. Coggins (from 6 Feb)
	Maj Benjamin W. Muntz (from 23 Jul)
	Capt Robert B. Carney (from 14 Sep)
Commanding Officer, Headquarters Battery	Capt Clarence E. Hixon (to 7 Apr 1951)
	1stLt Thomas C. Thompson (from 8 Apr)
	Capt Richard L. McDaniel (from 22 Aug)
	Maj Claudie "M" Hollingsworth (from 24 Sep)
	2dLt Chester E. Reese (from 17 Nov)
	1stLt Samuel S. Rockwood (from 9 Mar 1952)
Commanding Officer, Service Battery	Maj Thomas M. Coggins (to 5 Feb 1951)
	1stLt Fred Rea (from 6 Feb)
	1stLt John F. Gresham (from 21 May)
	2dLt Chester E. Reese (from 7 Nov)
	Capt Warren G. Hopkins (from 17 Nov)
	Capt William B. Tom (from 16 Dec)
Commanding Officer, Battery C, 1st 4.5 Inch Rocket Battalion	1stLt Eugene A. Busche (to 11 Jul 1951)
	1stLt Edward A. Bailey (from 12 Jul)
	1stLt Stephen R. Mihalic (from 2 Nov)
	1stLt Edward J. Pierson (from 30 Mar 1952)

1st Battalion, 11th Marines

Commanding Officer	LtCol Harvey A. Feehan (to 30 Mar 1951)
	Maj Thomas F. Cave, Jr. (from 31 Mar)
	Maj Gordon R. Worthington (from 8 Aug)
	LtCol Sherman W. Parry (from 13 Sep)
	LtCol James R. Haynes (from 30 Mar 1952)
Executive Officer	Maj Thomas F. Cave (to 30 Mar 1951)
	Maj Gordon R. Worthington (from 31 Mar)
	Maj George J. Kovich, Jr. (from 8 Aug)
	Maj Harold E. Nelson (from 17 Sep)
Commanding Officer, Headquarters Battery	Capt Haskell C. Baker (to 2 Jan 1951)
	Capt Arnold C. Hofstetter (from 3 Jan)
	Capt Alonzo C. Thorson (from 3 May)
	Capt John McCaffrey (from 2 Jul)
	Capt Rodman E. Street (from 17 Oct)

1stLt Charles D. Branson (from 26 Dec)
1stLt Harley "B" Riley (from 1 Feb 1952)
1stLt Joseph P. McDermott, Jr. (from 26 Mar)

Commanding Officer, Service Battery .... Capt Arnold C. Hofstetter (to 1 Jan 1951)
1stLt Kenneth H. Quelch (from 2 Jan)
Capt Philip D. Higby (from 1 Mar)
Capt Mont G. Kenney (from 9 Jul)
Capt Mansfield L. Clinnick (from 9 Jan 1952)

Commanding Officer, Battery A .... Capt James D. Jordan (to 1 Apr 1951
Capt Mont G. Kenney (from 2 Apr)
Capt Philip D. Higby (from 10 Jul)
Capt Joseph A. Goeke (from 22 Jul)
1stLt Richard J. Randolph, Jr. (from 11 Sep)
1stLt Robert O. Martin, Jr. (from 3 Oct)
Capt Duane W. Skow (from 9 Nov)
Capt Rodman E. Street (from 24 Dec)

Commanding Officer, Battery B .... Capt Gilbert N. Powell (to 12 Jun 1951)
Capt Charles D. Corpening (from 13 Jun)
Capt Leslie C. Procter, Jr. (from 27 Aug)
1stLt Donald T. Clark (from 13 Dec)
1stLt Jefferson S. Smith (from 1 Feb 1952)

Commanding Officer, Battery C .... Capt William J. Nichols, Jr. (to 14 Feb 1951)
Capt Haskell C. Baker (from 15 Feb)
Capt Glenn L. Tole (from 14 Jul)
Capt Mansfield L. Clinnick (from 12 Sep)
1stLt Harold H. Ramsour (from 5 Jan 1952)
Capt James C. Gasser (from 26 Mar 1952)

2d Battalion, 11th Marines

Commanding Officer ...... Maj Francis R. Schlesinger (to 4 Mar 1951)
Maj Jack C. Newell (from 5 Mar)
LtCol Merritt Adelman (from 14 Mar)
LtCol Dale H. Heely (from 13 Jun)
LtCol George B. Thomas (from 1 Jan 1952)

Executive Officer .... Maj Neal C. Newell (to 15 Mar 1951)
Maj Bruce E. Keith (from 16 Mar)
Maj Horace W. Card, Jr. (from 12 May)
Maj Peter J. Mulroney (from 4 Aug)
Maj Claudie "M" Hollingsworth (from 14 Aug)
Maj Frank W. Keith (from 11 Sep)
Maj James R. Haynes (from 1 Nov)

Maj Peter J. Mulroney (from 29 Nov)
Maj James R. Haynes (from 15 Dec)
Maj Morris R. Snead (from 29 Mar 1952)

Commanding Officer,
Headquarters Battery .... Capt George J. Batson, Jr. (to 27 Jun 1951)
1stLt Howard A. Blancheri (from 28 Jun)
Capt Raymond D. Spicer (from 3 Oct)
1stLt John J. Scollay (from 29 Oct)
2dLt Arthur H. Westing (from 15 Jan 1952)
2dLt John E. Buynak (from 16 Feb)
1stLt Ivan B. Clevinger (from 13 Mar)

Commanding Officer,
Service Battery .... Capt Herbert R. Merrick, Jr. (to 24 Feb 1951)
Capt William D. Gibson (from 25 Feb)
1stLt Walter L. Blocker (from 30 Jun)
Capt Robert N. Kreider (from 20 Jul)
1stLt Robert E. Santee (from 6 Oct)
1stLt Donald F. Schaller (from 3 Feb 1952)
1stLt James W. Bell (from 16 Feb)

Commanding Officer,
Battery D .... Capt Richard E. Roach (to 18 Mar 1951)
Capt William D. Stubbs, Jr. (from 19 Mar)
Capt Walter L. Blocker, Jr. (from 4 Aug)
1stLt John M. Hoben (from 4 Nov)

Commanding Officer,
Battery E .... Capt Richard N. Aufmann (to 25 Feb 1951)
Capt Herbert R. Merrick, Jr. (from 26 Feb)
Capt Robt. E. Dawson (from 2 Apr)
Capt Herbert R. Merrick, Jr. (from 27 Apr)
Capt George J. Batson, Jr. (from 28 Jun)
1stLt Albert "G" Harris, III (from 7 Aug)
Capt Raymond D. Spicer (from 11 Dec)

Commanding Officer,
Battery F .... 1st Lt Howard A. Blancheri (to 20 Jan 1951)
Capt George J. Kovich, Jr. (from 21 Jan)
Capt Robert E. Dawson (from 3 May)
Capt William D. Gibson (from 30 Jun)
1stLt James F. Shea (from 13 Aug)
1st Lt James W. Bell (from 8 Nov)
Capt Robert E. Dawson (from 24 Nov)
Capt John S. Adamson (from 24 Dec)
1stLt Frederick A. Koch, Jr. (from 31 Dec)

3d Battalion, 11th Marines

Commanding Officer ...... LtCol Francis F. Parry (to 6 Feb 1951)
LtCol William McReynolds (from 7 Feb)
Maj James R. Haynes (from 6 Sep)

LtCol James F. Coady (from 23 Oct)
LtCol Henry E. Barnes (from 2 Mar 1952)

Executive Officer .... Maj Norman A. Miller, Jr. (to 14 Jul 1951)
Maj Stephen K. Pawloski (from 15 Jul)
Maj James R. Haynes (from 16 Aug)
Maj Carl A. Neilson (from 6 Sep)
Maj Richard H. Jeschke, Jr. (from 1 Dec)
Maj Charles A. Lipot (from 4 Mar 1952)

Commanding Officer, Headquarters Battery .... 1stLt John J. Brackett (to 20 Jan 1951)
1stLt Eugene H. Brown (from 21 Jan)
1stLt Robert C. Cameron (from 6 Apr)
Capt Donald H. Campbell (from 21 May)
1stLt Robert H. Maurer (from 2 Aug)
1stLt Thomas E. Driscoll (from 18 Aug)
1stLt Hugh W. Manning (from 6 Sep)
2dLt John B. Buynak (from 7 Oct)
Capt Thomas L. Sullivan (from 20 Nov)
2dLt Thomas P. McGeeney, Jr. (from 3 Jan 1952)
2dLt Albert E. Shaw, Jr. (from 19 Feb)
1stLt William A. Barton, Jr. (from 14 Mar)

Commanding Officer, Service Battery .... Capt Samuel A. Hannah (to 25 Feb 1951)
1stLt Lawrence T. Kane (from 26 Feb)
1stLt David D. Metcalf (from 4 Apr)
Capt Arthur S. Tarkington (from 10 Sep)
Capt Charles J. Small (from 27 Nov)

Commanding Officer, Battery G .... Capt Ernest W. Payne (to 14 Jul 1951)
Capt Arthur S. Tarkington (from 15 Jul)
1stLt Arthur H. Fugalsoe (from 6 Sep)
1stLt Mervyn E. Kerstner (from 11 Sep)
1stLt Arthur H. Fugalsoe (from 15 Sep)
1stLt Edward S. McCabe (from 1 Nov)
1stLt Joseph M. Vosmik (from 13 Mar 1952)

Commanding Officer, Battery H .... Capt Mason D. McQuiston (to 24 Aug 1951)
Capt David D. Metcalf (from 25 Aug)
1stLt William A. Barton, Jr. (from 1 Nov)
1stLt George E. Chambers, Jr. (from 21 Jan 1952)
1stLt Russell E. Blagg (from 17 Mar)

Commanding Officer, Battery I .... Capt Robert T. Patterson, Jr. (to 13 Jun 1951)
Capt Floyd R. Jaggears (from 14 Jun)
Capt Donald H. Campbell (from 2 Aug)

1stLt Homer C. Wright (from 12 Aug)
Capt Donald H. Campbell (from 25 Aug)
1stLt Homer C. Wright (from 9 Sep)
1stLt Charles R. Davidson, Jr. (from 19 Feb 1952)

4th Battalion, 11th Marines

Commanding Officer ...... Maj William McReynolds (to 6 Feb 1951)
Maj Maurice J. Coffey (from 7 Feb)
Maj Norman A. Miller, Jr. (from 16 Jul)
LtCol Louis A. Jones (from 6 Sep)
LtCol William M. Gilliam (from 24 Nov)

Executive Officer .... Maj Maurice J. Coffey (to 6 Feb 1951)
Maj Donald V. Anderson (from 7 Feb)
Maj Bernard W. Giebler (from 17 Aug)
LtCol Bruce F. Hillan (from 24 Feb 1952)

Commanding Officer, Headquarters Battery .... 1stLt Michael B. Wier (to 10 Jun 1951)
1stLt Frank P. Zarzeka (from 11 Jun)
1stLt Arthur Coburn (from 21 Aug)
1stLt Paul R. Joyce (from 28 Aug)
1stLt Thomas C. Thompson, Jr. (from 25 Nov)
1stLt Earl C. Senter (from 10 Feb 1952)

Commanding Officer, Service Battery .... Capt Aldor B. Elmquist (to 9 Jun 1951)
1stLt Matthew J. Dennin (from 10 Jun)
1stLt William A. Mazzarella (from 1 Jul)
Capt Matthew J. Dennin (from 2 Sep)
Capt Eugene A. Frank (from 8 Sep)
Capt Matthew J. Dennin (from 16 Oct)
1stLt Leland B. Elton (from 19 Nov)

Commanding Officer, Battery K .... Capt Arthur D. Challacombe, Jr. (to 4 Aug 1951)
1stLt Albert E. Coffeen (from 5 Aug)
1stLt Paul M. Rice (from 23 Dec)
1stLt William L. Jesse (from 17 Mar 1952)

Commanding Officer, Battery L .... Capt Armond G. Daddazio (to 15 Apr 1951)
Capt Eugene A. Frank (from 16 Apr)
Capt William M. Sigler, Jr. (from 7 Sep)
1stLt Dennis Manko (from 28 Nov)

Commanding Officer, Battery M .... Capt Vernon W. Shapiro (to 3 Feb 1951)
Capt Charles E. Walker (from 14 Feb)
Capt Walter E. Magon (from 18 Jun)

1stLt George C. Briggs, Jr. (from 28 Nov)
1stLt Louis M. Dunklin (from 10 Feb 1952)
1stLt Billy J. White (from 18 Mar)

1st Amphibian Tractor Battalion

Commanding Officer ...... LtCol Erwin F. Wann, Jr. (to 26 Sep 1951)
LtCol Michiel Dobervich (from 27 Sep)

Executive Officer .... Maj Arthur J. Barrett (to 14 Sep 1951)
Maj William L. Eubank (from 15 Sep)

Commanding Officer,
Headquarters Company .. Capt Frank E. Granucci (to 12 Jun 1951)
Capt Lawrence H. Woods (from 13 Jun)
Capt Thomas J. Melcher (from 15 Sep)
1stLt Richard R. Myers (from 9 Jan 1952)
1stLt William H. Gatlin (from 10 Mar)

Commanding Officer,
Company A .... Maj James P. Treadwell (to 6 Apr 1951)
Maj Thomas H. Boler (from 7 Apr)
Capt Harry A. Steinmeyer (from 1 May)
Capt Dudley F. McGeehan (from 17 May)
Capt Robert L. Stuford (from 10 Jan 1952)

Commanding Officer,
Company B .... Capt Russell Hamlet (to 11 Apr 1951)
Capt Dudley F. McGeehan (from 12 Apr)
Capt John C. Crawley (from 17 May)
Capt Carl L. Hill (from 10 Jun)
Capt Harold W. Stroschein (from 1 Jan 1952)
Capt Samuel L. Eddy (from 10 Jan)

Commanding Officer,
Company C .... Maj Arthur J. Noonan (to 8 Aug 1951)
Maj William L. Eubank (from 9 Aug)
Maj Edward C. Nelson (from 10 Sep)
Capt Samuel L. Eddy (from 19 Dec)
Capt Robert T. Johnson (from 9 Jan 1952)

1st Armored Amphibian Battalion

Commanding Officer ...... LtCol Francis H. Cooper (to 15 Jun 1951)
Maj George M. Warnke (from 16 Jun)
LtCol John T. O'Neill (from 2 Oct)

Executive Officer .... Maj Richard G. Warga (to 7 Apr 1951)
Maj George M. Warnke (from 8 Apr)
Maj Bernard G. Thobe (from 16 Jun)
Maj Robert J. Murphy (from 1 Oct)
Maj David Young (from 6 Jan 1952)
LtCol James L. Jones (from 29 Feb)

634040 O-62-22

Commanding Officer, Headquarters Company	Capt Roger B. Thompson (to 10 May 1951) 1stLt Jean T. Fox (from 11 May) Capt Richard P. Greene (from 18 Jun) 1stLt Edward J. Sullivan (from 12 Oct) 2dLt Newton C. Tullis (from 2 Dec)
Commanding Officer, Company A	Capt Bernard G. Thobe (to 25 Apr 1951) 1stLt Clyde P. Guy (from 26 Apr) Maj Rex Z. Michael, Jr. (from 5 Sep) Maj David Foos (from 3 Oct)
Commanding Officer, Company B	Capt Lewis E. Bolts (to 26 Jun 1951) Maj Ralph H. Platt (from 27 Jun) Maj John M. Scarborough (from 3 Oct) Capt John B. Harney (from 10 Feb 1952)
Commanding Officer, Service Company	Capt Rex Z. Michael, Jr. (to 4 Sep 1951) 1stLt Presley K. Saine (from 5 Sep) 2dLt John A. Boone (from 5 Nov) Capt William H. Chandler (from 16 Mar 1952)

1st Combat Service Group

Commanding Officer	Col John N. Cook, Jr. (to 10 Jun 1951) LtCol John M. Brickley (from 10 Jun) Col Joseph P. Sayers (from 9 Aug) Col Russell N. Jordahl (from 30 Sep)
Executive Officer	LtCol Edward A. Clark (to 17 Jan 1951) LtCol Randolph S. D. Lockwood (from 18 Jan) LtCol John H. Brickley (from 9 May) Maj Murray F. Rose (from 11 Jun) LtCol Robert K. McClelland (from 17 Aug) Maj John R. Blackett (from 1 Sep) LtCol Robert T. Stivers (from 22 Oct) LtCol James G. Kelly (from 6 Jan 1952)
Commanding Officer, Headquarters Company	Capt Francis L. Miller (to 11 Apr 1951) Capt Raymond E. Wase (from 12 Apr) Capt Billie G. Hagan (from 19 Apr) Capt George M. Zellick (from 22 Jul) 1stLt William P. Lacy (from 21 Sep) Capt James H. Shaw (from 15 Jan 1952)
Commanding Officer, Maintenance Company	Maj Edward H. Voorhees (to 19 May 1951) 1stLt Donald M. Dackins (from 20 May) Maj Berny L. Thurman (from 3 Sep) Capt Warren H. Allen (from 25 Nov) Maj John R. Blackett (from 31 Dec)

Commanding Officer, Supply Company .... Maj Robert W. Hengesbach (to 17 Apr 1951)
Capt Bernard L. Keiter (from 18 Apr)
1stLt John Spiropoulas (from 24 Nov)
Maj William D. Porter (from 29 Dec)

Commanding Officer, Support Company .... Maj Donald B. Cooley, Jr. (to 22 Jan 1951)
Maj James T. Breen (from 23 Jan)
Maj Mason H. Morse (from 10 Oct)
Maj Howard T. Pittman (from 4 Nov)

Commanding Officer, Truck Company .... Capt Jack W. Temple (to 10 Jun 1951)
1stLt Cecil C. Spencer (from 11 Jun)
1stLt Frank W. Dickel (from 7 Jul)
1stLt James H. Shaw (from 8 Sep)
Capt Jacob Stocker (from 24 Sep)

Commanding Officer, 1st Fumigation and Bath Platoon .... 1stLt James L. Dumas (to 14 Aug 1951)
1stLt Raymond S. Eason (from 15 Aug)
1stLt Roger B. Meade (from 6 Sep)

Commanding Officer, 1st Air Delivery Platoon..... Capt Hersel D. C. Blasingame (to 10 Jun 1951)
2dLt Robert C. Morton (from 11 Jun)
CWO John T. Eakes (from 26 Jun)
1stLt William A. Reavis (from 30 Dec)
2dLt William S. Daniels (from 7 Feb 1952)

1st Engineer Battalion

Commanding Officer ...... LtCol John H. Partridge (to 10 Jun 1951)
LtCol John V. Kelsey (from 11 Jun)

Executive Officer .... Maj Richard M. Elliott (to 1 Feb 1951)
Maj Emile P. Moses, Jr. (from 2 Feb)
Maj Grover C. Williams (from 4 Aug)

Commanding Officer, Headquarters Company .. Capt Edward D. Newton (to 24 Mar 1951)
1stLt Gerald W. Wade (from 25 Mar)
1stLt Lee A. Kirstein (from 16 Jun)
Capt Leonard L. Schultz (from 22 Aug)
Capt Donald F. Draeger (from 24 Nov)
Capt Robert W. Hurley (from 20 Dec)

Commanding Officer, Service Company .... Capt Phillip A. Terrell, Jr. (to 25 Mar 1951)
Maj Richard M. Elliott (from 26 Mar)
Maj Louis L. Ball (from 6 Sep)
Capt Thirl D. Johnson (from 10 Jan 1952)
1stLt Arthur L. Rourke (from 9 Mar)

Commanding Officer, Company A	Capt William B. Gould (to 20 Apr 1951) Capt Harold R. Gingher (from 21 Apr) 1stLt George L. Bowman (from 15 Jun) 1stLt Floyd L. Vuillemot (from 1 Oct) Capt Walter L. Hill (from 5 Nov)
Commanding Officer, Company B	Capt Orville L. Bibb (to 25 Mar 1951) Capt Phillip A. Terrill, Jr. (from 26 Mar 1951) 1stLt Gerald W. Wade (from 17 Oct) 1stLt Clyde R. Kolahan (from 1 Mar 1952)
Commanding Officer, Company C	Capt Lester G. Harmon (to 15 Aug 1951) 1stLt Robert L. Brown (from 16 Aug) 1stLt Robert J. Hickson (from 4 Nov)
Commanding Officer, Company D	Capt Byron C. Turner (to 30 May 1951) Capt Edward D. Newton (from 31 May) Capt Thirl D. Johnson (from 29 Jun) 1stLt Lee A. Kirstein (from 23 Sep) 1stLt John J. Killelea (from 23 Dec)

1st Medical Battalion

Commanding Officer	Cdr Howard A. Johnson, USN (to 22 Jan 1951) Cdr Clifford A. Stevenson, USN (from 23 Jan) Cdr Richard Lawrence, Jr., USN (from 23 Sep)
Executive Officer	Cdr William S. Francis, USN (to 8 Jan 1951) LtCdr Gustave T. Anderson, USN (from 9 Jan) Cdr George A. Schlesinger, USN (from 4 Jul) Cdr Lewis E. Rector, USN (from 9 Aug) LtCdr Merrill W. Rusher, USN (from 28 Oct) Cdr James C. Luce, USN (from 28 Feb 1952)
Commanding Officer, Headquarters and Service Company	Cdr William S. Francis, USN (to 8 Jan 1951) LtCdr Gustav T. Anderson, USN (from 9 Jan) Cdr Lewis E. Rector, USN (from 7 Jun) Cdr George C. Schlesinger, USN (from 4 Jul) Cdr Lewis E. Rector, USN (from 9 Aug) LtCdr Merrill W. Rusher, USN (from 28 Oct) Lt Edgar F. Bechtel, USN (from 16 Dec) Lt(jg) Charles P. Richardson, USN (from 21 Mar 1952)
Commanding Officer, Company A	Cdr Byron E. Bassham, USN (to 3 Mar 1951) Cdr Philip L. Nova, USN (from 4 Mar) Cdr James A. Addison, USN (from 18 Apr) LtCdr Arvin T. Henderson, USN (from 22 Sep)

Commanding Officer, Company B .... LtCdr James A. Kaufman, USN (to 12 Jun 1951)
LtCdr Francis M. Morgan, USN (from 13 Jun)
Lt James F. Mumma, USN (from 11 Aug)
Lt Robert Fahrner, USN (from 17 Sep)
Lt John T. St. Mary, USN (from 20 Sep)
Lt(jg) Leroy F. Von Lackum, USN (from 15 Oct)
LtCdr Merrill W. Rusher, USN (from 8 Nov)
CWO William R. Lipscomb, USN (from 27 Nov)
WO Clarence B. Mohler, USN (from 7 Dec)
WO William R. Stanberry, USN (from 22 Jan 1952)

Commanding Officer, Company C .... Cdr Harold A. Streit, USN (to 8 Jan 1951)
Cdr Lewis E. Rector, USN (from 9 Jan)
LtCdr Merrill W. Rusher, USN (from 6 Jun)
Lt John P. McDonald, USN (from 28 Oct)
LtCdr Merrill W. Rusher, USN (from 27 Nov)
Lt(jg) Thaddeus H. Doggett, USN (from 26 Dec)
LtCdr James A. McLaughlin, USN (from 11 Jan 1952)
Lt(jg) Thaddeus H. Doggett, USN (from 7 Feb)

Commanding Officer, Company D .... LtCdr Gustave J. Anderson, USN (to 7 Jan 1951)
LtCdr Daniel M. Pino, USN (from 8 Jan)
Lt(jg) Hermes C. Grillo, USN (from 10 Aug)
Lt(jg) Powell H. Perkins, USN (from 8 Dec)
LtCdr James A. McLaughlin, USN (from 6 Feb 1952)

Commanding Officer, Company E .... LtCdr Charles K. Holloway, USN (to 8 Jan 1951)
LtCdr John H. Cheffey, USN (from 9 Jan)
LtCdr Robert G. Allen, USN (from 13 Jun)
Lt Robert J. Fahrner, USN (from 9 Sep)
LtCdr Clifford R. Hall, USN (from 17 Oct)

1st Motor Transport Battalion

Commanding Officer ...... LtCol Olin L. Beall (to 15 Mar 1951)
LtCol John R. Barreiro, Jr. (from 16 Mar)
LtCol Howard E. Wertman (from 18 Aug)

Executive Officer .... Maj John R. Barreiro, Jr. (to 15 Mar 1951)
Maj Edward L. Roberts (from 16 Mar)
Maj Eero Nori (from 6 Aug)
Capt Howard Dismeier (from 3 Feb 1952)
Maj Raymond L. Luckel (from 7 Mar)

Commanding Officer, Headquarters and Service Company .... Capt George B. Loveday (to 4 May 1951)
1stLt John C. O'Connell (from 5 May)
2dLt Walter R. Gustafson (from 21 Jul)
1stLt John C. O'Connell (from 17 Aug)
Capt Seneker Woll (from 1 Sep)
1stLt Eldon F. Kennedy (from 9 Jan 1952)

Commanding Officer, Company A .... Capt Arthur W. Ecklund (to 3 May 1951)
1stLt Mildridge E. Mangum (from 4 May)
Capt Arnold T. Reed (from 4 Sep)
1stLt Walter A. Knopp (from 30 Mar 1952)

Commanding Officer, Company B .... Capt James C. Camp, Jr. (to 9 Aug 1951)
1stLt Marshall "A" Webb, Jr. (from 10 Aug)
1stLt Gerald W. Gruber (from 13 Sep)

Commanding Officer, Company C .... 1stLt Norman E. Stow (to 15 Aug 1951)
Capt Joe P. England (from 16 Aug)

Commanding Officer, Company D .... 1stLt William D. Pothoff (to 8 Oct 1951)
1stLt Eldon F. Kennedy (from 9 Oct)
Capt Leroy P. Oetter (from 17 Oct)

Commanding Officer, Automotive Support Company .... 1stLt Mildridge E. Mangum (to 16 Feb 1951)
Capt Walter J. Desel, Jr. (from 17 Feb)
1stLt Marshall "A" Webb, Jr. (from 14 May)
Capt Leon Serkin (from 1 Aug)
Capt Charles R. Godwin (from 4 Nov)

Commanding Officer, Automotive Maintenance Company .... Maj Edward L. Roberts (to 15 Mar 1951)
Capt Victor E. Sellers (from 16 Mar)
Capt Ira N. Hayes (from 10 Apr)
Capt Harold L. Mayfield (from 3 Aug)
Maj Marion D. Grush (from 5 Nov)

7th Motor Transport Battalion

Commanding Officer ...... LtCol Carl J. Cagle (to 1 Oct 1951)
Maj Walter R. O'Quinn (from 2 Oct)

	Maj Herbert E. Pierce (from 3 Jan 1952)
Executive Officer	Maj Vernon A. Tuson (to 26 Jul 1951)
	Capt Joseph L. Bunker (from 27 Jul)
	Maj Walter R. O'Quinn (from 19 Sep)
	Maj Ben Sutts (from 2 Oct)
Commanding Officer, Headquarters and Service Company	2dLt Henry F. Finney (to 13 Jan 1951)
	2dLt Palmer B. Fordham (from 14 Jan)
	1stLt Richard J. Keeling (from 10 Feb)
	1stLt Earl H. Johnson (from 10 Apr)
	1stLt Louis C. Tauber (from 13 Aug)
	1stLt Kenneth F. Smith (from 1 Sep)
	Capt John J. Wilkinson (from 1 Jan 1952)
Commanding Officer, Company A	Capt Ira N. Hayes (to 8 Apr 1951)
	1stLt Landon E. Christian (from 9 Apr)
	Capt Robert B. Stone (from 8 Aug)
	Capt John J. Wilkinson (from 1 Sep)
	Capt Kenneth F. Smith (from 1 Jan 1952)
Commanding Officer, Company B	Capt Clovis M. Jones (to 11 Mar 1951)
	1stLt Lawrence C. Norton (from 12 Mar)
	1stLt John B. Wilson (from 1 Sep)
	1stLt Clyde H. Loveday, Jr. (from 15 Jan 1952)
Commanding Officer, Company C	Capt Fred B. Rogers (to 16 Apr 1951)
	1stLt Oscar A. Bosma (from 17 Apr)
	1stLt Richard C. O'Dowd (from 6 Jun)
	Capt Roscoe C. Hibbard (from 23 Nov)
	Capt Clifton G. Moore (from 28 Dec)
Commanding Officer, Company D	Capt Joseph L. Bunker (to 26 Jul 1951)
	1stLt Hubert J. Thomas (from 27 Jul)
	1stLt Clyde H. Stratton (from 1 Sep)
	Capt Clyde H. Stratton (from 1 Jan 1952)

1st Ordnance Battalion

Commanding Officer	Maj Lloyd O. Williams (to 31 Aug 1951)
	Maj Harold C. Borth (from 1 Sep)
Executive Officer	Maj Samuel A. Johnstone, Jr. (to 5 Jul 1951)
	Capt Theodore Tunis (from 6 Jul)
	Capt Gordon H. Moore (from 1 Aug)
	Maj Harold C. Borth (from 13 Aug)
	Maj Eugene Anderson (from 1 Sep)
	Capt Thomas J. Belt, Jr. (from 1 Jan 1952)
	Capt Frederick V. Osborn (from 6 Feb)

Commanding Officer, Headquarters Company . . Capt Gordon H. Moore (to 9 Sep 1951)
2dLt Willie B. Hayter, Jr. (from 10 Sep)
1stLt Henry "H" Best, Jr. (from 7 Jan 1952)

Commanding Officer, Ordnance Supply Company . . . . 1stLt Victor F. Brown (to 10 Aug 1951)
Capt Simon W. Vevurka (from 11 Aug)
Capt Thomas J. Belt (from 1 Nov)

Commanding Officer, Ammunition Company . . . Capt Richard W. Sinclair (to 6 Apr 1951)
Capt Robert C. Holder (from 7 Apr)
Capt David A. Malinsky (from 12 Sep)
Capt Chester D. Brown, Jr. (from 4 Jan 1952)
Capt Cecil B. Smith (from 21 Feb)

Commanding Officer, Ordnance Maintenance Company . . . . Capt George L. Williams (to 15 Aug 1951)
Maj James H. Pierce (from 16 Aug)
1stLt Charles B. Haslam (from 10 Dec)
Capt William E. L. Donner (from 20 Jan 1952)
Capt Dwight H. Sawin, Jr. (from 17 Mar)

1st Service Battalion

Commanding Officer . . . . . . LtCol Charles L. Banks (to 11 Jan 1951)
Col Gould P. Groves (from 12 Jan)
LtCol Horace E. Knapp (from 27 Mar)
LtCol Woodrow M. Kessler (from 6 Jul)
LtCol Bernard W. McLean (from 3 Mar 1952)

Executive Officer . . . . Maj John R. Stone (to 18 Jun 1951)
Capt Victor E. Johnson, Jr. (from 19 Jun)
Maj Louis G. Monville (from 3 Jul)
Maj George E. Allison (from 18 Feb 1952)

Commanding Officer, Headquarters Company . . Capt Morse "L" Holladay (to 20 Jan 1951)
1stLt Robert E. Follendorf (from 21 Jan)
1stLt James B. Lichtenberger (from 3 Sep)
1stLt Peter N. Pappas (from 10 Oct)
Capt John E. Welch (from 31 Dec)
1stLt Joseph D. Walker (from 10 Jan 1952)
1stLt Harry H. Saltzman (from 10 Mar)

Commanding Officer, Supply Company . . . . Capt Robert A. Morehead (to 13 Apr 1951)
Capt George K. Reid (from 14 Apr)
Capt Hayward M. Friedrich (from 27 May)
Capt Milton W. Magee (from 6 Jun)

Maj James R. Fury (from 13 Aug)
Capt Warren G. Hopkins (from 26 Dec)
Capt John H. Tomlinson (from 11 Mar 1952)

Commanding Officer, Support Company ...... Capt Thomas M. Sagar (to 22 Jan 1951)
Capt Morse "L" Holladay (from 23 Jan)
1stLt Victor E. Johnson (from 30 Mar)
Capt Hayward M. Friedrich (from 7 Jun)
1stLt Glenn P. Gasaway (from 2 Jul)
1stLt Robert W. Blum (from 3 Sep)
Capt Robert E. Moyer (from 1 Oct)
1stLt Jack A. Mackenzie (from 13 Nov)
1stLt Carlton R. Appleby (from 21 Dec)
1stLt Barry D. Diamond (from 8 Jan 1952)
Capt Seneker Woll (from 10 Mar)

1st Shore Party Battalion

Commanding Officer ...... LtCol Henry P. Crowe (to 10 May 1951)
LtCol Horace H. Figuers (from 11 May)
LtCol Harry W. Edwards (from 17 Jul)
LtCol George G. Pafford (from 29 Sep)
LtCol Franklin B. Nihart (from 20 Dec)
LtCol Warren S. Sivertsen (from 9 Mar 1952)

Executive Officer .... LtCol Horace H. Figuers (to 10 May 1951)
Maj John G. Dibble (from 11 May)
Maj Frederick F. Draper (from 7 Aug)
Maj Joseph T. Smith, Jr. (from 6 Sep)
Maj Frederick F. Draper (from 7 Nov)

Commanding Officer, Headquarters and Service Company .... Maj James I. Glendinning, Jr. (to 2 Jan 1951)
Maj George A. Smith (from 3 Jan)
Maj Burt A. Lewis (from 19 May)
Maj William T. Miller (from 20 Jun)
1stLt Robert H. During (from 20 Aug)
Maj Edson W. Card (from 29 Aug)
Maj Paul R. Nugent (from 12 Sep)
Capt Quentin H. Kravig (from 19 Jan 1952)

Commanding Officer, Company A .... Maj Charles E. Ingram (to 1 Jul 1951)
Maj Orville L. Bibb (from 2 Jul)
Capt Calvin Wall (from 10 Aug)

Commanding Officer, Company B .... Maj Henry Brezinski (to 17 Jun 1951)
Capt William A. Reno (from 18 Jun)
Maj Charles E. Ingram (from 3 Jul)
Maj George W. Ellis, Jr. (from 29 Jul)
Capt Francis V. Clifford (from 8 Dec)

Commanding Officer,
Company C .... Maj Murray F. Rose (to 9 Jun 1951)
Capt Henry J. Jadrich (from 10 Jun)
Maj Burt A. Lewis, Jr. (from 21 Jun)
Maj Edson W. Card (from 4 Aug)
Capt William A. Reno (from 29 Aug)
Maj Edson W. Card (from 8 Sep)
Capt Robert T. Weis (from 12 Dec)

1st Signal Battalion

Commanding Officer ...... LtCol Robert L. Schreier (to 6 Apr 1951)
Maj Richard A. Glaeser (from 7 Apr)
Maj Alton L. Hicks (from 31 Aug)
LtCol John E. Morris (from 20 Oct)

Executive Officer .... Maj Elwyn M. Stimson (to 9 Mar 1951)
Maj Richard A. Glaeser (from 10 Mar)
Capt Marion J. Griffin (from 7 Apr)
Maj Robert W. Nelson (from 20 Apr)
Maj Alton L. Hicks (from 20 Oct)
Maj Ernest C. Bennett (from 12 Feb 1952)

Commanding Officer,
Headquarters Company .. 2dLt Merle W. Allen (to 1 Mar 1951)
1stLt Raymond B. Spicer (from 2 Mar)
2dLt Richard D. Alexander (from 18 Jun)
1stLt Frank J. Cerny (from 16 Aug)

Commanding Officer,
Signal Company .... Maj Richard A. Glaeser (to 8 Mar 1951)
Capt John H. McGuire (from 9 Mar)
Maj Harold S. Hill (from 17 Aug)
Maj Bolish J. Kozak (from 1 Mar 1952)

Commanding Officer,
ANGLICO .... Maj Frederick N. Steinhauser (to 24 Oct 1951)
Maj Walter R. Miller (from 25 Oct)
LtCol Alton L. Hicks (from 13 Feb 1952)

1st Tank Battalion

Commanding Officer ...... LtCol Harry T. Milne (to 21 Apr 1951)
LtCol Holly H. Evans (from 22 Apr)
Maj Walter E. Reynolds, Jr. (from 9 Feb 1952)

Executive Officer .... Maj Philip C. Morell (to 2 Sep 1951)
Maj Walter E. Reynolds, Jr. (from 3 Sep)
Maj Edward C. Nelson, Jr. (from 9 Feb 1952)

Commanding Officer,
Headquarters Company .. 1stLt John B. Lund (to 21 Sep 1951)
Capt Robert S. Grether (from 22 Sep)
1stLt Jack D. Sheldon (from 10 Mar 1952)

Commanding Officer, Service Company . . . . Maj Douglas E. Haberlie (to 3 Jul 1951)
Maj George W. Bubb (from 4 Jul)
Maj Edward C. Nelson (from 27 Dec)
Capt Robt. H. Vogel (from 9 Feb 1952)

Commanding Officer, Company A . . . . 1stLt Robert J. Craig (to 20 Jan 1951)
Maj Arthur M. Hale (from 21 Jan)
Capt Robert M. Krippner (from 31 Mar)
Capt John E. Scanlon (from 17 Apr)
Capt Joseph W. Luker (from 14 Jun)
Capt Robert S. Grether (from 3 Sep)
Capt Albert W. Snell (from 21 Sep)
1stLt William E. Young (from 19 Feb 1952)
Capt Milton L. Raphael (from 10 Mar)

Commanding Officer, Company B . . . . Capt Bruce F. Williams (to 1 Jul 1951)
Capt Paul F. Curtis (from 2 Jul)
Capt John E. Lund (from 2 Oct)
1stLt Paul A. Wood (from 5 Nov)
Capt Jack J. Jackson (from 29 Dec)

Commanding Officer, Company C . . . . Capt Richard M. Taylor (to 5 Aug 1951)
Maj Walter Moore (from 6 Aug)
Capt Thomas W. Clark (From 21 Nov)

Commanding Officer, Company D . . . . Capt Joseph W. Malcolm, Jr. (to 2 Sep 1951)
Capt James L. Carey (from 3 Sep)
Capt Charles A. Sooter (from 28 Nov)

Marine Observation Squadron 6

Commanding Officer . . . . . . Maj Vincent J. Gottschalk (to 31 Mar 1951)
Capt Clarence W. Parkins (from 1 Apr)
Maj David W. McFarland (from 5 Apr)
Maj Allan H. Ringblom (from 6 Oct)
Maj Edward R. Polgrean (from 1 Nov)
Maj Kenneth G. Smedley (from 1 Feb 1952)
Maj William G. MacLean, Jr. (from 11 Feb)
LtCol William T. Herring (from 27 Feb)

Executive Officer . . . . Capt Andrew L. McVicars (to 13 Jan 1951)
Capt Clarence W. Parkins (from 14 Jan)
Capt Kenneth C. Smedley (from 21 Jul)
Maj William G. MacLean, Jr. (from 21 Nov)

Marine Helicopter Transport Squadron 161

Commanding Officer . . . . . . LtCol George W. Herring (to 17 Dec 1951)
Col Keith B. McCutcheon (from 18 Dec)

Executive Officer . . . . Maj William P. Mitchell (to 19 Mar 1952)
Maj James R. Dyer (from 20 Mar)

FIRST MARINE AIRCRAFT WING (1st MAW)

1 January 1951–31 March 1952

Commanding General . . . . . . MajGen Field Harris (to 28 May 1951)
BrigGen Thomas J. Cushman (from 29 May)
MajGen Christian F. Schilt (from 27 Jul)
Asst Commanding General. . BrigGen Thomas J. Cushman (to 28 May 1951)
BrigGen William O. Brice (from 29 May)
BrigGen Frank H. Lamson-Scribner (from 29 Sep)
Chief of Staff. . . . . Col Caleb T. Bailey (to 18 Aug 1951)
Col Arthur F. Binney (from 19 Aug)
Col Carson A. Roberts (from 2 Jan 1952)
Col Arthur F. Binney (from 26 Mar)
Asst Chief of Staff, G–1. . . . Col Raymond E. Hopper (to 10 Feb 1951)
Col Alexander G. Bunker (from 11 Feb)
LtCol Owen M. Hines (from 1 Nov)
Col Robert O. Bisson (from 27 Feb 1952)
Asst Chief of Staff, G–2. . . . Col Roger T. Carleson (to 18 Feb 1951)
LtCol Winson V. Crockett (from 19 Feb)
Capt John E. Buckle (from 21 Jun)
Capt William G. Redel (from 1 Aug)
LtCol Chester A. Henry, Jr. (from 1 Sep)
LtCol John W. Stage (from 12 Jan 1952)
Asst Chief of Staff, G–3. . . . Col Edward C. Dyer (to 28 Feb 1951)
LtCol Howard A. York (from 1 Mar)
LtCol Neil R. MacIntyre (from 12 Mar)
Col Rivers J. Morrell, Jr. (from 26 Jun)
Col Stanley W. Trachta (from 19 Aug)
Col Rivers J. Morrell, Jr. (from 7 Sep)
Col Guy M. Morrows (from 14 Sep)
Col Stanley W. Trachta (from 21 Jan 1952)
Asst Chief of Staff, G–4. . . . Col Thomas J. Noon (to 14 May 1951)
Col Wallace T. Breakey (from 15 May)
LtCol Carl M. Longley (from 21 Jul)
Col Luther S. Moore (from 5 Sep)
Col Elmer T. Dorsey (from 7 Jan 1952)
Col Robert E. Galer (from 12 Mar)

Marine Aircraft Group 33 (MAG–33)

Commanding Officer . . . . . . Col Frank G. Dailey (to 29 Dec 1950)
LtCol Radford C. West (from 30 Dec)
LtCol Paul J. Fontana (from 15 Jan 1951)
LtCol Richard A. Beard, Jr. (acting) (from 2 Apr)

	Col Guy M. Morrow (from 9 Apr)
	Col Carson A. Roberts (from 31 Jul)
	Col Arthur F. Binney (from 2 Jan 1952)
	Col Martin A. Severson (from 27 Mar)
Executive Officer	LtCol Richard A. Beard, Jr. (to 18 May 1951)
	LtCol James B. Moore (from 19 May)
	LtCol Nathan T. Post, Jr. (from 14 Jul)
	LtCol John W. Stage (from 2 Sep)
	LtCol Nathan T. Post, Jr. (from 12 Jan 1952)
	LtCol Vernon O. Ullman (from 6 Feb)

Marine Air Base Squadron 33 (MABS–33)

Commanding Officer	LtCol Nathan T. Post (to 10 Jan 1952)
	LtCol Finley T. Clarke, Jr. (from 11 Jan)
	Maj Frank P. Barker, Jr. (from 27 Mar)
Executive Officer	Maj George K. Harshbarger (to 24 Apr 1952)

Marine Aircraft Maintenance Squadron 33 (MAMS–33)

Commanding Officer	LtCol Joseph W. Kean, Jr. (2 Dec 1951 to 21 Jan 1952)
	Maj Zadik Collier (from 22 Jan)
Executive Officer	Maj Alton C. Bennett (to 4 Dec 1951)
	Maj Zadik Collier (from 5 Dec)
	Maj Alton C. Bennett (from 22 Jan 1952)

Headquarters Squadron 33 (HQSQ, MAG–33)

Commanding Officer	Capt Grover C. McClure, Jr. (to 14 Apr 1951)
	Maj William D. Armstrong (from 15 Apr)
	Maj Raymond F. Scherer (from 28 Jul)
	Maj Morgan C. Webb, III (from 27 Aug)
	Capt Allen R. Schutter (from 27 Mar 1952)

Marine Service Squadron 33 (SMS–33) [1]

Commanding Officer	LtCol James C. Lindsay (to 23 Jan 1951)
	Maj Edward J. Montagne (from 24 Jan)
	Maj William M. Lundin (from 26 Jan)
	Maj Elmer P. Thompson, Jr. (from 1 Apr)
	LtCol Allen T. Barnum (from 2 Jul)
	LtCol Joseph W. Kean, Jr. (from 12 Nov)
Executive Officer	Maj Edward J. Montagne, Jr. (to 13 Mar 1951) [2]
	Maj Elmer P. Thompson, Jr. [3] (from 2 Jul)
	Maj George K. Harshbarger (from 7 Aug)

[1] SMS disestablished 1 Dec 1951—concurrently MABS–33 and MAMS–33 formed.

[2] No Exec listed after Montagne was detached sometime in March 1951 until July 1951. Thompson came aboard 13 Mar 1951 which may well be date that Montagne was detached as Exec—however, nothing is recorded to this effect.

[3] It is quite possible and logical that Thompson was Exec from 13 Mar–2 Apr 1951—when he became CO.

Marine Aircraft Group 12 (MAG–12)

Commanding Officer ...... Col Boeker C. Batterton (to 28 May 1951)
Col Stanley W. Trachta (from 29 May)
Col Richard C. Mangrum (from 1 Aug)
Col Luther S. Moore (from 2 Jan 1952)
Col Elmer T. Dorsey (from 1 Apr)
Executive Officer .... LtCol Donald K. Yost (to 24 Feb 1951)
LtCol Rivers J. Morrell, Jr. (from 25 Feb)
LtCol Richard W. Wyczawski (from 26 Jun)
LtCol William G. Thrash (from 18 Jul)
LtCol Hugh M. Elwood (from 8 Aug)
LtCol Jens C. Aggerbeck, Jr. (from 17 Nov)
LtCol Robert J. Hoey (from 27 Feb 1952)

Headquarters Squadron, (HQSQ, MAG–12)

Commanding Officer ...... Maj John E. Hays (to 31 Dec 1950)
Capt William E. Lesage (from 1 Jan 1951)
Maj Bradley K. Schwarz (from 4 Apr)
Maj David P. John (from 2 Sep)
Capt Joseph E. Givens (from 9 Oct)
Capt George Byers, Jr. (from 1 Feb 1952)

Marine Service Squadron 12 (SMS–12) [4]

Commanding Officer ...... LtCol Charles E. McLean, Jr. (to 28 Jul 1951)
Maj Perry L. Shuman (from 29 Jul)
Executive Officer .... Maj Joseph W. Mackin (to 2 Apr 1951)
Maj Howard W. Bollmann (from 3 Apr)
Maj Raphael Ahern (from 8 Aug)
Maj Robert E. Wall (from 3 Oct)

Marine Air Base Squadron 12 (MABS–12) (Commissioned 1 Dec 1951)

Commanding Officer ...... Maj Perry L. Shuman (to 5 Jan 1952) [5]
Maj Robert L. Bryson (from 6 Jan)
LtCol Carl M. Longley (from 1 Mar)
Executive Officer .... Maj Floyd C. Kirkpatrick (to 18 Dec 1951)
Maj Robert L. Bryson (from 19 Dec)
Maj Floyd C. Kirkpatrick (from 6 Jan 1952)
Maj Robert A. Collett (from 1 Mar)

Marine Aircraft Maintenance Squadron 12 (MAMS–12)
(Commissioned 1 Dec 1951)

---

[4] SMS–12 disestablished 1 Dec 1951—concurrently MABS–12 and MAMS–12 formed and commissioned.

[5] Narrative of Jan 1952 CD MABS–12 states Shuman det 4 Jan 1952 and Bryson on same date took over as CO. Assumption of command order states that 6 Jan 1952 was date Bryson became CO.

Commanding Officer ...... Maj Robert E. Wall (to 10 Feb 1952)
LtCol Carl M. Longley (from 11 Feb)
LtCol Joseph A. Gray (from 1 Mar)
Executive Officer .... Capt Kenneth A. Anderson (to 26 Dec 1951)
Maj "S" "D" G. Peterson (from 27 Dec) [a]
Maj Robert E. Wall (from Feb/Mar 1952) [a]

Marine Wing Service Squadron 1 (MWSS-1) (Decommissioned 1 Jul 1953)
and
Marine Wing Service Group 17 (MWSG-17) (Commissioned 1 Jul 1953)

Commanding Officer ...... CWO Aubrey D. Taylor (to 23 Jan 1951)
LtCol James C. Lindsay (from 24 Jan)
Col Roger T. Carleson (from 19 Feb)
Col Elmer T. Dorsey (from 9 Sep)
Col John Wehle (from 7 Jan 1952)
Executive Officer .... None shown prior to 19 Feb 1951.
LtCol James C. Lindsay (to 16 Jul 1951)
LtCol Alton D. Gould (from 17 Jul)
Maj Edward J. McGee (from 13 Nov)
LtCol Robert M. Haynes (from 2 Dec)
LtCol Birney B. Truitt (from 15 Mar 1952)

Marine Ground Control Intercept Squadron 1 (MGCIS-1)

Commanding Officer ...... Maj Harold E. Allen (to 10 Jun 1951)
LtCol Manual Brilliant (from 11 Jun)
Maj Edward R. Polgrean (from 18 Aug)
LtCol William T. Herring (from 18 Sep)
Maj Milton M. Cook (from 1 Feb 1952)
LtCol Herbert D. Raymond, Jr. (from 16 Feb)
Maj Fred A. Steele (from 28 Mar)
Executive Officer .... Maj Richard Hey, Jr. (to 3 Apr 1951)
Maj Casper F. Hegner (from 4 Apr)
Maj Edward R. Polgrean (from 31 Jul)
Maj William T. Porter (from 21 Nov)
Maj Milton M. Cook, Jr. (from 11 Dec)
Maj Marvin R. Bridges, Jr. (from 2 Feb 1952)
Maj Fred A. Steele (from 16 Feb)
Maj Marvin R. Bridges, Jr. (from 28 Mar)

Marine Transport Squadron 152 (VMR-152)

Commanding Officer ...... Col Deane C. Roberts (to 15 Jul 1951)
LtCol John S. Carter (from 16 Jul)
Col William B. Steiner (from 27 Jul)

---

[a] These dates are those from the Station Lists—the diary records nothing (except in the case of Beatty (20 Feb 1952)) that would either prove or disprove these dates as being correct.

Marine Fighter Squadron 212 (VMF–212) redesignated Marine Attack Squadron 212 (VMA–212) on 10 Jun 1952

Commanding Officer ...... LtCol Richard W. Wyczawski (to 9 Mar 1951)
LtCol Claude H. Welch (from 10 Mar)
LtCol Manual Brilliant (from 21 Aug)
LtCol Joseph A. Gray (from 11 Dec)
LtCol Robert L. Bryson (from 1 Mar 1952)
Executive Officer .... Maj Elmer P. Thompson, Jr. (to 18 Mar 1951)
Maj Edward J. Montagne, Jr. (from 19 Mar)
Maj Joseph W. Mackin (from 13 Apr)
Maj Floyd C. Kirkpatrick (from 16 Jul)
Maj William H. Rankin (from 20 Sep)
Maj Robert A. Collett (from 11 Dec)
Maj Richard B. Elliott (from 23 Feb 1952)

1st 90mm AAA Gun Battalion Arrived Pusan, Korea—29 Aug 1951

Battalion Commander ...... LtCol Charles W. May (KIA) (to 21 Dec 1951)
LtCol Kenneth P. Dunkle (from 22 Dec)
Col John F. Dunlap (from 30 Jan 1952)
Col Max C. Chapman (from 23 Mar)
Executive Officer .... Maj Kenneth P. Dunkle (to 21 Dec 1951)
None shown 22–25 Dec 1951.
Maj David H. Simmons (from 26 Dec)
LtCol Kenneth P. Dunkle (from 30 Jan 1952)

Marine Fighter Squadron 311 (VMF–311)

Commanding Officer ...... LtCol Neil R. MacIntyre (to 10 Mar 1951)
LtCol John F. Kinney (from 11 Mar)
Maj Frank S. Hoffecker (from 28 Jul)
LtCol James B. Moore (from 1 Aug)
LtCol John S. Payne (from 1 Dec)
LtCol Darrell D. Irwin (from 27 Feb 1952)
Executive Officer .... Maj John R. Stack (to 20 Feb 1951)
Maj Samuel Richards, Jr. (from 21 Feb)
Maj Samuel B. Folsom, Jr. (from Apr) [7]
Maj Frank S. Hoffecker, Jr. (from 1 Jun) (KIA)
Maj Frank C. Drury (from 25 Aug)
Maj Carroll E. McCullah (from 1 Jan 1952)
Maj Jay E. McDonald (from 16 Feb)

Marine Night-Fighter Squadron 513 (VMF(N)–513)

Commanding Officer ...... LtCol David C. Wolfe (to 22 Feb 1951)
LtCol James R. Anderson (from 23 Feb)
LtCol Robert R. Davis (from 1 Jul)

---

[7] The absence of a specific date indicates that no specific date of assignment is shown in unit records.

LtCol Allen T. Barnum (from 22 Nov)
Maj Frank H. Simonds (from 1 Feb 1952)
LtCol John R. Burnett (from 1 Mar)

Executive Officer .... Maj Albert L. Clark (to 18 Dec 1950)
Maj George B. Herlihy (from 19 Dec)
Maj William G. Johnson[8] (from Feb 1951)
Maj Evans C. Carlson (from 23 Apr)
Maj John E. Reynolds (from 7 May)
Maj Leo F. Tatro, Jr. (from 25 Aug)
Maj Judson C. Richardson, Jr. (MIA) (from 4 Oct)
Maj Frank H. Simonds (from 14 Dec)
Maj Leroy T. Frey (from 1 Feb 1952)
Maj Frank H. Simonds (from 1 Mar)

Marine Night-Fighter Squadron 542 (VMF(N)–542)

Commanding Officer ...... LtCol Max J. Volcansek, Jr. (to 5 Feb 1951)
LtCol James R. Anderson (from 6 Feb)
Maj Albert L. Clark (from 23 Feb)[9]
LtCol Peter D. Lambrecht (from 24 Mar)

Executive Officer .... Maj Robert T. Whitten (to 23 Jan 1951)
LtCol James R. Anderson (from 24 Jan)

Marine Fighter Squadron 323 (VMF–323) redesignated Marine Attack Squadron 323 (VMA–323) on 30 Jun 1952

Commanding Officer ...... Maj Arnold A. Lund (to 24 Jan 1951)
Maj Stanley S. Nicolay (from 25 Jan)
Maj Donald L. Clark (from 1 Mar)
Maj Charles M. Kunz (from 3 May)
LtCol George F. Vaughan (from 25 Sep)
Maj John L. Dexter (from 26 Oct)
LtCol Richard L. Blume (from 16 Jan 1952)

Executive Officer .... Maj Robert E. Johnson (to 31 Jan 1951)
Maj Donald L. Clark (from 1 Feb)
Maj Wilbur F. Evans, Jr. (from 1 Mar)
Maj John L. Dexter (from 7 Jul)
Maj Floyd C. Kirkpatrick (from 25 Oct)
Maj Andrew J. Voyles (from 22 Nov)
Maj Howard E. Cook (from 18 Dec)
Maj Herbert D. Raymond, Jr. (from 13 Jan 1952)
Maj Howard E. Cook (from 14 Feb)
Maj William A. Weir (from 16 Mar)

---

[8] The absence of specific dates indicates that no specific assignment dates can be found in existing records.

[9] VMF(N)–542: At sea bound for United States 12–21 Mar 1951—arrived El Toro, 24 Mar 1951.

Marine Air Control Group 2 (MACG–2) (Arrived Korea 11 Apr 1951)

Commanding Officer ...... LtCol Manual Brilliant (from 10 Apr 1951)
Col Edwin P. Pennebaker, Jr. (from 30 Apr)
Col Martin A. Severson (from 1 Jan 1952)
Col Frederick R. Payne, Jr. (from 1 Mar)

Executive Officer .... None shown during period LtCol Brilliant was CO.
LtCol Manual Brilliant (from 30 Apr 1951)
LtCol Joseph W. Kean (from 10 Jun)
LtCol Robert R. Davis (from 4 Dec)
LtCol Russell D. Rupp (from 6 Feb 1952)

Marine Tactical Air Control Squadron 2 (MTACS–2)

Commanding Officer ...... Maj Christian C. Lee (to 30 Apr 1951)
Maj James A. Etheridge (from 1 May)
Maj Milton M. Cook, Jr. (from 6 May)
Maj Wade W. Larkin (from 28 May)
LtCol Henry W. Bransom (from 25 Jun)
LtCol Hensley Williams (from 1 Dec 1951)

Executive Officer .... Maj Harlen E. Hood (to Mar/Apr 1951)
Maj James A. Etheridge (from 26 Apr) [10]
Maj Wade W. Larkin (from 1 May)
Maj Milton M. Cook, Jr. (from 28 May)
Maj Clinton E. Jones (from 23 Sep)

Marine Ground Control Intercept Squadron 3 (MGCIS–3)

Commanding Officer ...... Maj Raymond H. George (to 15 Feb 1951)
Maj Jack R. Moore (from 16 Feb)
LtCol Hoyle R. Barr (from 1 Nov)
LtCol Owen W. Hines (from 2 Mar 1952)

Executive Officer .... Maj David M. Hudson (to 15 Aug 1951)
Maj Daniel L. Cummings (from 16 Aug)
Maj James H. Foster (from 17 Feb 1952)

Marine Attack Squadron 121 (VMA–121)

(Departed El Toro—2 Oct 1951 for Korea; 21 Oct 1951 reported to CG, 1stMAW, for duty; 22 Oct 1951 CO arrived Pohang (K–3), Korea.)

Commanding Officer ...... LtCol Alfred N. Gordon (KIA) (to 17 Nov 1951)
Maj Frank P. Barker, Jr. (from 18 Nov)
LtCol Phillip B. May (from 1 Dec)
LtCol William A. Houston, Jr. (from 15 Mar 1952)

[10] His date of attachment is vague.

Executive Officer ....Maj Frank P. Barker, Jr. (to 17 Nov 1951)
Maj Edward B. Harrison (from 18 Nov)
Maj Frank P. Barker, Jr. (from 1 Dec)
Maj Edward B. Harrison (from 1 Jan 1952)
Maj Richard J. Flynn, Jr. (from 15 Feb)
Maj Henry W. Horst (from 26 Mar)

Marine Fighter Squadron 214 (VMF–214)

Commanding Officer ......Maj William M. Lundin (to 25 Jan 1951)
Maj James A. Feeley, Jr. (from 26 Jan)
Maj Edward Ochoa (from 5 May)
LtCol James W. Poindexter (from 16 May)
Maj Charles M. Kunz (from 4 Nov) [11]

Executive Officer ....Maj Edward Ochoa (to 31 Jan 1951)
Maj Hugh B. Calahan (from 1 Feb)
Maj Herbert C. Langenfeld [12] (from 1 Jun)

Marine Fighter Squadron 115 (VMF–115) (Arrived Pohang (K–3), Korea on 25 Feb 1952)

Commanding Officer ......LtCol Thomas M. Coles (25 Feb–20 May 1952)

Executive Officer ....Maj Conrad G. Winter (25 Feb–26 Apr 1952)

Marine Fighter Squadron (VMF–312) redesignated Marine Attack Squadron (VMA–312) on 1 Mar 1952

Commanding Officer ......LtCol "J" Frank Cole (to 28 Jan 1951)
Maj Donald P. Frame (KIA) (from 29 Jan)
Maj Frank H. Presley (from 4 Apr)
Maj Edward J. McGee (from 20 Jun)
LtCol Harry W. Reed (KIA) (from 22 Jul)
Maj Edward J. McGee (from 31 Jul)
LtCol Russell D. Rupp (from 15 Aug)
LtCol Joe H. McGlothlin, Jr. (from 8 Jan 1952)

Executive Officer ....Maj Frank H. Presley (to 3 Apr 1951)
Capt Phillip C. DeLong (from 4 Apr)
Maj Robert J. Shelley, Jr. (from 22 Jun)
Maj Edward J. McGee (from 22 Jul)
Maj Robert J. Shelley, Jr. (from 31 Jul)
Maj Edward J. McGee (from 14 Aug)
Maj James H. Crutchfield (KIA) (from 25 Oct)
Maj Jay W. Hubbard (from 4 Nov)
Maj Richard J. Webster (from 19 Dec) [13]
Maj Fred A. Steele (from Jan 1952) [13]

---

[11] VMF–214 departed Korea for Itami on 4 Nov 1951—en route to USA (El Toro) aboard the *Lenawee*, 8–27 Nov 1951.

[12] Records do not indicate specific date.

[13] Records do not indicate specific date.

	Maj Alexander S. Walker, Jr. (from 28 Jan)
	Maj Edmond P. Hartsock (from 30 Mar)

Photographic Unit—commissioned Marine Photographic Squadron 1 (VMJ-1) on 25 Feb 1952

Commanding Officer	Maj Donald S. Bush (to 14 Jun 1951)
	Maj Edgar L. Smith (from 15 Jun)
	Maj James W. Dougherty (from 27 Jul)
	Capt Edward A. Fitzgerald (from 29 Oct)
	LtCol Alton D. Gould (from 12 Nov)
	Maj Robert R. Read (from 26 Mar 1952)
Executive Officer	Maj Robert R. Read (to 25 Mar 1952)
	Maj Albert E. James (from 26 Mar)

HQSQ, 1st MAW

Commanding Officer	Capt Earl B. Sumerlin, Jr. (to 12 Jan 1951)
	Maj John A. Reeder (from 13 Jan)
	Capt Edwin H. McCaleb, III (from 17 Jun)
	Maj Herbert C. Langenfeld (from 11 Oct)
	Maj Earl C. Miles (from 2 Dec)

APPENDIX D

Unit Citations

THE SECRETARY OF THE NAVY

WASHINGTON

The President of the United States takes pleasure in presenting the PRESIDENTIAL UNIT CITATION to the

FIRST MARINE DIVISION, REINFORCED

for service as set forth in the following CITATION:

"For extraordinary heroism in action against enemy aggressor forces in Korea during the periods 21 to 26 April, 16 May to 30 June, and 11 to 25 September 1951. Spearheading the first counteroffensive in the spring of 1951, the First Marine Division, Reinforced, engaged the enemy in the mountainous center of Korea in a brilliant series of actions unparalleled in the history of the Marine Corps, destroying and routing hostile forces with an unrelenting drive of seventy miles north from Wonju. During the period 21 to 26 April, the full force of the enemy counteroffensive was met by the Division, north of the Hwachon Reservoir. Although major units flanking the Marine Division were destroyed or driven back by the force of this attack, the Division held firm against the attackers, repelling the onslaught from three directions and preventing the encirclement of the key center of the lines. Following a rapid regrouping of friendly forces in close contact with the enemy, the First Marine Division, Reinforced, was committed into the flanks of the massive enemy penetiation and, from 16 May to 30 June, was locked in violent and crucial battle which resulted in the enemy being driven back to the north with disastrous losses to his forces in the number of killed, wounded and captured. Carrying out a series of devastating assaults, the Division succeeded in reducing the enemy's main fortified complex dominating the 38th Parallel. In the final significant offensive of the action in Korea, from 11 to 25 September 1951, the First Marine Division, Reinforced, completed the destruction of the enemy forces in Eastern Korea by advancing the front against a final desperate enemy defense in the 'Punch Bowl' area in heavy action which completed the liberation of South Korea in this locality. With the enemy's major defenses reduced, his forces on the central front decimated, and the advantage of terrain and the tactical initiative passing to friendly forces, he never again recovered sufficiently to resume the offensive in Korea. The outstanding courage, resourcefulness and aggressive fighting spirit of the officers and men of the First Marine Division, Reinforced, reflect the highest credit upon themselves and the United States Naval Service."

The following reinforcing units of the First Marine Division participated in operations against enemy aggressor forces in Korea during the cited periods:

FLEET MARINE FORCE UNITS AND DETACHMENTS: "C" Battery, 1st 4.5 Rocket Battalion; 1st Combat Service Group; 1st Amphibian Tractor Battalion; 7th Motor Transport Battalion; 1st Armored Amphibian Battalion; "A" Company, 1st Amphibian Truck Battalion (Redesignated 1st Amphibian Truck Company 18 July 1951); Team #1, 1st Provisional Historical Platoon; 1st Fumigation and Bath Platoon; 1st Air Delivery Platoon; Radio Relay Team, 1st Signal Operations Company; Detachment, 1st Explosive Ordnance Disposal Company; 2nd Platoon, Auto Field Maintenance Company; 1st Provisional Truck Company; Detachment, 1st Air Naval Gunfire Liaison Company.

UNITED STATES ARMY UNITS: (For such periods not included in Army Unit Awards) 1st Bn, 32d Regt, 7th Inf Div; 7th Inf Div; 74th Truck Co; 513th Truck Co; 1st Ord Medium Maint Co, USA; 3d Plt, 86th Engr Searchlight Co (passed to operational control of 11th Marines); 558th Trans Truck Co (Amphibious, was attached to 7th MT Bn, FMF); 196th Field Arty Bn; 92d Army Engr Searchlight Plt; 181st CIC Det USA; 163d MIS Det USA; TLO Det USA; UNMACK Civil Affairs Team USA; 61st Engr Co; 159th Field Arty Bn (155 Howitzer); 623d Field Arty Bn; 17th Field Arty Bn "C" Btry; 204th Field Arty Bn "B" Btry; 84th Engr Construction Bn; 1st Bn, 15th US Inf Regt; 1st Bn, 65th US Inf Regt; 1st Bn, 9th Regt, 2d US Div (attached to KPR); Recon Co, 7th US Inf Div; 461st Inf Bn; Heavy Mortars, 7th Inf Div; 204th Field Arty Bn "A" Btry; 69th Field Arty Bn; 64th Field Arty Bn; 8th Field Arty Bn; 90th Field Arty Bn; 21st AAA-AW Bn; 89th Tank Bn; 441st CIC Det, USA; Prov Bn, USA (Dets 31st and 32d RCTS); Co D, 10th Engr (C) Bn, USA; Tank Co, 31st Inf, USA; Hqr Co, 31st Inf, USA; Co B, 1st Bn, 31st Inf, USA; 2d Bn, 31st Inf, USA (less Co E).

For the President,
CHARLES S. THOMAS
Secretary of the Navy

THE SECRETARY OF THE NAVY

WASHINGTON

The President of the United States takes pleasure in presenting the PRESIDENTIAL UNIT CITATION to the

FIRST MARINE AIRCRAFT WING, REINFORCED

for service as set forth in the following CITATION:

"For extraordinary heroism in action against enemy aggressor forces in Korea from 8 March to 30 April, 18 May to 30 June, and 3 August to 29 September

1951. Carrying out 'round-the-clock' combat flights during these periods, often under hazardous conditions of weather and terrain, the First Marine Aircraft Wing, Reinforced, provided unparalleled close air support for friendly ground forces, effectively reducing the enemy's power to resist and contributing materially to the sweeping victories achieved by our ground forces. Operating continuously in the most advanced areas under fire, the Wing consistently maintained a high degree of combat readiness and struck savage blows to inflict tremendous damage and heavy casualties upon the enemy. Individually capable and determined, the gallant officers and men of this indomitable team achieved a distinctive combat record during a period of vital operations against a stubborn foe. This record is a lasting tribute to the courage and fighting spirit of all members of the First Marine Aircraft Wing, Reinforced, and reflects the highest credit upon the United States Naval Service."

All organic units (excepting Marine Fighting Squadrons 214 and 323 for the periods 8 March to 30 April 1951 and 18 May to 30 June 1951, and Marine Observation Squadron 6 for the entire three periods) and the following reinforcing units of the First Marine Aircraft Wing participated in operations against enemy aggressor forces in Korea during one or more of the above cited periods: 1st 90mm Anti-Aircraft Artillery Gun Battalion and Ground Control Approach Unit 41M.

For the President,
Charles S. Thomas
Secretary of the Navy

Bibliography

DOCUMENTS

Department of the Army. Joint Daily Situation Reports, December 1950–March 1952. Reports and Orders (1950–1952) R&O File, HQMC Historical.

Smith, Oliver P. MajGen, USMC. Chronicle of the Operations of the First Nine Months of the Korean War, 1950–1951. MS. Manuscript File, HQMC Historical.

———. Notes on the Operation of the 1st Marine Division During the First Nine Months of the Korean War, 1950–1951. MS. Manuscript File, HQMC Historical.

U.S. Air Force. U.S. Air Force Historical Study Number 72 (Continuation of Army Historical Study Number 71), "U.S. Air Force Operations in Korean Conflict 1 November 1950 to 30 June 1952." July 1955. UHR File, HQMC Historical.

U.S. Marine Corps. Interviews with participants in the Korean War, 1950–54. Interviews (Korea) File, HQMC Historical.

———. Letters and comments from participants in the Korean War. Comments File, HQMC Historical.

———. Letters, memoranda, narratives, and statements received by Historical Branch, G–3, concerning Korean operations. Monograph and Comments File, HQMC Historical.

———. Commandant of the Marine Corps letter to Distribution List: "Analysis of Close Air Support Systems," 19 August 1952. HQMC Historical.

———. Report of Joint Army–Navy Mission at Headquarters U.S. Marine Corps, 9 November 1951. G–4 Files, HQMC.

———. *Instructional Information, Vest, Armored,* M–1951. G–4 Files, HQMC.

Commander in Chief, U.S. Pacific Fleet, Interim Evaluation Report Number 1, 25 June to 15 November 1950. 20 January 1951. 6 vols. UHR File, HQMC Historical.

———. U.S. Pacific Fleet. Interim Evaluation Report Number 2, 16 November 1950 to 30 April 1951. 6 vols. UHR File, HQMC Historical.

———. U.S. Pacific Fleet. Interim Evaluation Report Number 3, 1 May 1951 to 31 December 1951. 6 vols. UHR File, HQMC Historical.

———. U.S. Pacific Fleet. Interim Evaluation Report Number 4, 1 January to 30 June 1952. 6 vols. UHR File, HQMC Historical.

Far East Command. Allied Translator and Interpreter Service. Enemy Documents: Korean Operations. Intelligence File, HQMC Historical and Document File Section.

———. Operations Branch, Theater Intelligence Division, Military Intelligence Section. Order of Battle Information Chinese Communist Third Field Army. Intelligence File, HQMC Historical.

Chief, Army Field Forces Headquarters, Tactical Air Command. Joint Training Directive for Air-Ground Operations. UHR File, HQMC Historical.

Fleet Marine Force, Pacific. Chinese Communist Forces Tactics in Korea. 22 March 1951. UHR File, HQMC Historical.

———. Staff Study: The Establishment of a Balanced Fleet Marine Force Air-Ground Force in the Western Pacific. 19 October 1950. R&O File, HQMC Historical.

———. Historical Diaries, December 1950–March 1952. Diary File, HQMC Historical.

Eighth U.S. Army in Korea. War Diaries, Command Reports, and supporting documents, December 1950–March 1952. Departmental Records Branch, The Adjutant General's Office. Alexandria, Va. (DRB, TAGO).

———. Order of Battle Branch, Office of the Assistant Chief of Staff, G–2. CCF Army Histories. 1 December 1954. Copy at OCMH.

Fifth Air Force. Reports, orders, and supporting documents, May 1951–March 1952. Copies at HQMC Historical.

X Corps. War Diaries, Command Reports, and supporting documents, December 1950–March 1952. DRB, TAGO.

———. G–2 Section. Periodic Intelligence Reports, December 1950–March 1952. UHR File, HQMC Historical.

1st Marine Air Wing, FMF. Historical Diaries, December 1950–March 1952. Command Diary (Korea), Type B Report File (Diary File), HQMC Historical.

1st Marine Division, FMF. Historical Diaries, December 1950–March 1952. Diary File, HQMC Historical.

———. Periodic Intelligence Reports, December 1950–March 1952. Correspondence File, 1stMarDiv (Korea), HQMC Historical.

———. Periodic Operations Reports, December 1950–March 1952. Correspondence File, 1stMarDiv (Korea), HQMC Historical.

———. Reports, messages, journals, correspondence, orders, and miscellaneous matter, December 1950–March 1952. Correspondence File, 1stMarDiv (Korea), HQMC Historical.

———. Historical Diaries, December 1950–March 1952. Diary File, HQMC Historical.

Commander Amphibious Group One (CTF 90). Action Report Hungnam Operation; Period 9 December 1950 through 25 December 1950. 21 January 1951. R&O File, HQMC Historical.

Commander Amphibious Group Three (CTG 90.1). War Diary, January 1951. UHR File (Navy), HQMC Historical.

1st Marines. Historical Diaries, December 1950–March 1952. Diary File, HQMC Historical.

———. Unit Reports, December 1950–March 1952. Correspondence File, 1stMarDiv (Korea), HQMC Historical.

5th Marines. Historical Diaries, December 1950–March 1952. Diary File, HQMC Historical.

———. Unit Reports, December 1950–March 1952. Correspondence File, 1stMarDiv (Korea), HQMC Historical.

7th Marines. Historical Diaries, December 1950–March 1952. Diary File, HQMC Historical.

———. Unit Reports, December 1950–March 1952. Correspondence File, 1stMarDiv (Korea), HQMC Historical.

11th Marines. Historical Diaries, December 1950–March 1952. Diary File, HQMC Historical.

———. Unit Reports, December 1950–March 1952. Correspondence File, 1stMarDiv (Korea), HQMC Historical.

Marine Air Group 12. Historical Diaries, December 1950–March 1952. Diary File, HQMC Historical.

Marine Air Group 33. Historical Diaries, December 1950–March 1952. Diary File, HQMC Historical.

1st Korean Marine Corps Regiment. Unit Reports, Intelligence Summaries, Periodic Operation Reports, Periodic Intelligence Reports, Dispatch Summaries, Patrol Orders, Special Action Reports, 1951–1952. HQMC Historical.

Headquarters Battalion, 1st Marine Division. Unit Reports, December 1950–March 1952. Correspondence File, 1stMarDiv (Korea), HQMC Historical.

1st Battalion, 1st Marines. Historical Diaries, December 1950–March 1952. Diary File, HQMC Historical.

2d Battalion, 1st Marines. Historical Diaries, December 1950–March 1952. Diary File, HQMC Historical.

3d Battalion, 1st Marines. Historical Diaries, December 1950–March 1952. Diary File, HQMC Historical.

1st Battalion, 5th Marines. Historical Diaries, December 1950–March 1952. Diary File, HQMC Historical.

2d Battalion, 5th Marines. Historical Diaries, December 1950–March 1952. Diary File, HQMC Historical.

3d Battalion, 5th Marines. Historical Diaries, December 1950–March 1952. Diary File, HQMC Historical.

1st Battalion, 7th Marines. Historical Diaries, December 1950–March 1952. Diary File, HQMC Historical.

2d Battalion, 7th Marines. Historical Diaries, December 1950–March 1952. Diary File, HQMC Historical.

3d Battalion, 7th Marines. Historical Diaries, December 1950–March 1952. Diary File, HQMC Historical.

1st Engineer Battalion. Historical Diaries, December 1950–March 1952. Diary File, HQMC Historical.

1st Tank Battalion. Historical Diaries, December 1950–March 1952. Diary File, HQMC Historical.

VMO–6. Historical Diaries, December 1950–March 1952. Diary File, HQMC Historical.

VMF–212. Historical Diaries, December 1950–March 1952. Diary File, HQMC Historical.

VMF–214. Historical Diaries, December 1950–March 1952. Diary File, HQMC Historical.
VMF–311. Historical Diaries, December 1950–March 1952. Diary File, HQMC Historical.
VMF–312. Historical Diaries, December 1950–March 1952. Diary File, HQMC Historical.
VMF–323. Historical Diaries, December 1950–March 1952. Diary File, HQMC Historical.
VMF(N)–513. Historical Diaries, December 1950–March 1952. Diary File, HQMC Historical.
VMF(N)–542. Historical Diaries, December 1950–March 1952. Diary File, HQMC Historical.
Tactical Air Control Squadron One. War Diary, December 1950–March 1952. Diary File, HQMC Historical.
Marine Wing Service Squadron One. Historical Diary, December 1950–March 1952. Diary File, HQMC Historical.
Mobile Construction Battalion Two. Report of Activities, January 1951. UHR File, HQMC Historical.
USS *Bataan* (CVL–29). War Diary, January 1951. Unit Report File, HQMC Historical.
USS *Bataan* (CVL–29). Action Report, "Operations off the West Coast of Korea," 15 January–7 April 1951. UHR File, HQMC Historical.
USS *Consolation* (AH). Commanding Officer's Report to Commander, Naval Forces Far East dated 26 January 1952. UHR File, HQMC Historical.

OTHER SOURCES

Almond, Edward M., Gen, USA. Quoted in *U.S. News and World Report,* 34:40–2 (13 February 1953).
Barclay, Cyril N., Brigadier, British Army. *The First Commonwealth Division.* Aldershot: Gale and Polden, Ltd., 1954.
Berger, Carl. *The Korean Knot.* Philadelphia: University of Pennsylvania Press, 1957.
Burchette, Wilford G. *This Monstrous War.* Melbourne: J. Waters, 1953.
Cartier, Raymond. "Top Reporters Forum," *UN World,* v. 5, no. 10 (October 1951).
Fowler, Delbert E., Capt, USA. "Operations at the Hwachon Dam, Korea," *The Military Engineer,* v. 44, no. 297 (Jan–Feb 1952).
Gavin, James M., Gen, USA, (Ret.). *War and Peace in the Space Age.* New York: Harper and Brothers, 1958.
Gugeler, Russell A., Capt, USA, Editor. *Combat Actions in Korea.* Washington: Combat Forces Press, 1954.
Harrison, Charles W., Col, USMC, "KMC Attack on Taeu-san, 8–11 July 1951." Manuscript, HQMC Historical.
Joy, C. Turner, Adm, USN, (Ret). *How Communists Negotiate.* New York: Macmillan, 1955.

Kihss, Peter. "One Year in Korea," *United Nations World*, v. 5, no. 7 (July 1951).

Metcalf, Clyde H., LtCol, USMC. *A History of the United States Marine Corps*. New York: G. P. Putnam's Sons, 1939.

Montross, Lynn. *Cavalry of the Sky*. New York: Harper and Brothers, 1954.

Ridgway, Matthew B., LtGen, USA, (Ret.) as told to H. M. Martin. *Soldier, The Memoirs of Matthew B. Ridgway*. New York: Harper and Brothers, 1956.

Stewart, James T. *Airpower, The Decisive Force in Korea*. Princeton: D. Van Nostrand Company, Inc., 1957.

Taylor, Maxwell D., Gen, USA, (Ret.) *The Uncertain Trumpet*, New York: Harper and Brothers, 1959.

U.S. Army. FM 70–10, *Mountain Operations*. Washington: The Adjutant General's Office, 1947.

———. TM 9–1980, *Bombs For Aircraft*. Washington: The Adjutant General's Office, 1950.

U.S. Department of State. *U.S. Relations with China with Special Reference to the Period 1944–1949* [China White Papers]. State Department Publications 3573, Far East Series 30. Washington, 1949.

U.S. Marine Corps. Landing Force Bulletin Number 6, *Night Vision and Night Combat*, HQMC, Washington, D.C., 5 December 1953.

———. Landing Force Bulletin Number 18, *Battlefield Illumination*, HQMC, Washington, D.C. 4 June 1956.

U.S. Military Academy. *Operations in Korea*. West Point: Department of Military Art and Engineering, U.S. Military Academy, 1956.

U.S. Navy, Office of Chief of Naval Operations. "Korean Air War," *Naval Aviation News* (April 1951).

Van Fleet, James A., Gen, USA (Ret). "The Truth About Korea," *Life*, 34:126–8 (11 May 1953).

Vatcher, William H. Jr. "Inside Story of Our Mistakes in Korea," *U.S. News and World Report*, 34:35–6 (23 January 1953).

Vatcher, William H. Jr. *Panmunjom, The Story of the Korean Military Negotiations*. New York: F. Praeger, 1958.

Walker, Richard L. *China Under Communism: The First Five Years*. New Haven: Yale University Press, 1955.

Weintal, E. "What Happened at Kaesong and What is in Prospect," *Newsweek*, 38:38 (23 July 1951).

Index

Adelman, LtCol Merritt, 122*n*, 134, 149
A-frames, 86, 163
Agan, Capt Alfred H., 47
Air attack, enemy, 65
Air battle, 103
Air control, 97
  1st MAW, 89
  System, USAF, 18
  System, USMC, 18
Aircraft, 46, 47, 63, 71, 83, 89, 96, 117, 125, 131, 137, 159, 188–190, 208. *See* Helicopters and Communist Forces.
  AD (Skyraiders), 116, 122, 125
  B–26, 33
  B–29, 90
  F3D (SkyKnight), 89
  F4U (Corsair), 13, 27, 33, 49, 63, 71, 78, 89, 96, 107, 108, 116, 125, 137, 142, 186
  F7F (Tigercat), 49, 50, 63, 89, 131
  F9F (Panther), 63, 96, 108, 125, 137, 148
  F–51 (Mustang), 28, 33, 116, 125, 137, 206
  F–80 (Shooting Star), 15, 97, 125
  F–82, 89
  F–84, 125
  Jets, 15, 49, 63, 96
  Marine, 63, 78, 103, 108, 120
  Mars, 31
  Mosquito, 18, 26–28, 47, 50, 51, 97, 107
  Observation planes
    OEs, 224
    OYs, 1, 14, 49, 63, 104, 107, 116, 125, 134, 137, 142, 164, 187, 188, 224
  PB4Y–2, 138
  PO–2, 170
  R4D, 55, 63, 138
  R5D, 31, 32, 63, 89
  Sabre jets, 28
  SNB (Beechcraft), 63
  SNJ, 18
  TBM (Avenger), 50, 63
  T–6, 18
  Transports, 31
Air drop, 73, 93, 116, 181
Air Fields
  Bofu, repair of, 33
  Kimpo, 28, 29, 59, 64
  K–1, 29, 32, 33, 35, 46, 48, 56, 62, 63, 89, 96, 108, 137, 171
  K–2, 29, 32
  K–3, 32, 62, 63, 89, 108, 171
  K–4, 29
  K–9, 1, 15, 28, 29, 32, 46, 48–50, 52, 56, 90
  K–10, 29
  K–16, 28, 108, 116, 137, 254
  K–18, 171, 185, 186
  K–46, 137, 138, 142, 170
  K–50, 215, 254
  K–51, 215
  X–77, 214
  X–83, 189, 208, 219, 220, 241, 242
  Seoul, 29
  Suwon, 28, 29
  Tsuika, 33*n*
  Wonsan, 29
  Yonpo, 29
Air Force, 28, 31, 33, 34, 63, 76, 170, 224
  Eighth Air Force, 167
  Far East Air Forces (FEAF), 14, 29, 32, 33, 45, 47, 95
    Commander, 33
    Combat Air Command, 73
    Military Air Transport Service, 31
  Fifth Air Force, 14, 15, 18, 49, 64, 67, 69, 78, 135, 142 143, 170, 185, 257
    Control, 15
    543d Tactical Support Group, 64
    8th Fighter Bomber Squadron, 15
    606th Aircraft Control and Warning Squadron, 90
Airheart, 1stLt William C., 151
Air Liaison officer, 17
Air lift, 31, 32, 215
  Troop, 32
Air mattresses, 94
Air observation, 182
Air operations
  Control, 26
  United Nations, 33

Air reconnaissance, 49, 50, 64, 78, 84, 101, 143, 207
Armed, 96
Patrols, 28
United States, 47
Air strike, 17, 69, 70, 75, 91, 92, 119, 125, 130, 142, 148, 149, 159, 179, 183, 185, 186, 195, 206, 219, 221
Control, 17, 142
Interdiction, 27, 144
Tactics, 64
United States, 13, 52
U. S. Marine Corps, 65, 70, 74, 115
Air support, 27, 28, 45, 69, 76, 78, 81, 86, 96, 97, 108, 130, 135–138, 141, 142, 148, 150, 178, 185, 191, 206, 210, 220, 258, 260
Close, 14, 18, 26, 27, 49, 71, 76, 78, 96, 108, 114, 125, 136, 142–144, 169, 170, 185, 186, 224, 258, 260
Control of, 14, 15, 17, 136
Air Force, 17
Marine Corps, 17
Deep, 49, 50
Interdictory, 18
Naval (British), 28
Reconnaissance, 64
United States, 56, 71, 83
ALBANY, Phase Line, 80, 84, 85
Allen, BrigGen C. R., 246*n*
Allen, Maj H. E., 90, 90*n*
Allert, 1stLt William J., 111
Almond, MajGen Edward M., 5, 19, 121, 128, 133, 133*n*, 137, 149, 154, 157, 162
Alston, LtCol William P., 194, 204, 242
Ambushes, 70, 217, 229, 231, 242
Chinese Communist Forces, 44
Marine, U.S., 53
North Korean People's Army, 49
American Civil War, 165
Americans, Fighting for Korea, 7
Ammunition, 72, 122, 144, 170, 179, 181, 187, 190, 193, 210, 214, 221, 243, 244, 260
Artillery, 185
Mortar, 180
11th Marines, 122
Dump, 104, 210
Enemy, 130
Mortars, 114, 244
Shells, 104
Shortage, 244
Small-arms, 114
Supply Point (ASP) 60–B, 181
United States, 64, 93, 122
U. S. Marine Corps, 113
Amphibious
Assault, 257
Attack, 250
Landing, 45, 47
Operation, 250
Tactics, 250
Anderson, LtCol James R., 46, 63, 89
Andong, 29, 42, 44, 49, 51, 55, 62, 66
Andong-Taegue area, 45
Andrewes, VAdm Sir William G., RN, 15, 28, 46
Antiaircraft
Defenses, 103
Fire, enemy, 47
ARIZONA, Phase Line, 73, 75
Armor, body, 165, 166, 168, 169, 237–239, 251, 257. *See* Vest, Armored.
Army, U. S., Strength figures, 11
Army, U. S. Units
Eighth U. S. Army in Korea (EUSAK), 5, 8, 10, 14, 18, 19, 21, 22, 25–27, 30, 35, 37, 38, 41, 44, 49, 51, 54, 58–60, 63, 65, 67, 69, 84, 86, 92, 93, 109, 119, 121, 122, 127, 128, 132, 133, 156–158, 163, 176, 185, 201, 203, 215, 218, 225, 242, 243, 250–253, 258, 259
Commander, 7, 11, 12, 19, 41, 50, 58, 60, 67, 68, 72, 73, 99, 120, 199, 202, 261
Fortifications, 11
Morale, 7, 8, 156
Tactics, 8
Transportation Section, 163
Units, supply of, 12
Tenth Army, 72
I Corps, 22, 24, 25, 44, 50, 66, 79, 84, 108, 116, 119, 121, 131, 154, 163, 201, 251, 253, 256, 259
Headquarters, 50
IX Corps, 22, 24–26, 44, 50, 58, 60, 62, 65–67, 72, 79, 80, 84, 86, 93–95, 99–102, 107–109, 118, 121, 126, 128, 130–132, 154, 201, 214, 253, 259
Commander, 60, 71, 79, 215
Command Post, 72
X Corps, 1, 5, 19, 21, 22, 24–26, 29, 45, 65, 66, 68, 71, 72, 79, 93, 94, 108, 120, 121, 125, 128, 131, 146, 149, 154, 158, 159, 161, 181, 199, 201, 206, 210, 212, 216, 218, 221, 224, 227, 253, 256, 259

Army, U. S. Units—Continued
X Corps—Continued
Commander, 162
Command post, 19
Evacuation, 2, 5
XVIII Airborne Corps, 7
1st Cavalry Division, 8, 22, 65, 80, 86, 94–97, 102, 119, 121, 201
2d Infantry Division, 8, 19, 22, 25, 26, 62, 65, 66, 71, 72, 80, 119–121, 123, 125, 126, 128, 146, 159, 161, 168, 171, 173, 176, 180, 202, 253
Morale, 199
3d Infantry Division, 8, 14, 19, 22, 25, 121, 123, 144, 201, 253
7th Infantry Division, 8, 19, 22, 25, 66, 71, 72, 95, 121, 130, 144, 201, 217, 253
24th Infantry Division, 8, 22, 58, 59, 65, 107, 121, 201
25th Infantry Division, 8, 22, 41, 107, 121, 144, 201, 253, 254
40th Infantry Division, 253
45th Infantry Division, 253
2d Logistical Command, 12, 254
5th Cavalry Regiment, 119
7th Cavalry Regiment, 95, 97
8th Cavalry Regiment, 95, 97
9th Infantry Regiment, 126
17th Regimental Combat Team, 253
23d Regiment, 62, 168
35th Regiment, 51
38th Regiment, 126, 168
65th Regimental Combat Team, 253
187th Airborne Regimental Combat Team, 22, 25, 66, 68, 86, 121, 128, 131, 133
17th Field Artillery Battalion, 122
92nd Armored Field Artillery Battalion, 66, 109
96th Armored Field Artillery Battalion, 122
196th Field Artillery Battalion, 171
987th Armored Field Artillery Battalion, 111, 114
92d U. S. Army Searchlight Company, 227
4th Ranger Company, 102
74th Truck Company, 66
U. S. Army Searchlight Company, 245
Artillery, 159, 176, 195
Army, 11, 59, 107, 116, 121, 210
Chinese Communist Forces, 70, 90, 134
Marine, 52, 69, 70, 74–76, 86, 92, 107, 114, 116, 125, 180, 183, 210, 218, 244
North Korea People's Army, 180
Support, 78, 92, 93, 114, 130, 142, 148, 149, 178, 191, 195, 206, 220, 227
Atomic bomb, 37, 187
Australia, 24
Autrey, Maj Robert L., 103, 182, 199*n*
Averill, Maj Gerald P., 194*n*

BADGER Line, 157
Badoeng Strait (CVE), USS, 1, 2, 15
BAKER, Phase Line, 86, 91, 92
Banning, LtCol Virgil W., 74, 91, 111, 117, 118
Barclay, Brig C. N., 116*n*
Barbed wire, 197, 232
Bataan (CVL), USS, 1, 2, 14, 28, 35, 41, 46, 47, 89, 103, 146
Bates, Maj William L., 111
Batterton, Col Boeker C., 48
Baugh, Maj William E., 223
Bay of Masan, 2
Bayonet assault, 150
Bean Patch, 2, 3
Belgium, 24
Bell. *See* Helicopters.
Belleau Wood, 118, 152
Berger, Carl, 160*n*, 171*n*
Bethel, Col Ion M., 235
Black, Capt Albert A., 215
Blakeney, Jane, 193*n*
Bofu, 33, 35, 45, 47, 48, 50, 56, 62
Bohannon, Capt Thomas J., 52, 111, 113, 116
Booby traps, 206
Borth, Maj Harold C., 181
Bougainville, 118
Bowser, Col Alpha L., 7, 18, 19, 60*n*, 62*n*, 94, 94*n*, 102, 102*n*, 107*n*, 120*n*
Boyd, Maj Vernon D., 235
Brice, BrigGen William O., 136
Bridges, 64, 102, 103
Enemy, 143
Floating, 102
Improvised, 74
Korean, 15
Sections, 64
Swiss bent, 74
Brown, Capt Leslie E., 15
Brown, Col Wilburt S., 126, 134, 134*n*, 148–150, 150*n*, 152, 152*n*, 157, 158
BROWN Line, 150, 151, 157
Brunelli, Col Austin R., 255
Buckingham, CWO Robert C., 170
Buckner, LtGen Simon Bolivar, 72
BUFFALO, Phase Line, 85, 92
Bugles, NKPA, 192, 217

Bunkers, 184, 191, 192, 194, 210, 217, 219, 221, 224, 230, 232, 242, 244, 245, 261
  Chinese Communist Forces, 70, 75, 83, 91, 92
  Enemy, 149
  North Korean People's Army, 141, 142, 182, 183, 191, 216, 220
Burchett, Wilford G., 247, 247n, 248
Burton, Col Custis, Jr., 179, 210, 232, 255
Bush, 2dLt Clayton O., 83, 83n
Bush, Maj Donald S., 63, 64, 134
Bushe, Capt Eugene A., 119
Butterfly bombs, USMC, 88
Byers, MajGen Clovis E., 162, 208

C Rations, 175
Cagle, LtCol Carl J., 181
CAIRO, Phase Line, 92, 93
Camel, 64
Cameras, 92
  K–17 camera, 134
Camp Lejeune, 31, 167, 168, 234, 236
Camp Pendleton, 31
Camp Tripoli, 233, 243, 255
Canada, 24
Cargadores, 163, 173, 181, 190
Cargo ship, USN, 30
Carney, Maj Edgar F., Jr., 192, 193n
Cartier, Raymond, 155
Casualties, 210, 213, 220, 221, 238, 239, 243
  Chinese Communist Forces, 38, 70, 75, 76, 79, 83, 84, 91, 116, 118, 121, 125–128, 131, 259
  Eighth Air Force, 167
  Enemy, 153, 202, 206, 210, 216, 220, 246, 259
  Evacuating of, 165, 189, 190, 214
  Korean Marine Corps, 55, 147, 159
  IX Corps, 84
  Non-Battle, 234
  North Korean (NKPA), 51, 52, 57, 146, 180, 184, 194, 198, 210, 217, 220, 225, 228
  X Corps, 84
  United Nations, 34, 161, 202
  United States, 60, 156
  U. S. Marine Corps, 30, 49, 51, 57, 69, 70, 75, 76, 81, 83, 86, 91, 111, 116, 118, 125, 131, 133, 143, 150–152, 159, 173, 178, 180, 184, 186, 193, 194, 196, 199, 206, 210, 216, 218, 220, 246
    Evacuation of, 164
Cates, Gen Clifton B., 232
Chachon-dong, 52
Chaegok-tong, 54
Chang, Gen, 109
Changhang, 216
Changwon, 3
Chechon, 26
Chiang Kai-shek, 36
Chidi San mountain mass, 3
China, Red, 261
  Funds for war, 37
  "Hate America" Campaign, 35, 36
Chinandong, 54
Chinese Civil War, 155
Chinese Communist Forces (CCF), 6, 58, 59, 85, 123, 128, 154
  Air strike, 15
  Armies, 45
  Attack, 26
  Bugle calls, 113
  Concentration, 27
  Counterstroke, 63
  Entrenchments, 15, 248
  Equipment, 39, 153
  Gun positions, 15
  Infiltration, 35
  Logistical, 144
  Morale, 155
  Offensive, 11, 15, 19, 33, 107, 120–122, 126, 152, 164, 259
    Fifth Phase, 103, 105, 121, 126, 155, 164
    January, 19, 34, 41, 154
    Sixth Phase, 178
  Penetration, 21
  Road block, 111
  Shore batteries, 156
  Soldier, 35
  Strategy, 35, 80, 101
  Strength figures, 218
  Supply
    Dumps, 19
    Lines, 41
  Tactics, 35–38, 100, 248
    Human Sea, 35
  Troops, 27, 41, 50, 51, 170
    Shelters, 15
  Uniform, 35
  Yak fighter, 103
  Units
    20th Army, 24, 101
    24th Army, 24
    26th Army, 24, 101
    27th Army, 24
    30th Army, 24
    32d Army, 24
    37th Army, 24
    38th Army, 24
    39th Army, 24, 66, 93, 101
    40th Army, 24, 60, 66, 101, 106, 115

Chinese Communist Forces—Continued
- Units—Continued
  - 42d Army, 24, 101
  - 48th Army, 24
  - 49th Army, 24
  - 50th Army, 24
  - 65th Army, 24
  - 66th Army, 24, 60, 66, 81, 93, 101
  - 44th Division, 126
  - 58th Division, 101
  - 59th Division, 101
  - 60th Division, 101
  - 76th Division, 101
  - 77th Division, 101
  - 78th Division, 101
  - 115th Division, 101
  - 116th Division, 101
  - 117th Division, 101
  - 118th Division, 101
  - 119th Division, 101
  - 120th Division, 101, 106, 115
  - 124th Division, 101
  - 125th Division, 101
  - 126th Division, 101
  - 196th Division, 66, 81
  - 197th Division, 81
  - 198th Division, 81
  - 358th Regiment, 106
  - 359th Regiment, 115
  - 360th Regiment, 115

Chinese Nationalists, 36
Chinhae, 2, 54
Chinju, 3
Chiso-dong, 52
Chodo-ri, 154
Choe, MajGen Am Lin, 146
Chogutan, 171
Chongja-dong, 52
Chon-San, 90
Chorwon, 95, 154, 201
- Hwachon area, 27

Chosin, 70
- Breakout, 3, 14, 101, 109, 234
- Campaign, 6, 32, 38, 42, 45, 52, 57, 118, 257
- Fighting, 8
- Operation, 2, 5, 152, 168
- Reservoir, 12, 66, 136, 152, 257
- Withdrawal, 120

Christmas in Korea, 5
Chumunjin, 80
Chunchon, 8, 80, 94, 117, 119, 181, 215
Chunchon-Hongchon highway, 123
Chunchon-Wonju-Chechon corridor, 26
Chungju, 22, 26, 58, 60, 66, 67
Chipyong-ni, 71
Clark, Maj Albert L., 89
Clothing
- Camouflage, 228
- Cold weather, 163, 221, 222

Coffey, Capt John, 109, 109*n*, 111, 111*n*, 113
Cole, LtCol Frank J., 46
Commander Naval Forces, Far East (ComNavFE), 5
- *See* VAdm C. Turner Joy, USN.

Communications, 137, 208
- Air-ground, 260

Communist
- Cause, 3
- Delaying tactics, 249
- Photographers, 156, 157
- Truce Team, 160

Consolation (AH), USS, 57, 165
Cornwell, Maj Charles E., 240
Correspondents, at Hagaru, 6
Cotton, Capt James T., 215
Cowan, Cdr John S., 236
Craig, BrigGen Edward A., 2, 3*n*, 7, 8*n*, 12*n*, 21, 49, 57, 164
Cronin, Capt James T., 106, 106*n*
Cronk, Capt Welby D., 91
Culhane, Col Thomas A., Jr., 254*n*
Cushman, BrigGen Thomas J., 136

Daigh, 1stLt Harold D., 103
Davis, Maj Daniel H., 71
Davis, LtCol Raymond G., 107*n*, 123, 133
Delegates
- Communist, 171, 220, 225, 229, 233
  - Delaying tactics, 160
- United States, 248, 249

DeLong, Capt Philip C., 103
Demolitions, 102
Devastate BAKER, 96
Doriot, Col George F., 166, 167
Doyle, RAdm James H., 13, 14
DUCK, Line, 224, 229, 230

Edwards, LtCol Harry W., 189*n*, 199*n*, 207, 213, 222*n*
El Toro, California, 88, 188
England, 250
Equipment
- Eighth Army, 12
- Shortages, 30
- USMC, 12
  - Shortages, 11, 12
  - Worn, 13

Ethiopia, 24
Evacuation, 115
- Casualities, 164, 165, 250
- Hungnam, 5

Evans, LtCol Holly F., 114, 191, 210
Everest, MajGen Frank E., 136

Feeney, Maj James A., Jr., 46
Fenton, Capt Donald L., 170
Ferenbaugh, MajGen Claude F., 215
Fink, Maj Gerald, 223*n*
Fisher, Capt Don H., 52
Fisher, 1stLt Joseph R., 91
Flak traps, 143
Flares
  CCF, 113
  Trip, 216
Flournoy, Col Walter N., 204, 232
Flying Dutchman (legendary ship), 57
Fontana, LtCol Paul J., 46, 47, 62
Ford, Capt John, USNR, 6
Forward Air Controller (FAC), 17, 52, 71, 96, 97, 116, 142, 186, 206, 230
Forward Observers, 144
Forward OPs, 210
Fowler, Capt Delbert M., 102*n*
Frame, Maj Donald P., 46
France, 24, 65, 68, 246
Franklin, Benjamin, 251
Freeman, Col Paul, 62
French, 202
Fresh Rations, Marines, 5
Frostbite, 234, 235
  Casualty, Medical, 57
Fulton, Capt Floyd K., 52

Gall, Maj Walter, 60, 83, 168
Garvin, BrigGen Crump, 12
Gases, 134
Gasoline, 32, 34, 72, 170, 190, 222
Gavin, Gen James M., 38, 38*n*
Gayle, LtCol Gordon D., 199*n*, 232, 254
Geiger, MajGen Roy S., 72
General Darby, USNS, 31
General W. H. Gordon, USS, 254
Glendinning, Maj James I., 74
Gomez, Pfc Edward, 193, 193*n*
Gorman, LtCol John E., 193, 193*n*, 204, 240
Gottschalk, Maj Vincent J., 14
Greece, 24, 99
Griffin, LtCol Louis C., 175, 183, 184
Groff, Capt Goodwin C., 91
Groves, Col Gould P., 158*n*, 159, 176*n*, 199*n*
Guadalcanal campaign, 118, 152
Guam, 32
Guerrillas, 36, 44, 203, 213, 222, 257, 258
  Anti-, 241
  Communist, 3
  Communist Chinese Forces, 35, 44, 51
  Enemy, 53, 57, 58
  North Korean People's Army, 26, 44, 45, 48, 51, 52, 55, 56
Gugeler, Capt Russell A., 111*n*, 116*n*

Haffey, Capt Eugene H., 107
Hagaru-ri, 6
Hager, Col Frank P., Jr., 199*n*, 227
Haiti, 118, 232
Hamhung, 6
Han River, 11, 25, 28, 50, 58, 59, 71, 108, 119, 122
Hangye, 130, 171
Hardwick, LtCol Gordon A., 166*n*, 234*n*, 235, 237*n*, 238*n*
Harper, 2dLt William E., 230
Harris, MajGen Field, 14, 15, 46, 48, 49, 50, 62, 78, 88, 89, 90, 135, 136
Harrison, LtCol Charles W., 54, 54*n*, 92, 93*n*, 102*n*, 146*n*, 158*n*
Hart, LtGen Franklin H., 232
Hawaii, 89
HAYS Line, 180, 182, 192, 204, 213
Hayward, Col Richard W., 130, 146, 148
Heely, LtCol Dale H., 191
Helicopter, 1, 14, 47, 72, 79, 111, 115, 135, 164, 165, 187–190, 207, 208, 211, 212, 214, 219, 220, 223, 240, 241, 250, 251, 257
  Evacuating of casualties, 49, 56
  Laying wire, 49
  Lift, 211, 213
    Troop, 207, 212
  HO3S–1, 63, 164, 188
  HTL, 59, 63, 164, 188
Helmets, steel, 166
Hemostat, 115
Hemphill, Col Bruce T., 199*n*, 227, 242*n*, 246*n*, 251*n*
Hering, Capt Eugene R. (MC), USNR, 57
Herring, LtCol George W., 188, 189, 207, 211, 213, 223
Hewitt, Maj Roy R., 96, 96*n*, 97
Hickman, Maj William T., 152*n*
Hicks, 1stLt Norman W., 115, 115*n*, 223*n*
Higgs, 1stLt Richard C., 207
Highways, 80
  Coastal, 27
  Japanese, 33
  Traffic jams, 32
Hilian, Lt Donald L., (MC), USN, 215
Hill 201, 75
Hill 208, 75
Hill 246, 91
Hill 313, 105–107
Hill 321, 75
Hill 330, 92
Hill 333, 75
Hill 335, 75
Hill 356, 90
Hill 381, 92
Hill 399, 91

Hill 420, 151
Hill 428, 91
Hill 509, 105, 106
Hill 516, 144, 146
Hill 536, 75
Hill 549, 83
Hill 602, 173, 178, 179, 183
Hill 610, 141, 142
Hill 651, 141
Hill 673, 182–184, 194
Hill 680, 142, 173
Hill 692, 142
Hill 702, 173, 178, 211, 214
Hill 749, 182–184, 187, 190–194
Hill 751, 190, 191, 193, 194, 196, 216
Hill 755, 176
Hill 761, 157, 158
Hill 802, 150
Hill 812, 183, 194–198, 204, 230, 232, 243, 245. *See* Kanmubong Ridge
Hill 854, 204, 206, 243, 245
Hill 883, 130
Hill 884, 203, 204, 206, 208, 210, 220, 240, 241. *See* "Mount Helicopter"
Hill 902, 111
Hill 924, 173, 176, 178–180
Hill 930, 173, 176
Hill 951, 210
Hill 975, 93
Hill 980, 183, 194–197
Hill 1000, 173
Hill 1001, 159
Hill 1026, 173, 176, 179, 180
Hill 1042, 196
Hill 1051, 130
Hill 1052, 182, 183, 196–198, 216, 219
Hill 1100, 159
Hill 1122, 146, 147
Hill 1218, 146
Hill 1316, 146
Hire, LtCol Homer E., 144, 148
Hiroshima, 187
Hoengsong, 28, 60, 63, 66, 70, 72–74, 80–83, 89, 137, 170
Hoengsong-Hongchon road, 80, 81
Hoge, MajGen William H., 79, 111*n*
Holliday, Maj Morse L., 126
Hongchon, 72, 80, 86, 88, 90, 91, 94, 96, 120, 128, 176, 181
Hongchon-Hangye road, 171
Honsowetz, Col Russell E., 255
Hopkins, LtCol John L., 106*n*, 141
Hospital
  Clearing station, 190
  Japan, 12
  Ships, 165
Horseshoe Ridge, 113, 115
Hovatter, 1stLt Eugenous M., 83
Howard, LtCol Harold C., 241
Hughes, James J., 230
Hungnam, 2
  Beachhead, 2
  Evacuation of, 2, 5
  Redeployment, 1, 8, 13, 14, 65
Hwachon, 45, 102, 105, 109, 122, 201
  Reservoir, 27, 96, 102, 119, 128, 131, 148, 260
Hwanggi, 182

Illinois Wesleyan University, 118
Imjin, 26, 27, 107, 108, 116, 154, 201
  River, 73, 79
Inchon, 27, 28, 30, 47, 50, 51, 59, 63, 80, 254, 256
  Evacuation of, 28
  Landing, 1, 63, 257
Inchon-Seoul, 118
  Operation, 2, 44, 135, 168
Indianapolis (CA), USS, 232
Inje, 128, 133, 147, 161, 168, 171, 175, 181, 187, 215
Inje-Kansong road, 128
Innovations, 257
  Body armor, 239
  Helicopter, 164, 240
  Tactical, 165, 251
Iron Triangle, 95, 101, 107, 143, 154
Irwin, Capt John R., 242
Itami, 2, 28, 29, 32, 48, 49, 63, 89
Itazuke, 1, 33, 49, 50, 63
Iwo Jima, 152, 187

Japan, 1, 12, 33, 50
JAMESTOWN, Line, 253, 256
Johnson, Capt Horace L., 111
Johnston Island, 32
Jones, Capt Jack R., 52, 68
Jones, Maj Jack R., 53*n*
JOC (FAF–EUSAK Joint Operations Center), 17, 27, 45, 47, 64, 65, 67, 69, 70, 78, 81, 96, 135–138, 258, 260
  Air Control, 67, 76, 97, 135
Joy, VAdm C. Turner (ComNavFE), 5, 14, 155, 157, 157*n*, 160, 160*n*, 161, 161*n*, 221*n*, 223*n*, 248*n*, 249*n*
Junks, enemy, 15
Jutlandia, Danish Hospital Ship, 85, 158

Kaesong, 154, 156, 157, 160, 171, 180, 220, 222, 247, 248
Kangnung, 171, 185
Kanmubong Ridge, 183, *See* Hill 812.
KANSAS Line, 94, 95, 99, 108, 115, 116, 118, 141, 147, 148, 150–153, 162, 168, 175, 176, 178, 179, 182, 233

Kansong, 128, 187
Kelly, LtCol Bernard T., 123, 125n, 135, 135n, 175, 175n, 176n, 179, 183
Kelly, LtCol James G., 176, 183, 184, 199n, 216
Kelly, Capt John L., 141
Kelsey, LtCol John V., 191
Kennedy, LtCol Donald R., 91, 194
Kerrigan, 1stLt William E., 68, 68n, 141
Kihss, Peter, 153n
Kim, Col Dae Shik, 204
Kim, Col Dong Ha, 106, 227
Kim, 1st Lt SikTong, 93
Kim, Col Sung Eun, 54
Kimpo, 253
  Peninsula, 119, 201, 254, 255
Kingsley, J. Donald, 153
Kirby-Smith, Maj Ephraim, 207
Kirkpatrick, Capt Louis P., (MC), USN, 166n
Knapp, LtCol Horace E., Jr., 193n
Kobe, 30, 62, 143
Kopas, Capt William T., 141
Korea, 1
  North
    North Korean People's Army (NKPA), 52, 53, 55, 85, 154
      Irregulars, 258
      Morale, 53
      Soldiers, 55
      Strategy, 196
      Strength figures, 218
      Troops, 132, 147, 152, 170, 184, 185, 260
      Units
        I Corps, 24
        II Corps, 51, 173, 182
        III Corps, 182, 217
        V Corps, 24, 60, 132
        1st Division, 24, 182, 217, 218, 228
        2d Division, 182, 217
        3d Division, 24
        10th Division, 44, 45, 48, 56, 57
        12th Division, 132, 146
        15th Division, 24, 217, 218, 224, 228
        19th Division, 218, 224
        45th Division, 228
        47th Division, 224, 228
        6th Regiment, 173
        25th Regiment, 44, 55
        27th Regiment, 44, 55
        29th Regiment, 44
        41st Regiment, 151
        91st Regiment, 244
Korea,
  South
    Republic of Korea, 3, 11
    Army (ROKA), 3, 19, 26, 54, 73, 108, 116, 163, 206
      Units
        I Corps, 22, 108, 121, 123, 128, 154, 202, 215, 253
        II Corps, 22
        III Corps, 22, 25, 26, 121, 123
        Civil Transport Corps, 86, 162, 163, 181
        Korean Service Corps, 181
        Capital Division, 22, 58, 121, 202, 253
        1st Division, 22, 24, 84, 121 201, 251, 253
        2d Division, 22, 25, 27, 57, 201, 253
        3d Division, 22, 72, 121, 202, 216, 253
        5th Division, 22, 25, 121, 123, 180, 202, 211, 253
        6th Division, 22, 65, 66, 74, 97, 99, 101, 105–109, 116, 120, 121, 201, 253
        7th Division, 22, 108, 121, 123, 148, 202, 253
        8th Division, 22, 25, 176, 182, 201–204, 253, 255
        9th Division, 22, 121, 253
        11th Division, 121, 202, 203, 213, 253
        5th Regiment, 148, 149
        10th Regiment, 204
        11th Regiment, 207
        21st Regiment, 204, 206
        National Guard, 86, 163
        Soldiers, Lack of training, 24
    Marine Corps (KMC), 54, 93, 94, 102, 105, 106, 109, 113, 117, 126, 146–148, 150–152, 158, 159, 162, 171, 176, 178–181, 183, 243, 254, 260
      1st Regiment, 54, 58, 73, 92–95, 101, 102, 106, 120, 126, 130, 146, 147, 158, 159, 162, 171, 175, 176, 178, 182, 198, 204, 210, 218, 224, 227, 246, 253, 254
        1st Battalion, 54, 55, 105, 146, 159, 176, 178
        2d Battalion, 55, 93, 146, 176, 178–180, 216
        3d Battalion, 54, 93, 146, 176, 178

Korea—Continued
South—Continued
Marine Corps—Continued
1st Regiment—Continued
5th Battalion, 55, 201
1st Korean Artillery Battalion, 227
Diet, 54
Police, 3, 52, 53
Korean entertainers, 5
Korean War, police action, 7
Koto-ri, 6
Koutsandreas, 2dLt John D., 230
Krisky, Corpsman R. E., 57
Krulak, Col Victor H., 189, 199*n*, 213, 250*n*
Kumhwa, 95 143, 154, 201
Kumsong, 154
Kunsamma Pass, 81
Kurdziel, LtCol Edward G., 184*n*, 241
Kyongju, 19, 22, 29, 30, 42
Kyongju-Yungchon-Uisong road, 42

La Hue, LtCol Foster C., 190, 206
Lavoie, LtCol Leon F., Jr., 66, 109, 115
Lawrence, Capt George E., 216
Lee, MajGen Ban Nam, 45, 51, 53
Lejeune, MajGen John A., 72
Lewis, LCdr Frederick J. (MSC), USN, 165, 167, 168, 236, 237, 237*n*
ICELAND, Line, 233
Litters, 188
Litzenberg, MajGen Homer L., 64*n*
Logistics, 72, 73, 180, 182, 215, 258. *See* Supplies.
Long, Capt Edwin B., 170
Los Angeles (CA), USS, 219
Lund, Maj Arnold A., 15, 27, 46
Lundin, Maj William M., 15, 46

Mabry, Maj Clarence J., 148
MacArthur, General of the Army Douglas, 5, 8, 8*n*, 14, 33, 50, 68, 91, 92
Recall of, 99
MacDonald, Maj J. Angus, 222*n*, 223*n*
MacIntyre, LtCol Neil R., 15
Mago-ri, 218
Main line of resistance (MLR), 21, 122, 157–159, 180, 203, 204, 211–218, 220, 221, 223, 224, 229, 231, 232, 238, 240, 243, 245, 250, 253, 259–261
Main Supply Route (MSR), 29, 30, 38, 42, 44, 67, 80, 91, 123
Hongchon-Chunchon, 119
Malik, Jacob, Foreign Minister of the USSR, 154
Manchuria, 85, 160
Mao Tse-tung, 36, 128, 155
Marine Corps, U. S.
Role in Korea, 6, 32
Units
Fleet Marine Force, Pacific, 12, 31, 241,
1st Air Delivery Platoon, 241
1st Combat Service Group, 3, 241
1st 4.5 Rocket Battalion, 119, 227
Battery C, 119
7th Motor Transport Battalion, 3, 30, 60, 181, 255
1st Marine Aircraft Wing, 1, 14, 29, 32, 33, 33*n*, 34, 45, 46, 50, 56, 62, 63, 67, 71, 78, 84, 88, 89, 96, 103, 116, 120, 125, 131, 134–138, 143, 146, 164, 169, 170, 185, 203, 219, 249, 260, 261
Commanding Officer, 257
Headquarters Squadron, 63
Marine Air Control Group–2, 90
Marine Ground Control Intercept Squadron–1, 1, 89, 90
Marine Ground Control Intercept Squadron–3, 90
Marine Tactical Air Control Squadron–2, 1, 2, 89, 96
Air Defense Section, 89
Marine Aircraft Group 12, 33, 35, 46, 48, 56, 60, 81, 83, 137, 170, 171, 185
Marine Aircraft Group 33, 2, 33*n*, 34, 45–48, 56, 62, 81, 96
Service Squadron 33, 46
HMR–161, 187–190, 207, 208, 213–215, 217, 222, 240–242
HMX–1, 164, 187, 188, 223
HRS–1, 188, 189, 197, 211, 212, 214
MASRT–1, 186
VMF–152, 31, 32
VMF–212, 1, 13, 14, 27, 28, 35, 41, 46, 47, 89, 108
VMF–214, 1, 13, 15, 28, 29, 46, 48, 62, 91, 92, 96, 107, 137, 141, 142, 144, 150, 185
VMF–311, 1, 15, 28, 29, 49, 63, 64, 95, 96, 148
VMF–312, 46, 47, 50, 62, 71, 89, 103, 146, 185
VMF–323, 1, 14, 15, 27–29, 46, 48, 52, 55, 62, 92, 107, 108, 137, 144, 146
VMF–352, 31
VMF(N)–513, 1, 49, 52, 55, 63, 89, 131, 138, 150, 170
VMF(N)–542, 1, 46, 49, 50, 63, 88

Marine Corps, U. S.—Continued
  Units—Continued
    VMO-6, 1, 14, 49, 50, 56, 104, 107, 111, 115, 116, 125, 134, 135, 142, 164, 165, 187–189, 206, 223, 229, 258
    VMR-152, 31, 32, 46, 89
    VMR-352, 31
    I Marine Amphibious Corps, 118, 232
    1st Provisional Marine Brigade 2, 95, 160, 164, 188
    1st Marine Division, 1, 6, 8, 12, 14, 18, 19, 21, 22, 29–31, 34, 35, 38, 42, 44, 48–50, 54, 56–60, 65–67, 72, 76, 78, 80, 83, 84, 86, 88, 92–96, 99, 101, 105, 108, 109, 111, 113, 117, 118, 120–123, 125, 126, 128, 130, 132–135, 137, 138, 142–144, 146, 147, 152, 157, 159, 161, 164, 168, 169, 171, 173, 176, 180–182, 184–186, 192, 194, 198, 202–204, 208, 211, 212, 215–219, 221, 227, 228, 231–233, 236, 242, 249, 253, 254, 256–261
      Assembly area at Masan, 2, 17
      Administrative headquarters at Masan, 1, 22
      Commander, 22, 260
      Command Posts, 21, 22, 34, 42, 68, 78, 79, 244, 251, 255, 256
      EUSAK Control, 5
      Equipment, Shortage of, 12
      Fresh rations, 5
      Headquarters, 3, 5
      Hospital, 3
      Military Police Company, 3
      Morale, 7, 199
      Personnel, Shortage of, 11
      Reconnaissance Company, 3, 56, 60, 66, 83, 103, 105, 108, 161, 168, 176, 182, 192, 207, 208, 210, 213, 217
      Security Measures, 3
      Strength figures, 11, 12
    1st Marines, 30, 34, 42, 51, 54, 60, 66, 68–75, 81, 83, 86, 90–94, 101, 105, 117, 120, 125, 126, 130, 134, 144, 146–152, 157, 171, 176, 182, 186, 187, 190, 191, 193, 194, 204, 210, 211, 215, 218, 219, 224, 227, 229, 243, 255
      Commanding Officer, 255
      Command Post, 157, 208
      Antitank Company, 149
      1st Battalion, 51, 69, 70, 105, 107, 109, 111, 113–117, 119, 148–150, 187, 192, 193, 197, 204, 210, 216, 217, 240
        Command Post, 114
        Company A, 111
        Company B, 111
        Company C, 51, 70, 111, 113, 115, 216
        Weapons Company, 111
      2d Battalion, 55, 69, 75, 83, 90, 117, 119, 144, 148–150, 187, 189–193, 197, 204, 211, 213, 215, 220
        Commanding Officer, 150, 189
        Command Post, 190
        Company D, 91
        Company E, 91, 92, 211
        Company F, 91, 92, 193
      3d Battalion, 44, 55, 74, 75, 81, 91, 109, 111, 114, 117, 118, 144, 148, 150, 157, 158, 186, 191–194, 204, 206, 210, 217, 229
        Company G, 83, 111, 114
        Company H, 111, 114, 206, 229–231
        Company I, 91, 111, 114, 230, 231
        Weapons Company, 74, 111
    5th Marines, 2, 5, 42, 48, 51, 52, 54, 55, 62, 66–70, 72, 73, 81, 92–95, 101, 105, 106, 109, 113, 117, 120, 126, 130, 133, 137, 141, 146–148, 161, 168, 171, 175, 176, 179, 182, 192, 194, 204, 213–215, 218, 219, 224, 227, 232, 236, 243, 255, 260
      1st Battalion, 54, 68–70, 94, 106, 141, 175, 194, 195, 197, 204, 220, 228, 241
        Command Post, 105
        Company A, 141, 142
        Company B, 107, 141, 228
        Company C, 52, 68, 141
      2d Battalion, 54, 70, 94, 130, 141, 148, 194–197, 223
        Company D, 130, 194, 196
        Company F, 106, 194–197
        Weapons Company, 196
      3d Battalion, 69, 70, 91, 94, 119, 126, 130, 148, 178, 194–196

Marine Corps, U. S.—Continued
Units—Continued
7th Marines, 42, 51, 52, 54, 55, 62, 66, 72–75, 81, 83, 86, 91, 92, 94–97, 101, 106, 107, 109, 111, 114, 117, 120, 123, 125, 126, 130, 133, 137, 141, 144, 146, 151, 152, 161, 171, 175, 176, 180, 182, 183, 186, 192, 204, 213–215, 218, 224, 227, 243, 255
Commanding Officer, 255
Command Post, 175
Outpost, 133
1st Battalion, 52, 75, 88, 91, 106, 107, 111, 117, 151, 176, 178, 179, 183, 184, 187, 213, 216, 241
Command Post, 51, 91
Company A, 51, 52, 83
Company C, 107, 216
2d Battalion, 52, 74, 75, 78, 90, 111, 117, 130, 151, 175, 176, 180, 183, 184, 187, 190, 191, 208, 213, 238, 241
Company F, 190
3d Battalion, 52, 74, 75, 78, 90, 96, 97, 111, 113–115, 117, 119, 123, 125, 135, 151, 157, 173, 175, 176, 178, 179, 183, 184, 187, 213, 214, 244
Company G, 123, 151, 179, 184
Company H, 175, 179, 183, 184
Company I, 125, 151, 183, 184
11th Marines, 3, 42, 54, 56, 68, 75, 81, 84, 114, 122, 149, 171, 179, 182, 189, 191, 192, 210, 215, 218, 225, 227, 242, 245, 254, 260
1st Battalion, 141, 179, 228, 254
2d Battalion, 70, 134, 149, 171, 191
3d Battalion, 92, 161, 171, 179
1st Amphibian Tractor Battalion, 1, 3
1st Armored Amphibian Battalion, 254
1st Engineer Battalion, 3, 34, 42, 102
Company C, 191
Company D, 88
1st Motor Transport Battalion, 3, 30, 181, 255
1st Ordnance Battalion, 3, 34, 42, 181, 230
1st Signal Battalion, 3
1st Service Battalion, 3, 34, 42
1st Shore Party Battalion, 3, 189, 207, 211, 213, 241
1st Tank Battalion, 3, 34, 42, 62, 176, 210, 218, 245, 255
Company A, 81, 114
Company B, 62, 92, 114
Company C, 133, 141
H & S Company, 62
See also Replacement Drafts and Rotation Drafts.
2d Marine Division, 167
Marks, LtCol Alfred H., 227
Marquez, Cpl Leo, 113, 114
Marshall, Col S. L. A., 6
Martin, LtCol Glen E., 106
Martin, Cpl Paul G., 103, 104*n*
Masan, 1–3, 5–7, 12–14, 18, 19, 22, 29, 30, 34, 89, 257
Massacre Valley, 70
Matthewson, 1stLt Robert E., 115
Mayer, Capt Franklin B., 70, 70*n*
McAlister, Col Francis M., 42*n*, 52*n*, 66, 105, 114, 117, 126
McClelland, LtCol Robert K., 90, 91, 148
McCutcheon, Col Keith B., 199*n*, 223, 240, 241, 242
McDonald, Capt Charles W., 228
McElroy, Capt John W., USNR, 165
McFarland, Maj David W., 107, 134, 134*n*, 223
McGahn, 2dLt Patrick T., 106, 106*n*
McGill, Col Robert A., 254
McGrew, Capt David G., Jr., 166*n*, 235, 238, 238*n*, 239
McGuire's University School, 231
McMahon, Capt John R., 216
McRay, Capt H. G., 115
McReynolds, LtCol William, 92, 179
MELLOW, Code Name, 17. *See* JOC.
Meuse-Argonne offensive, 118
Meyerhoff, Col Wilbur F., 74*n*, 107*n*, 109*n*, 131*n*, 151
Michener, James, 257*n*
MIG Alley, 65
Mills, 1stLt Niel B., 109*n*, 115
Milne, LtCol Harry T., 62*n*, 114
Mine fields, 130, 191, 230, 261
NKPA, 184
Mines, 130, 151, 178, 197, 206, 211, 224
Antipersonnel, 211, 236
Enemy, 206
Friendly, 206
MINNESOTA Line, 213, 217, 219, 227, 229, 232, 251
Miryang, 22
Missouri (BB), USS, 27
Mitchell, 1stLt Weldon R., 64

Mitchell, LtCol William P., 188, 199*n*, 207, 213, 214, 223, 241
Mojin, 119
Mongolian horses, 64
Montross, Lynn, 164*n*, 187*n*, 207*n*,
Moore, MajGen Bryant E., 60, 65, 66, 68, 71, 72
Morae-Kogae, 123
Moscow, 99
Mount Helicopter, 220. *See* Hill 884.
Mount McKinley (AGC), USS, 14
Mountain warfare, 138
Mukkye-dong, 51
Munsan, 8, 220
Munsan-ni, 201, 255
Murray, Col James C., 155*n*, 160*n*
Myers, Maj Reginald R., 117, 118

Naktong, Battle of, 2
Nam Il, Gen, 157, 160
Napoleon, 73
Native laborers, 11
Naval Medical Field Research Laboratory, 167, 169, 234, 236, 237
Naval gunfire, 13, 50, 119, 143, 210, 219, 220, 244
Navy, U. S., 5, 28, 34, 135, 138, 167, 234
  Role in Korea, 6
  Units
    Naval Forces Far East, 5
      Commander, (ComNavFE). *See* VAdm C. Turner Joy.
    Fleet Logistics Air Wing, 31, 32
    Seventh Fleet, 142
      Tactical Air Control Squadron–1, 14
      Tactical Air Control Squadron–3, 27, 28
    Task Force 77, 14, 27
    Task Force 90, 13
      Commander, 13. *See also* RAdm J. H. Doyle.
    Task Group 95.1, 15
    Escort Carrier Task Group 96.8, 2, 27, 28
    Western Deployment Group, 28
      See RAdm Lyman A. Thackrey.
    Mobile Construction Battalion 2 (Seabees), 34, 35, 45, 46
    VR–5, 31
    VR–9, 31
    VR–21, 32
Netherlands, 24, 202
New Britain operation, 232
New Jersey (BB), USS, 210, 217, 220
New Zealand, 24
Nicaragua, 150
Nickerson, Col Herman, Jr., 107*n*, 109*n*, 114, 114*n*, 123, 133, 144, 151, 183
Nicolay, Maj Stanley S., 46
Nihart, LtCol Franklin B., 187, 190, 191, 199*n*, 204, 211, 246*n*, 257*n*
Nolan, 2dLt Harvey W., 106
NO NAME Line, 120, 122, 125, 126
Noon, Col Thomas J., 33*n*
Norton, LtCol Kirt W., 220, 223, 228, 241

Objective 1, 176, 178
Objective 2, 176, 179, 230, 231
Objective 3, 176, 180, 231
Objective ABLE, 182, 183. *See also* Hill 673.
Objective BAKER, 182–184, 192, 193. *See also* Hill 749.
Objective CHARLIE, 182, 198
Objective DOG, 194, 196
Objective YOKE, 176, 180
Okinawa, 72, 167
O'Neill, LtCol John T., 254
Operation BLACKBIRD, 211, 212
Operation BUMBLEBEE, 214, 241
Operation BUSHBEATER, 217
Operation CHANGIE-CHANGIE, 241
Operation CLAM-UP, 242–244
Operation FAREWELL, 223
Operation HOUSEBURNER I, 217
Operation HOUSEBURNER II, 217
Operation KILLER, 65, 67, 69, 71, 73, 76, 79, 258
Operation LITTLE SWITCH, 223
Operation MIXMASTER, 251, 256
Operation MOUSETRAP, 241
Operation MULETRAIN, 240
Operation ROTATE, 241
Operation ROUNDUP, 51, 59
Operation RIPPER, 73, 79, 80, 84, 85, 88, 93, 258
Operation RUGGED, 94
Operation SEA LION, 250
Operation STRANGLE, 143, 144, 169, 185, 186, 249, 260, 261
Operation SUMMIT, 207, 208, 213
Operation SWITCH, 219
Operation THUNDERBOLT, 50, 51, 59
Operation WEDGE, 215
Operation WINDMILL I, 189, 190
Operation WINDMILL II, 190, 197
Operation WOLFHOUND, 41
Operations Research Office of Johns Hopkins University, 6
Outpost, 244
  Line of Resistance, 122, 238
Ord, Capt James B., 229, 229*n*, 230, 231
Osan, 26

Oum Mountain, 80
Oum San, 83
Ozuki, 33

Paekcha-dong, 55
Pafford, LtCol George G., 213
Parka, 69, 75
Parkins, Capt Clarence W., 57
Parry, LtCol Sherman W., 228
Partridge, MajGen Earle E., 14, 49, 78, 88*n*, 135, 136, 136*n*,
Partridge, Col John H., 73*n*, 102, 102*n*
Patrols, 91, 102, 104, 117, 122, 179, 182, 210, 213, 218–220, 222, 224, 225, 228, 229, 232, 238, 242–244, 249, 258
  Aggressive tactics, 11
  Air, 49, 50
  Ambush, 220, 228, 244
  Bases, 157–159
    Concept, 158
  Chinese Communist Forces, 11
  Coastal, British, 28
  Enemy, 224
  1st Marines, 70
  1st ROK Division, 84
  Foot, 215, 224
    C/1/5, 52
  Helicopter, 222
  Jungle, 232
  Marine, 53, 56, 70
  Motor, 49, 88, 103
  Reconnaissance, 67, 148, 158, 244
  Rice paddy, 257
  RCT–7, 53
  Sea, 50
  3/1, 44
  United States, 66, 70, 83
  United States Army, 3, 26
Panel markings, 115
Panmunjom, 220, 221, 224, 225, 229, 233, 247, 249
Patterson, Capt Russell G., Jr., 47
Pearson, Capt John A., 130
Pearl Harbor, 166
Peleliu, 232
PENDLETON, Phase Line, 109, 111
Pentagon, 11
People's Tribunal, China, 36
Pershing, Gen John J., 166
Philippines, 24
Photo laboratory, 135
Pierce, Maj Herbert E., 256
Piner, Sgt William, 106
Plans and Orders
  EUSAK
    Operation Plan 20, 25
    Operation Instruction (OI)–272, 253
  IX Corps
    Operation Plan 17, 120
  X Corps
    Operation Instruction 235, 203
  1st Marine Division
    Operation Order 1–51, 30, 34
    Operation Order 2–51, 30, 34
    Operation Order 3–51, 42, 44
    Operation Order 4–51, 54
    Operation Order 5–51, 60
    Operation Order 6–51, 66
    Operation Order 22–51, 176
    Operation Order 23–51, 182
    Operation Order 25–51, 192, 194
    Operation Order 26–51, 198
    Operation Order 27–51, 204, 206
    Operation Order 50–51, 219
    Operation Order 2–52, 253
    Operation Order 8–52, 255
    Training Order 2–51, 161
Planes, Marine, 219. *See also* Aircraft.
Pohang, 21, 30–34, 42, 44, 45, 49, 51, 54, 62, 66
Pohang-Andong, 50
Pohang-dong, 19
Pohang-Kyongju road, 42
Pohang-Kyongju-Andong MSR, 42, 44, 49
Police action, Korean War, 7
Political commissars, NKPA, 132
Political indoctrination of the CCF soldier, 36
Polgrean, Maj Edward R., 223
Port of debarkation, Pusan, 18
Pratt, LtCol Spencer H., 229–231
Princeton (CV), USS, 122
Prisoners, 56, 222, 229, 243
  Chinese Communist Forces, 100, 101, 103, 131, 249, 260
  Interrogations, 19, 22
  NKPA, 44, 45, 57, 132
  UN, 222, 223
  of War (POWs), 151, 176
    American, 222
    Exchange, 249
    Interrogations, 53, 100, 134, 182
Press correspondents, 6, 35, 239, 247
Privateers, 138
Propaganda, 156, 157, 222, 233
  Chinese, 104, 249
  Chinese Communist Forces, 134
Psychological warfare, 215, 233
Pukhan, 102, 109, 111, 115, 119
  River, 117, 122, 201
  Valley, 102

Puller, Col Lewis B., 52*n*, 57, 60, 67, 68, 72, 76, 78, 79
Punchbowl, 141, 154, 158, 173, 176, 180, 182, 185, 211, 228, 260
Pusan, 2, 8, 12, 15, 19, 22, 46, 49, 62, 90, 95, 108, 185, 215
  Air Force Field at, 1
  Perimefer, 164
  Port of debarkation, 18
  Withdrawal to, 8
Pyongchang, 51
Pyonggang, 143
Pyongtaek, 26
Pyongyang, 95, 248
Pyongyang-Chinnanpo, 103

Quantico, Virginia, 164, 223, 234
QUANTICO, Phase Line, 95, 99
Quilici, LtCol Clifford E., 220
Quinlan, John F., 237

Radford, Adm Arthur H., 31, 32
Radio, 36, 68, 116, 168, 190, 195, 231
  CCF/NKPA, 51
  Trouble in Jets, 28
Rail, 89
  Traffic jams, 32
  Transportation, 60
Railhead at Munsan-ni, 254
Railroads, enemy, 143
Railway system, South Korea, 33
  Tunnels, North Korea, 15
Randazzo, TSgt Carmelo J., 74
Rations, 170, 181, 222
  USMC, 54
Reavis, 1stLt William A., 241
Reconnaissance, 229
Redalen, Capt Dwail L., 164
Refrigeration ship, UN, 5
Refugees, 13, 25, 153
Reisler, 2dLt Joseph M., 105, 105*n*, 113
Replacement Drafts, USMC, 6, 31
  14th Replacement Draft, 219
  15th Replacement Draft, 224
  18th Replacement Draft, 254
  19th Replacement Draft, 254
Replacements, 258
  Airborne, 32
  Chinese Communist Forces, 38
  U. S., 42
  U. S. M. C., 18, 31, 32, 48, 49, 57, 89, 214
Reusser, Maj Kenneth L., 27
Reynolds, Maj Walter E., Jr., 245
Rhee, President Syngman, 11, 25
Rice paddy patrols, 48, 52, 57. *See also* Patrols.

Ridgway, LtGen Matthew B., 6, 7, 7*n*, 8, 10, 11, 11*n*, 19, 22, 24, 25, 25*n*, 26, 29, 30, 38, 38*n*, 41, 42, 45, 51, 58, 59, 59*n*, 60, 62*n*, 65, 68, 72, 73, 79, 84, 86, 91, 93–95, 99, 108, 115, 136, 160, 258, 261, 261*n*
  Declaration of Faith, 10
  Moral Leadership, 10
  Strategy, 35
  Tactics, 11
Ringblom, Maj Allan H., 223
Roach, Maj Maurice E., 74, 113, 117
Road
  Andong-Yongdok, 30
  Center at Yanggu, 133
  Networks, 143
  Traffic, 143
  Transportation, 60
Roadblocks, U. S., 24
Roads, 48, 66, 73, 86, 95, 123, 143, 163, 181
Roberts, Col Deane C., 32, 46, 89
Rockey, 2dLt William K., 211
Rooney, Col John T., 107*n*, 109*n*, 151
Rotation draft, USMC, 57
  10th Rotation Draft, 219
  11th Rotation Draft, 219
  12th Rotation Draft, 224
  27th Rotation Draft, 219
Rubber boats, 102
Ruble, RAdm Richard W., 27
Russia, 160
  Soviet, 37, 261

Saipan, 187
Samchok, 26, 58
Samgo-ri, 55
Sangyong, 54
Sapyong-ni, 99
Sawyer, Maj Webb D., 88, 88*n*, 91, 106
Schening, 1stLt Richard J., 53
Schilt, MajGen Christian F., 219
Schmuck, LtCol Donald M., 70, 149
Scott, 1stLt John L., 56
See, 2dLt Milo J., 229
Selden, MajGen John T., 199*n*, 231, 234*n*, 236, 243, 255
Seoul, 19, 22, 24, 28, 47, 50, 51, 59, 66, 73, 79, 84, 107, 108, 119, 123, 137, 170, 242, 253, 254, 259
  Evacuation of, 25
  Historic invasion route, 253
  UN Airfield at, 29
Sexton, Capt Martin J., 19
Shepherd, Gen Lemuel C., Jr., 31, 76, 76*n*, 78, 78*n*, 96, 96*n*, 135, 136, 208, 232

Ships
LSD, 256
LST, 254–256
LST 898, 34, 42
LST 914, 34, 42
See also Ship by name.
Shoes, 114
Lace, 68
Show, Gen Shiu Kwai, 81
Sicily (CVE), USS, 2, 15 146
Sikorsky. *See* Helicopters.
Simmons, Maj Edwin H., 74, 74*n*, 111, 113*n*
Sinhung, 34, 42
Smedley, Maj Kenneth C., 223, 229
Smith, Corpsman E. N., 115
Smith, Capt Jack A., 91
Smith, Capt Samuel S., 130
Smoke, 116
Screen, 93
Shells, 134
Smith, MajGen Oliver P., 3*n*, 5, 5*n*, 6, 12*n*, 18, 18*n*, 19, 19*n*, 21, 22*n*, 29, 30, 30*n*, 45, 45*n*, 50, 54*n*, 56, 56*n*, 57*n*, 58, 58*n*, 60*n*, 66, 67, 68, 68*n*, 72, 79, 91, 92, 92*n*, 94, 94*n*, 99, 106, 106*n*, 118, 118*n*, 257
Snedeker, Col Edward W., 8*n*, 21, 109*n*
Sniper, 215
Fire, 92, 229
Sniping, 196, 232
NKPA, 49
Snows, 69, 228, 230, 234
So-chon River, 151, 158
Sohwari, 175
Soissons, 118
Sokcho-ri, 215, 254, 256
Som River, 66, 74
South Africa, 24
Soyang, 119, 128, 130, 175
Bridgehead, 128
River, 94, 117, 146, 150, 173, 197, 216, 217, 230
Valley, 175, 181, 182
Stamm, Col John F., 199*n*, 235, 246*n*
Steiner, Col William B., 31
Stewart, James T., 131*n*, 169*n*
Stewart, LtCol Joseph L., 69, 69*n*, 94, 95*n*
Stiff, LtCol Houston, 194, 195
Stoyanow, Capt Victor, 125
St. Paul (CA), USS, 50, 224, 244
Strategy
Effect of the truce talks, 161
Limited war, 36, 37, 38
Nuclear war, 36, 37, 38
Stratemeyer, LtGen George E., 14, 33, 136
Strongholds, CCF, 71
Struble, VAdm Arthur D., 27
Supplies, 72, 95, 144, 147, 154, 170, 171, 176, 181, 187, 189, 190, 220, 222, 241, 243
See Logistics.
Dumps, 119, 219
U. S., 30, 62, 95
USMC, 48, 113
Surrender message, 55,
Survival suits, 47
Sutter, LtCol Allan, 75
Suwon, 22, 41, 50, 51
UN Airfield at, 29
Swanson, 1stLt William, 111

Tactical Air Control Center (TACC), 17
Tactical Air Control Party (TACP), 17, 26, 137
Tactical Air Controller (Airborne) (TACA), 97
Tactical Air Co-ordinator, Airborne (TACA), 17, 116
Tactics, 10, 258
Advantage, 169
Aggressive patrolling, 11
CCF, 83, 138
Delaying, 258
Eighth Army, 8
Firepower, 11
Marine Corps, 138, 260
NKPA, 192
Plan of Attack, 8
Ridgway, 11
Taeam-san, 141, 146, 147
Taegu, 14, 29, 44, 58, 60, 64
Taejon, 44, 90
Taeu-san, 158, 159
Tanks, 83, 151, 191, 245
British Centurion, 25
Column, USA, 62
Fake, CCF, 103
-Infantry patrols, 151
Support, 138
US, 121
USMC, 68, 70, 75, 83, 92, 104, 114, 117, 122, 125, 130, 133, 219
Tanyang, 44
Task Force Puller, 52
Taylor, Gen Maxwell D., 38, 38*n*
Taylor, Capt Richard M., 133, 141
Telephone, 158, 208
Wire, 36, 71, 74
Thailand, 24
Thackrey, RAdm Lyman A., 27, 28
Thermal boots, 231, 234–236, 240, 250, 251
Innovations, 234
Thermopylae, 128
Theseus, HMS, 28, 29, 46

Thirty-eighth Parallel, 8, 15, 45, 79, 220, 258
Thirty-ninth Parallel, 15
Thomas, MajGen Gerald C., 118, 120, 133, 133*n*, 137, 147, 150, 151, 158, 158*n*, 186, 186*n*, 189, 197, 199*n*, 207, 208, 211, 217, 228, 231, 246*n*
Tierney, Elizabeth L., 164*n*
Timberlake, MajGen Edward J., 136
Tim, Col Tai Shik, 159
Tinsley, LtCol James H., 119*n*, 232, 246*n*
Todun-ni, 111
Toledo (CA), USS, 217
Tongchon, proposed landing at, 132
Tongchon-Kumhwa road, 132
Tonpyong, 176
Topyong-dong, 52, 55
Transportation, 36, 46, 89
  Highways, 47
  Human, 181
  Motor, 42, 163, 181
  Motor lift, 42
  Railroads, 47
  Roads, 36
  Trains, 44
  Water lift, 42
  See Logistics.
Treasure Island, 31
Trenches, 217, 232, 246, 248
  Marine, 243
  Warfare, 232
Trompeter, Maj Joseph D., 117, 118
Truce, 160
  Talks, 154–156, 161, 163, 171, 180, 222, 233, 247–249, 261
Truman, President Harry S., 5, 37, 99
Tsuika, 33
Tucker, LtCol Roy A., 111*n*
Tundong-ni, 171
Turkey, 24
Typhus, 53

Uihung, 30
Uijongbu, 84
Uisong, 30, 42, 44, 49, 52, 54
Uisong-Andong road, 42
Ulsan, 8
United Kingdom, 24
United Nations (UN), 29, 156
  Delegates, 156, 160, 220, 229, 233
  Forces, 24, 35, 60, 80, 126, 154, 161, 163
    Commander, 257
    Strength figures, 85, 218
  Truce Team, 160
Units
  British Commonwealth
    1st Commonwealth Division, 71, 116, 119, 201, 253
    27th Brigade, 22, 65, 107, 116, 121
    29th Brigade, 22, 116, 121
    Royal Australian Regiment, 116
    Canadian Battalion, 22
    Canadian Light Infantry, Princess Patricia, 116
    Gloucestershire Battalion, 116
    Independent Commandos, Royal Marines, 11, 12
      41 Independent Commando, 3
    New Zealand Field Artillery Battalion, 22
    Royal commonwealth naval forces, 15
  Belgian Battalion, 201
  Colombian Battalion, 201
  EUSAK. *See* Army Units.
  Ethiopian Battalion, 201
  French Battalion, 176
  French Units, 15
  Greek Battalion, 201
  60th Indian Ambulance Group, 85
  Philippine 20th BCT, 201
  Swedish Evacuation Hospital Unit, 85
  Thai units, 15
  Thailand Battalion, 22, 201
  Turkish Brigade, 22, 201

Van Fleet, LtGen James A., 99, 109, 109*n*, 115, 118, 121–123, 123*n*, 126, 127, 127*n*, 133 142, 154, 161, 161*n*, 199*n*, 201, 202, 218, 219, 221, 248, 248*n*, 249, 251
Van Kueren, RAdm Alexander H., 166
Van Ryzin, BrigGen William J., 136*n*
Vatcher, William H., Jr., 155*n*, 221*n*
Vehicles, 142, 175, 186, 244
  DUKWs, 94, 102, 175
  EUSAK, 12
  Enemy, 15
  Jeep, 70, 91, 94, 131, 256
    Trailers, 256
  Ox cart, CCF, 143
  Tank cars, 32
  Tracked, USMC, 62
  Trucks, 105, 119, 131, 175, 181, 215, 256
    Convoy, 54, 70
    Enemy, 131, 143
    Fake, CCF, 103
    USMC, 30, 147
  U. S., 13, 62

Vests, armored, 165, 167, 168, 236–238, 240, 251. *See* Armor, body.
Vittori, Cpl Joseph, 193, 193*n*
Volcansek, LtCol Max J., Jr., 46

Wade, Col Sidney S., 119*n*, 227, 246*n*, 255
Walker, Maj Carl E., 211
Walker, Richard L., 36*n*
Walker, LtGen Walton H., 7
Ward, 1stLt Alfred J., 47
Warehouses, Korean, 15
Warfare
  Land, 250, 257
  Movement, of, 198, 201, 224, 229
  Position, of, 198, 201, 218, 229, 231
  Static, 247
  Trench, 232
Warner, 1stLt Robert E., 141
Warships, UN, 123. *See* Ships.
Washington, D. C., 7, 8, 28, 156, 161
Watson, 2dLt John E., 229
Weapons
  CCF, 35
    Antiaircraft guns, 103
    Artillery, 35, 248
    Automatic weapons, 96
    Booby traps, 74
    Burp guns, 156, 160
    Enemy mines, 70
    Grenades, 35, 96, 125
    Land mines, 70
      Explosions, 44
    Machine gun, 47, 70, 106, 125, 130 138
    Mines, 74
    Mortars, 35, 74, 81, 83, 125, 130, 138, 144, 248
      Fire, 92
      120mm, 96
    Rifle, 47
      Recoilless, 125
    Russian 76mm guns, 125
    Satchel charges, 125
    Small arms, 35, 81, 83, 92, 96
      Fire, 74
  NKPA
    Artillery, 44, 149, 151, 191
    Automatic, 151
    Burp, 197
    Grenades, 141, 197
    Machine guns, 55, 141, 178, 180, 192
    Mortars, 44, 51, 55, 142, 149, 151, 179, 180, 184, 187, 191, 192, 217
      76mm, 187, 192, 196
      81mm, 55
      82mm, 192
      105mm, 192
      120mm, 192
      122mm, 192
    Rifle, 141
    Small arms, 55, 142, 184, 217
  ROKA
    105 mm Howitzer, 227
    155 mm Howitzer, 227
  United States, 69
    Army Artillery, 11
    Grenade, 10
    Howitzer, 277
    Napalm, 17, 27, 51, 55, 71, 83, 132, 142, 171, 183, 195
    Pistol, 168
    Thompson submachine gun, 168
    USMC
      Air, 217. *See* Aircraft.
      Antitank guns, 149
      Artillery, 148, 217
      Bayonet, 149, 150, 168
      Bombs, 142, 143
        Napalm, 27
      Booby traps, 216
      Flame throwers, 217
      Grenades, 83, 91, 92, 149, 150, 168, 184, 195, 217
      105mm Howitzer, 49, 233
      Machine guns, 135, 179, 216, 217
      Mines, 176, 216. *See also* Mines.
      Mortars, 148, 206, 217
        4.2" Mortars, 191, 195
        81mm Mortars, 90, 113, 168, 195
      Rifles, 115
        M–1, 12
        75mm Recoilless, 148, 195
        90mm, 210
        Tank, 149
      Rockets, 195
Weather, 25, 33, 39, 69, 131, 233–235, 255
Weede, Col Richard G., 149, 192, 196, 199*n*, 204, 232
Wehrmacht, 250
Weintal, E., 155*n*
Wermuth, LtCol John J., 204, 227
Welch, LtCol Claude H., 89
Wertman, LtCol Howard E., 181, 255
West Point, 72
West, Col Radford C., 46
West, LtCol Robley E., 105, 148, 149
Weyland, MajGen Otto P., 169
Whaling, BrigGen William J., 232
White, 1stLt Francis E., 230
Williams, BrigGen Gregon A., 57
Wilson, TSgt Harold E., 114
Winecoff, Col Joseph L., 149, 150

Winfrey, 1stLt Frank A., 151
Wire laying, 230
Wisconsin, (BB), USS, 244
Whistles, NPKA, 192
Wolfe, LtCol David C., 89
Wonju, 18, 19, 22, 29, 44, 45, 59, 60, 63, 66–68, 91
  Highway and rail center at, 26
Wonju-Hoengsong highway, 67
Wonsan, 143, 156, 168
  Landing, 63
  UN airfield at, 29
Wontong-ni, 187, 192, 204, 213, 233
Wornham, Col Thomas A., 186, 191, 204
Worthington, Maj Gordon R., 179
World War I, 65, 68, 72, 166, 231, 232, 246
World War II, 6, 7, 33, 44, 72, 99, 100, 160, 167, 194, 232, 250, 257, 261
Wound statistics, 166
Wray, Capt Robert P., 51 70, 70*n*, 109*n*, 111, 113, 115*n*
Wyczawski, LtCol Richard W., 14, 46, 46*n*, 89

Yanggu, 128, 133, 135, 141, 146, 161, 171, 202
  Road center at, 132
Yanggu-Inje, 132
Yangjimal, 91
Yangyang, 8, 202
Yoju, 26, 66, 72, 79
  Corridor, 58
YOKE Ridge, 173, 176, 178, 183. *See also* Objective YOKE
Yonchon, 201
Yongchon, 30, 44, 49
Yongdok, 42, 54, 55
Yongdok-Andong road, 54
Yongwol-Pyongchang road, 66
Yonpo Airfield, 14, 29
Yu, Maj Dung Nam, 53
Yudam-ni area, 6

U.S. GOVERNMENT PRINTING OFFICE: 1962—0 634040

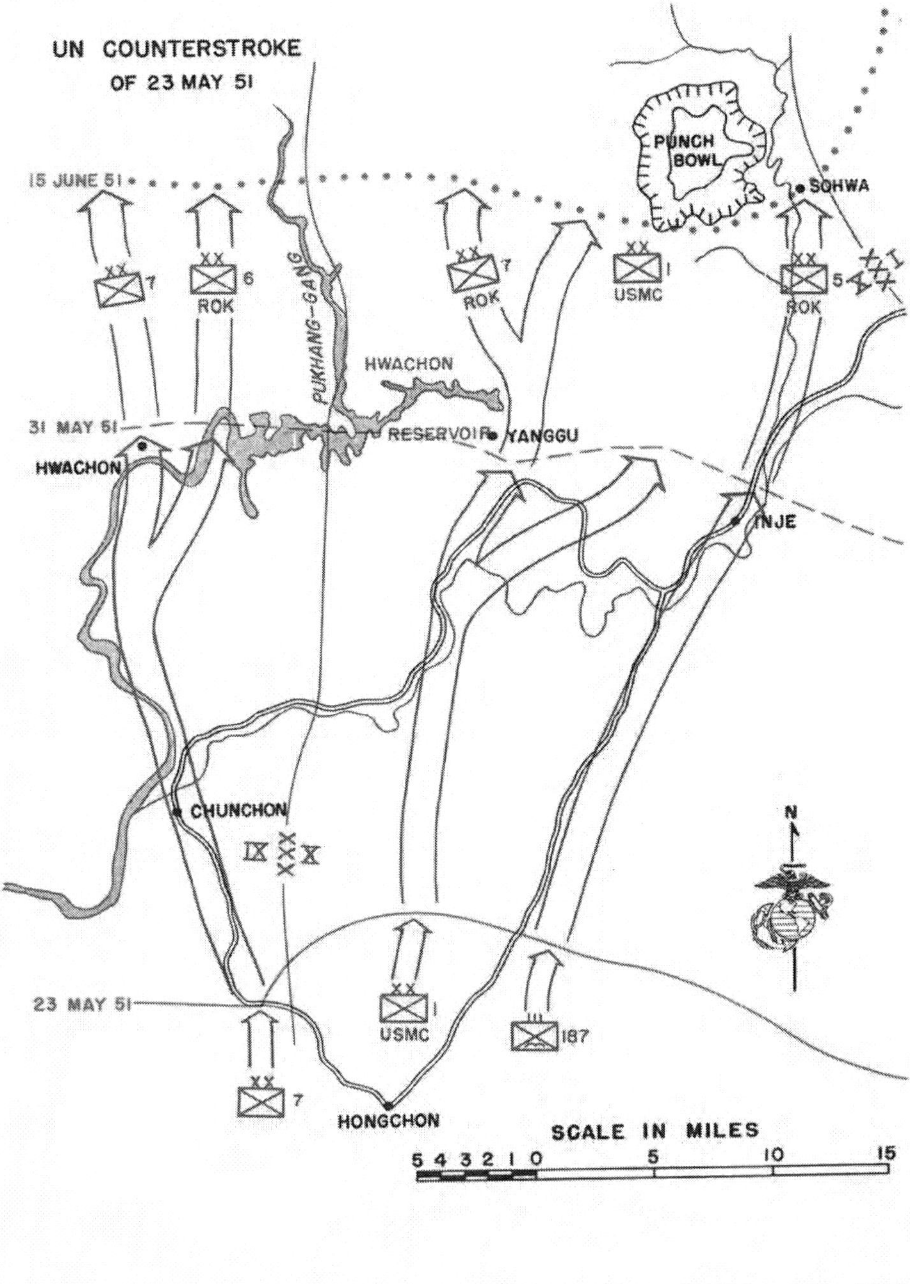
UN COUNTERSTROKE
OF 23 MAY 51
15 JUNE 51
31 MAY 51
23 MAY 51
PUNCH BOWL
SOHWA
PUKHANG-GANG
HWACHON
RESERVOIR
YANGGU
HWACHON
INJE
CHUNCHON
HONGCHON
7
6
ROK
7
ROK
1
USMC
5
ROK
IX
X
1
USMC
187
7
N
SCALE IN MILES
5 4 3 2 1 0 5 10 15

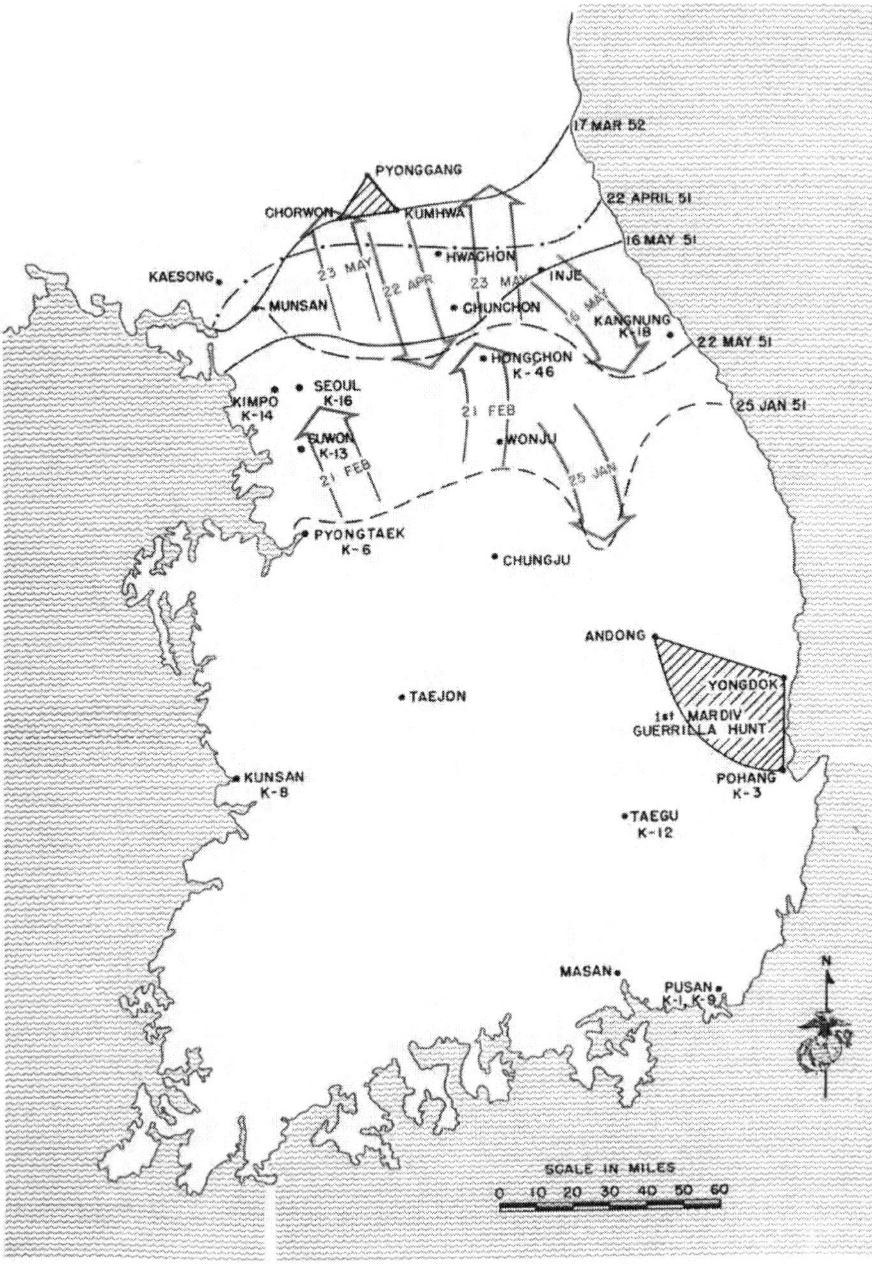

17 MAR 52
PYONGGANG
CHORWON
KUMHWA
22 APRIL 51
16 MAY 51
HWACHON
KAESONG
23 MAY
22 APR
23 MAY
INJE
MUNSAN
CHUNCHON
16 MAY
KANGNUNG
K-18
22 MAY 51
HONGCHON
K-46
KIMPO
K-14
SEOUL
K-16
21 FEB
25 JAN 51
SUWON
K-13
WONJU
21 FEB
25 JAN
PYONGTAEK
K-6
CHUNGJU
ANDONG
YONGDOK
TAEJON
1st MARDIV
GUERRILLA HUNT
KUNSAN
K-8
POHANG
K-3
TAEGU
K-12
MASAN
PUSAN
K-1, K-9
N
SCALE IN MILES
0 10 20 30 40 50 60